The Gateway to t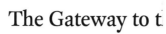

MW00635622

HISTORICAL STUDIES OF URBAN AMERICA

Edited by Lilia Fernández, Timothy J. Gilfoyle, Becky M. Nicolaides, and Amanda I. Seligman

James R. Grossman, Editor Emeritus

Recent titles in the series

Bulls Markets: Chicago's Basketball Business and the New Inequality
by Sean Dinces

Newsprint Metropolis: Newspapers and the Urbanization of Americans, 1880–1930
by Julia Guarneri

Evangelical Gotham: Religion and the Making of New York City, 1783–1860
by Kyle B. Roberts

Crossing Parish Boundaries: African Americans, Catholicism, and Sports in Chicago, 1914–1954
by Timothy Neary

The Fixers: Devolution, Development, and Civil Society in Newark, NJ, 1960–1990
by Julia Rabig

Chicago's Block Clubs: How Neighbors Shape the City
by Amanda I. Seligman

The Lofts of SoHo: Gentrification, Art, and Industry in New York, 1950–1980
by Aaron Shkuda

The Newark Frontier: Community Action in the Great Society
by Mark Krasovic

Making the Unequal Metropolis: School Desegregation and Its Limits
by Ansley T. Erickson

Confederate Cities: The Urban South during the Civil War Era
edited by Andrew L. Slap and Frank Towers

The Cycling City: Bicycles and Urban America in the 1890s
by Evan Friss

Making the Mission: Planning and Ethnicity in San Francisco
by Ocean Howell

A Nation of Neighborhoods: Imagining Cities, Communities, and Democracy in Postwar America
by Benjamin Looker

A World of Homeowners: American Power and the Politics of Housing Aid
by Nancy H. Kwak

Demolition Means Progress: Flint, Michigan, and the Fate of the American Metropolis
by Andrew R. Highsmith

Metropolitan Jews: Politics, Race, and Religion in Postwar Detroit
by Lila Corwin Berman

Blood Runs Green: The Murder That Transfixed Gilded Age Chicago
by Gillian O'Brien

A City for Children: Women, Architecture, and the Charitable Landscapes of Oakland, 1850–1950
by Marta Gutman

A World More Concrete: Real Estate and the Remaking of Jim Crow South Florida
by N. D. B. Connolly

Urban Appetites: Food and Culture in Nineteenth-Century New York
by Cindy R. Lobel

Crucibles of Black Empowerment: Chicago's Neighborhood Politics from the New Deal to Harold Washington
by Jeffrey Helgeson

The Streets of San Francisco: Policing and the Creation of a Cosmopolitan Liberal Politics, 1950–1972
by Christopher Lowen Agee

Harlem: The Unmaking of a Ghetto
by Camilo José Vergara

A complete list of series titles is available on the University of Chicago Press website.

The Gateway to the Pacific: Japanese Americans and the Remaking of San Francisco

MEREDITH ODA

The University of Chicago Press

CHICAGO AND LONDON

PUBLICATION OF THIS BOOK HAS BEEN AIDED
BY A GRANT FROM THE BEVINGTON FUND.

The University of Chicago Press, Chicago 60637
The University of Chicago Press, Ltd., London
© 2019 by The University of Chicago
Published 2019
Printed in the United States of America

28 27 26 25 24 23 22 21 20 19   1 2 3 4 5

ISBN-13: 978-0-226-59260-2 (cloth)
ISBN-13: 978-0-226-59274-9 (paper)
ISBN-13: 978-0-226-59288-6 (e-book)
DOI: https://doi.org/10.7208/chicago/9780226592886.001.0001

Library of Congress Cataloging-in-Publication Data

Names: Oda, Meredith, author.
Title: The gateway to the Pacific: Japanese Americans and the remaking of San
Francisco / Meredith Oda.
Other titles: Historical studies of urban America.
Description: Chicago: The University of Chicago Press, 2019. | Series: Historical
studies of urban America
Identifiers: LCCN 2018019707 | ISBN 9780226592602 (cloth: alk. paper) |
ISBN 9780226592749 (pbk: alk. paper) | ISBN 9780226592886 (e-book)
Subjects: LCSH: Japanese Americans—California—San Francisco—History—
20th century. | Urban renewal—California—San Francisco—History—
20th century. | Ethnic neighborhoods—California—San Francisco—History—
20th century. | San Francisco (Calif.)—History—20th century.
Classification: LCC F869.S39 J3586 2019 | DDC 979.4/6104956—dc23
LC record available at https://lccn.loc.gov/2018019707

♾ This paper meets the requirements of ANSI/NISO Z39.48–1992
(Permanence of Paper).

Contents

Acknowledgments • vii

Introduction • 1

CHAPTER ONE • 15
Japan and Japanese Americans in the Pacific Metropolis
through World War II

CHAPTER TWO • 41
Orienting the Gateway to the Pacific:
Reconsidering Japan and Reshaping Civic Identity

CHAPTER THREE • 73
Redeveloping Citizens: Planning a New Japanesetown

CHAPTER FOUR • 106
Pacific Crossings: Japan, Hawai'i, and
the Redefinition of Japanesetown

CHAPTER FIVE • 138
Intermediaries with Japan:
The Work of Professional Japanese Americans in the Gateway

CHAPTER SIX • 174
Local Struggles: Japanese American and African American Protest
and Cooperation after 1960

Conclusion • 213

Notes • 219
Index • 273

Acknowledgments

My scholarship and thinking for this book have been shaped by so many people and institutions along the way. There is no way I can possibly do justice to the tremendous debts I have accumulated, but I am grateful for the opportunity to at least acknowledge some here.

The University of Chicago has been a source of wonderful mentors and colleagues. Mae Ngai prodded a microstudy of redevelopment into a more expansive story of San Francisco's transpacific history. Her incisive feedback and her own model of scholarship improved this project in immeasurable ways and continue to inspire me. Kathleen Conzen was a kind and able guide through urban and ethnic history, while Thomas Holt's acute thinking on race and racism has shaped my understanding of race, difference, and power. George Chauncey, Kyeong-Hee Choi, Bruce Cumings, James Grossman, and Amy Dru Stanley have all helped make me a better historian. I also had the pleasure and privilege to learn among a cohort of excellent scholars and friends: Thomas Adams, Melissa Borja, Michael Carriere, Kornel Chang, Jessica Graham, Allyson Hobbes, Molly Hudgens, Gwennan Ickes, Kouslaa Kessler-Mata, Kelly King-O'Brien, Alison Lefkovitz, Jon Levy, Jason McGraw, Arissa Oh, Sarah Potter, Gautham Rao, David Spatz, Timothy Stewart-Winter, Ellen Wu, and Ryan Yokota.

A number of people have been instrumental in helping me tell this story. First, I would like to thank all those who generously shared their memories and time in sometimes quite lengthy interviews: Masao Ashizawa, Richard Hashimoto, Uta Hirota, Steve Nakajo, Noboru Nakamura, Victor Ona, Sam Seiki, Marshall Sumida, Tomoye Takahashi, Franklin Tokioka, Lionel Tokioka, Marvin Uratsu, George Yamasaki Jr., and Mas Yonemura.

Archivists and librarians are particular allies of historians, and my experience has been no exception. Marie Browning let me camp out at the San Francisco Redevelopment Agency Central Records for months of research; the agency has since closed, its records transferred to the San Francisco History Center at the San Francisco Public Library. Karl Matsushita lugged around boxes, provided invaluable information, and offered a little refreshing gossip at the Japanese American National Library. Tami Suzuki, Jeff Thomas, and the rest of staff at the San Francisco History Center at the San Francisco Public Library were incredibly gracious with their time and knowledge. Sherman Seki at the Archives and Manuscripts Department at the University of Hawai'i at Manoa Hamilton Library offered helpful suggestions and access. Rosalyn Tonai, Peter Yamamoto, Judy Yamaguchi, and Max Nihei at the National Japanese American Historical Society offered timely assistance. Kenji Taguma is not an archivist, but he was a cordial and informative host while I paged through decades of the newspaper he edited, the *Nichi Bei Times* (now reconfigured as the online *Nichi Bei*). I also thank the many archivists and librarians who have contributed to this project at the Bancroft Library, University of California, Berkeley; California Historical Society; University of California, Berkeley, Ethnic Studies Library; University of California, Los Angeles, Young Research Library Special Collections; Hawai'i State Archives; the Japanese American National Museum's Hirasaki Resource Center; Manuscripts and Records Collection at the Walter P. Reuther Library at Wayne State University; and National Archives and Records Administration at College Park, Maryland. Finally, I would like to thank Mark Lucas and the staff at the Mathewson-IGT Knowledge Center Interlibrary Loan, who tracked down my avalanche of requests with good humor.

I have found a wonderful and supportive home at the University of Nevada, Reno, in large part due to a brilliant bunch of friends and colleagues. Emily Hobson is a partner in crime, model of socially engaged scholar/teacher, and generous reader. Greta de Jong has hosted many a delicious summer solstice celebration and offered incredibly useful feedback and ideas. Jennifer Ng and Edward Schoolman have commiserated over food and drink. My chairs have been fantastic guides: Linda Curcio introduced me to the world of the tenure track, Bruce Moran was kind and supportive, and Dennis Dworkin—who took a fellow Maroon under his wing right away—offered crucial resources that helped me push to the finish line in a relatively timely manner. Jenni Baryol in the Department of History office is a lifesaver, time and time again. Chris Church, Marwan Hanania, Martha Hildreth, Renata Keller, Elizabeth Raymond, Bill Rowley, Hugh Shapiro,

Kevin Stevens, Cameron Strang, Charles Tshimanga-Kashama, and Barbara Walker have all helped make the department a collegial and smart department. Outside of history, Mikaela H. Rogozen-Soltar provided regular writing deadlines (she did what she could) and fun reminders that there's so much more than writing. Others, too, have made Reno a supportive, lively place: Debbie Boehm and Patrick Jackson, Katherine Fusco and Blake Watson, Jen Hill, Stephanie Gibson and David Rondel, Eunkang Koh, Jessica Nakamura, Amy Pason, and Daniel Enrique Pérez.

This book has also profoundly benefited from the feedback and assistance of a number of fellow historians. I cannot thank enough those who have read and commented upon versions of these chapters: Charlotte Brooks, Jason Chang, Grace Delgado, Mary Dudziak, Dennis Dworkin, Rámon Gutiérrez, Lisbeth Haas, Theresa Mah, Jennifer Miller, Amy Scott, and Naoko Shibusawa. Bill Issel generously read the entirety in an earlier version and offered sharp and constructive comments. Mariko Iijima and Yuki Oda kindly answered my questions about Japanese sources. Other scholars have also offered critical suggestions and ideas at crucial junctures: Eiichiro Azuma, Andrew Friedman, Madeline Hsu, Lon Kurashige, Scott Kurashige, Clement Lai, Meghan Mettler, Dennis Ogawa, Greg Robinson, Joel Tarr, Yujin Yaguchi, Henry Yu, and all the participants at the Race and Racial Ideologies Workshop at the University of Chicago and at conferences of the Urban History Association, the Pacific Coast Branch of the American Historical Association, Western History Association, Association for Asian American Studies, Organization of American Historians, American Historical Association, Society for the History of Foreign Relations, and the American Studies Association. Ayuko Takeda was an amazingly organized and diligent research assistant in the National Diet Library in Tokyo. Jennifer O'Donnell provided translation, as did my incredibly generous colleague Yoshie Kadowaki. Finally, the wonderfully careful and critical readers for Stanford University Press and University of Chicago Press helped me rethink and reshape the book in vital ways.

The University of Chicago Press has been absolutely marvelous to work with. My friend Charlotte Brooks and former professor Jim Grossman paved the way for me at the Press, for which I am truly grateful. Tim Mennel has been a gracious, supportive, and critical editor, while Rachel Kelly has been patient and responsive in moving the book toward publication. I thank the series editors, Lilia Fernández, Timothy J. Gilfoyle, and Amanda I. Seligman. I especially want to thank Becky M. Nicolaides, whose careful scrutiny of the manuscript produced insights, suggestions, and critiques that have vastly improved this book.

Finally, my friends and family have been my anchor throughout. My bestie, Michella Rivera-Gravage, is family, and her intelligence, creativity, and warmth are my consistent inspirations. Here in Reno, Emily H. and Felicia Perez first made Reno seem like an actual home. Lara Braff, Sandie Stringfellow, Heidi Duran, and Elizabeth Wayne are lifelong buddies. My late grandparents, Ethel and Takashi "Paul" Oda, Mildred and Masao Kato; my late Aunty Carol and Uncle Mel Teramoto; Uncle Roy and Aunty Lorraine; cousins Sherrie, Christine, and Elsa and their families; Aunty Laraine Oda; Uncle Cy, Aunty Jan, Angela, and Laurie Oda; Paul Teramoto; Don and Kathy Teramoto and the 5 Ts; Kurt, Dorothy, Chloe, and Haley Teramoto; Lance, Deana, Leah, and Caileen Teramoto; Aunty Linda Kalua; Aunty Amy and Uncle Wes Harano (who introduced me to foodie Honolulu!) and all of my aunties and uncles up and down California and in Hawai'i make up my own Pacific world. The Stanleys, Levins, and Clauders welcomed me into their family and made me feel at home: Sandra Clauder; Fay and Myron Levin; and Zac, Kim, Stella, and Townes Stanley. My partner, Jared Stanley, is everything I could have thought to hope for. There's no one who is as fun to be around. His wide-ranging intellectual curiosity, his own work, and his unwavering support of mine has sustained me for years and will for the rest of my life. Eleanor came into our life at the busiest time, just as I was going through the last series of revisions. She has been a hilarious, fun, troublesome, jaw-droppingly smart little addition to my life. While she has done nothing but hinder this book's progress, she just makes life right. Finally, I can't conceive of all the ways in which my parents, Frances and Terry Oda, have offered support and love over the years. There's almost no sense in trying, so to them, as to everyone else, I can only say: thank you.

INTRODUCTION

Tomoye and Henri Takahashi were doing quite well for themselves by 1959. They owned a flourishing business, the Takahashi Trading Company, which sold imported Japanese housewares to eager San Franciscans. This was quite a step up for them. The Takahashis and their growing family had been incarcerated during the war in the desolate Topaz detention center in Utah. When they returned to San Francisco, Henri, a minister's son and a journalist by trade, could find no newspaper willing to hire a Japanese American and so worked for a period as a dishwasher. Soon, however, the Takahashis started their business by sending care packages for Japanese Americans to relatives in war-devastated Japan and importing small, cheap, colorful trinkets such as origami paper, wooden toys, and brilliantly colored carp banners. From this, they moved on to custom-made lacquerware dishes, washi-paper lamps and screens, uniquely shaped bamboo baskets, and other items hungrily consumed by midcentury US shoppers enamored with Japanese design and seeking "modern . . . goods that will fit into any home."[1] Their business responded to and fostered a growing US fascination with Japan, thriving at the intersection of Japanese and US commerce. Tomoye—who, unlike Henri, was fluent in Japanese because of determined migrant parents, twelve full years of language study, and a major at the University of California, Berkeley—negotiated with suppliers in Japan. This involved convincing dubious Japanese manufacturers to accept her authority as a woman, and to depart from centuries of established styles to experiment with bold new colors and designs. Tomoye and Henri's work therefore brought innovative Japanese goods to US markets, made Japanese producers responsive to US consumer desires, and produced a highly successful company. By this point, the Takahashis' profits had also bought them a spacious home with sweeping views of the surrounding mountains and bay in San Francisco's residential Richmond district. The move had not been easy: they had desegregated a once lily-white neighborhood, leading to at least one racist episode in which a neighbor shot through a window one evening while Tomoye was at home alone. Shaken but undeterred, they remained. Indeed, Tomoye earned her real-estate license and assisted a veritable migration of Japanese and Chinese American families into the district. Their proceeds also eventually funded a charitable foundation to support Japanese American community organizations and Japanese cultural initiatives in the Bay Area.[2]

The Takahashis successfully maneuvered transpacific networks to wealth and prominence, but those same networks, put to use by others, also presented them with a serious challenge. In 1959, they were forced to relocate their business, ironically, for the Japanese Cultural and Trade Center, built as part of the city's first redevelopment project. This was a big change. It severed the business's deep Japanese American roots—roots that had been important even if their customer base was primarily white: their enterprise had begun with family care packages, and their original location was in the Japanese American residential and commercial enclave in the Western Addition district, generally called Japanesetown or Nihonjinmachi by its residents. This location had meant they had to work hard to attract white San Franciscans, who did not view the neighborhood with the same touristic interest as Chinatown. Furthermore, Japanesetown had been transformed by the war. Its Japanese American residents were evicted into inland detention centers, and African Americans had settled in. It was one of the few neighborhoods available to nonwhites and before the war already had a small black community, although for decades black San Franciscans had found far fewer residential restrictions than Asian Americans. By the late 1950s, the area was bustling, crowded, and multiracial, with black barbers and shops squeezed alongside Japanese groceries and hotels, and altogether far more residents than originally intended. These conditions helped make the neighborhood the city's first project of its new redevelopment program, and the Redevelopment Agency proposed a Japanese center as its crown jewel. As a result, the Takahashis, along with thousands of their neighbors, were once again uprooted by transpacific dynamics: this time, the close postwar US-Japanese alliance and the economic and political connections it generated.

Although the Takahashis would likely never have predicted it, US ties with Japan had moved to the heart of San Franciscan civic life. This was a far cry from their return from Topaz in the mid-1940s, when Henri was turned away from job after job because of lingering anti-Japanese sentiment. By the late 1950s, city boosters, municipal leaders, and ordinary San Franciscans had embraced San Francisco's self-styled identity as the Gateway to the Pacific, the link between the United States and the Pacific region. The "Gateway to the Pacific" was a ubiquitous phrase in the Cold War metropolis. It bedecked phone books, Chamber of Commerce publications, municipal reports, festival banners, and newspapers, and accented the speeches of city officials and business leaders alike. It was also much more than a nickname or motto. As expressed by Mayor George Christopher in 1958, it was a kind of optimistic geography: as "the gateway for trade, communication, and transportation with the Pacific nations," San Francisco was tied to, and could benefit from, "the very rapid economic growth of the Pacific Coast States and the Pacific area."[3] Anchoring San Franciscans' imagined Pacific was Japan. The nation was beginning its period of high-speed growth, having reached a 10 percent rate of economic increase that would remain in place for well over another decade. This "'unequalled' Japanese boom" was an irresistible lure to municipal and business leaders in a city long connected to Japanese and Asian ports.[4]

From its beginnings in the boosterism of a handful of civic leaders, the Gateway to the Pacific identity evolved into a transpacific urbanism that characterized San Francisco after World War II.[5] As sociologist Louis Wirth defined it decades ago, urbanism is that "complex of traits which make up the characteristic mode of cities," a "way of life" that distinguishes cities as a social, economic, cultural, and political whole.[6] San Francisco's Cold War iteration revolved around its relationship to the Pacific region and especially Japan. This was an urbanism—truly, an urbanity, according to the word's dual definitions of sophistication and urban life—that described a cosmopolitan municipality shaped by Asian influences, a newfound embrace of the city's Japanese and other Asian Americans, an economy naturally linked with Japan, and a physicality complementary to transpacific partnership. Expressions of this urbanity could be found in City Hall or a Western Addition apartment, downtown boardrooms or a small community shop, the modernized port or a neighborhood's bulldozed plot of land. This transpacific urbanity built on the metropolis's rich history of Pacific transportation and commerce as the West Coast artery between the United States and sought-after Asian markets, and it also responded to rising urban competition, especially from Los Angeles. Such competition encouraged boosters to develop all sorts of innovative, extra-market ways to naturalize and expand their relations with Japan. Amplifying a growing expanse of private and commercial Japanese-influenced institutions, these civic projects included informational organizations, business tours of Asia, a vibrant sister-city affiliation, a host of cultural activities, and, most significantly of all, transformations in the built environment. Together, these projects, institutions, events, and activities created a new "way of life" in San Francisco that contextualized and facilitated the city's economic and political transpacific connections with historical precedent and popular support. City leaders and ordinary San Franciscans alike shaped the metropolis's transpacific urbanity that framed a city oriented to Japan.

Japanese Americans, more than others, contributed to their city's transpacific urbanity in symbolic and concrete ways. This urbanity in fact hinged upon their relatively small population, whose migratory antecedents and enclave grounded and justified the metropolis's new relationship to the Pacific region. In ways that would have astounded the Takahashis as they returned from their inland incarceration, the development of San Francisco's urbanity centered the once-marginalized group in civic life and invested them with a celebratory significance previously unthinkable. Their presence provided the foundation for the redevelopment of their enclave, but also the larger rehabilitation of the city's transpacific economy, reorientation of municipal government, and celebration of their nation of origin.

This book traces the development of San Francisco's postwar transpacific urbanity in order to tell the story of the city's relations with and ideas about Japan in the decades following the Pacific War, and to argue that those were made within and remade the familiar, local sites of civic life, neighborhood, home, and identity. This urbanity took many forms, from attitudinal shifts among city leadership, new municipal institutions, celebrations, and expanded employment opportunities, but most intimately felt were transformations in the built environment. This, too, had varied expressions, but the axis was the construction in the Western Addition's Japanesetown of the Japanese Cultural and Trade Center that uprooted the Takahashis. The Japanese Center brought together the myriad other economic, cultural, civic, and racialized elements of San Francisco's urbanity into one institutionalized form. As such, the center and the urbanism it expressed was the result of choices and events in disparate places around the Pacific. In San Francisco, Pacific-trade businessmen rallied their city to revisit its long-standing Japanese commercial and cultural relations, while a sympathetic mayor and redevelopment officials turned to the tools of urban renewal to remake their cityscape along sympathetic lines. Japanesetown merchants, anxious to save their businesses from redevelopment's bulldozers, offered up proposals for a Japan-themed shopping center that lent officials the inspiration and the location for a hub of Japanese exchange. These officials secured a developer from Honolulu, Masayuki Tokioka, who came to the Japanese Center on a well-worn path of capital and sugar linking San Francisco and Hawai'i. Tokioka, a spectacled, slight Japanese American banking executive, exercised his connections with Japanese businesses and political figures—fostered in Hawai'i's shifting racial politics and pivotal location in transpacific travel—to build and tenant the center with Japanese companies. His ideas were considered in boardrooms and offices in Tokyo and Osaka, where Japanese executives strategized their own ways to secure international opportunities in the heady domestic context of rapid economic growth.

These decisions and partnerships precipitated transformations that reverberated throughout San Franciscans' everyday lives. Most directly, the Takahashis and thousands of their Western Addition neighbors and businesses were forced to move in order to make way for the Japanese Center and the surrounding developments. What places were open to them and what kinds of resources they could draw on for assistance, though, were also affected by the city's transpacific ambitions. Japanese Americans, who had struggled to make inroads into jobs long closed to them, found newly receptive employers and fresh business opportunities at the crux of white American interest in Japan. The context of friendly relations between Japan and the United States transformed Japanese American status, providing them new, if highly constrained, prospects in work and residence just as their African American neighbors struggled against rising barriers. Black San Franciscans reclaimed their homes in the Western Addition and their place in the city without the lubrication offered by valued migratory connections and while disproportionately burdened by renewal efforts. African Americans' strategies and proposals, even though they overlapped to a significant degree with those of Japanese Americans, therefore took on an opposing valence in the context of the city's transpacific priorities. The process of redevelopment and the broader regional context amplified racialized divergences while it drowned out commonalities in goals and strategies between the neighboring communities. The construction of San Francisco's transpacific urbanity—which shaped the homes and livelihoods of thousands of people living in the Western Addition, reshaped municipal priorities, reframed city business, and created new routes to prosperity and careers across the city while closing off others—was the product of events, policy, and decision-making across the Pacific as much as local conditions and ideas.

San Francisco's story is an inherently local one, in which planners, officials, and residents used new urban tools to carve out solutions to perceived problems in their streets, communities, and services and to connect their city to the Pacific world. These bifurcated goals were of entirely different scales yet were understood as inextricably linked. For San Francisco city officials, executives, and planners, the infrastructure of daily life and metropolitan functioning were expressions of and facilitators for their city's status, which in turn hinged upon exchanges with Japan and other places around the Pacific Ocean. These local actors therefore worked to deepen, extend, or at least publicize those transpacific exchanges. At times, this work meant aligning municipal goals with national foreign policy; at other times it meant departing from national agendas for more agreeable local priorities. It also required cultivating links that could elevate the city's economy while reallocating resources throughout the population. For Japanese American residents, the commercial, political, and even social links traversing the ocean were currents upon which they could navigate toward better homes, jobs, or civic position. Their activities both furthered the city's transpacific economic and cultural exchanges and reaffirmed their own vulnerability to international relations. In turn, the transnational networks cultivated by both elite and ordinary San Franciscans were informed by conditions close to home, such as the city's changing population, the shifting status of Chinatown in city life, and urbanist ideas produced in municipal offices and the academy.

This study of San Francisco's transpacific linkages parallels David Eltis's characterization of places in the Atlantic world, another oceanic region created in the movement of people, commerce, and ideas, which is far more studied than the Pacific: San Francisco was "fundamentally different from what [it] would have been without participation" in Pacific networks.[7] *The Gateway to the Pacific* therefore builds on emerging scholarship that has excavated these networks to show how disparate, multilingual, and multicultural countries, polities, and colonies, fractured by empire and only minimally unified by settlement, could be configured into, if not one, then a connected set of regions. Historians have begun to outline the Pacific Ocean in analytic terms analogous to the Mediterranean, Atlantic, and Indian Ocean worlds, terms that argue for the Pacific's long underexamined

importance in shaping the contours of US history.[8] Most of this historical work, however, focuses on earlier periods, from the eighteenth century
through the early twentieth, as trade, migration, and ideas consolidated
into established rhythms and the United States had not yet made the ocean
the postwar "American lake," as one British ambassador sardonically described it.[9] The Cold War period, in contrast, was a time of global US power
that was partly solidified through the definition and control of large parts
of the Pacific region. The Pacific world of this book, on one hand, was a
consequence of "movements and transits" that connected far-flung places
and, on the other, consisted of frameworks of knowledge and interpretation
through which the Cold War United States understood and therefore acted
in the region.[10]

   This, then, is a story of the making of the *US* Pacific after World War II,
an imperial process in which Americans selectively mapped the region,
magnifying the significance of some parts while suppressing the importance of others to create a Pacific world defined by foreign relations and
economic interests. This mapping was an "[invention] and [construction]
of the powerful" that helped justify increasing US intervention in Asia and
supported the consolidation of US power in the postwar years.[11] This process can be seen, for instance, in Lyndon B. Johnson's 1966 speech rallying
Americans behind the growing US involvement in Vietnam. Johnson argued
that the United States as a "Pacific power" had an obligation to foster the
"international trade," "free flow of peoples and ideas," and "full participation by all nations" that would sustain "the peace we seek in Asia."[12] In
many ways, his position was a culmination of the "Pacific Era" predicted in
1903 by Theodore Roosevelt—whom Johnson referenced in his speech—in
which the United States gained global status through "the commerce and
the command of the Pacific."[13] But clearly, Johnson's was also a specific
interpretation of a region articulated by a US president seeking to balance
Chinese and Soviet competition and prop up Japan, a critical US ally whose
booming economy knit economic relations together throughout Asia and
relied on resources from Indochina.[14] As a result, Johnson's Pacific was one
in which the thousands of Pacific islands were relegated to, at most, a poi-
gnant memory of World War II battlefields and stratagems, Pacific island-
ers had been all but forgotten, and the "Pacific" was now largely defined
by a handful of geopolitically important nations and colonies north of the
equator. San Franciscans were a part of this redefinition as they reimagined
Japan and Hawai'i, and other places around the ocean, in ways that made
them more "porous" to US trade, migration, politics, and ideas.[15]

Exploring the transpacific history of San Francisco requires attention to the everyday realms of streets and neighborhoods, sites underexplored in most studies of the United States in the world more broadly yet precisely telling of the ways in which movements across borders and around oceanic worlds reshape people's everyday lives.[16] As sociologist and historian Charles Tilly reminds us, cities have been the historic juncture for transborder mobilities in goods, people, and ideas and so are, in many ways, an ideal subject for investigation into the local, intimate consequences of these global processes.[17] In San Francisco, transpacific networks and goals both emerged from and reshaped where and how people lived, in elemental ways.

The crucial lever for these transformations was urban redevelopment. The program was conceived as a way to correct the city's postwar challenges of infrastructure, economy, and population, difficulties shared with postwar cities across the country. City leaders exuberantly responded to the legislation and money emerging from Washington as well as the ideas about the built environment and civic health sprouting up in new planning departments in cities in the United States and, indeed, around the world. This optimistic moment refined ideas about urban environments promoted at the turn of the century by Progressive reformers, and sharpened them into financial, legislative, and aesthetic tools to both define and address postwar problems. These tribulations included aging infrastructure, a shrinking tax base, changing local economies, overcrowding, overburdened social services, and suburbanization. As the encapsulation of them all, in many senses, officials focused on the issue of blight to direct intervention. Planners rendered blight as objective and quantifiable, easily distilled into statistics and visualized on maps. As historians have shown in other cities, however, its definition and measurement were suffused with assumptions about race and class. It was no coincidence that city planners' ideas met with newly honed federal tools and willing municipal implementation at the same time that northern and western US cities saw a massive migration of millions of African Americans from the South. This response was more complex than blunt racism, although certainly in San Francisco as elsewhere that could be found. More pervasive and ultimately more consequential was how race and class shaped ideas about who was deserving of social services, how tax burdens were understood, who was San Franciscan and who was alien, what financial pressure was debilitating and what was an accepted part of urban life, what economic growth meant and where to look for it, and what defined community. As elsewhere, rapid demographic change—in which a small population of about 5,000 African Americans grew to over 40,000 in just ten years, and continued to grow into the largest nonwhite population in the city—informed how urban problems and goals were conceived.[18] Furthermore, redevelopment programs were not only informed by ideas about race. They produced intensified if uneven racial segregation, heightened boundaries of class and income, and new definitions of urban desirability and cooperation that in turn shaped terms of inclusion and forms of racial thinking with lasting effects.[19]

San Francisco's racial terrain, however, was fundamentally different from that of the midwestern, southeastern, and northeastern cities that form the core of postwar urban scholarship. Historians have evocatively demonstrated that the color of California and other parts of the West was dissimilar in its multiracial hue from much of the country—indeed, prophetically so—and this condition shaped different kinds of politics.[20] In San Francisco, people of Asian descent had long been defined as the principal social problem, and the city had been the headquarters of the anti-Asian movement through World War II. Therefore, when the city sought to reshape its built environment, the legacy of Asian exclusion as much as black migration and transpacific ambitions imprinted the infrastructure, buildings, designs, and resultant social geography. City leaders as well as district residents understood urban problems and possibilities in the Western Addition in the context of Chinatown and historical Asian segregation, the ethnic diversity of the white population, the in-migration of black southerners, and even the growing Latino community in the Mission district.

San Francisco's multiracial context informed the development of the city's transpacific urbanism and Japanese Americans' place in it. This is a story of the built environment in which Asian Americans play a pivotal role.[21] In planning discussions, ideas about design, and choices in residence, they reshaped what the city looked like and how civic inclusion functioned. Japanese American small-scale proprietors, working-class renters, property owners, developers, and wealthy financiers from the mainland and Hawai'i not only reinterpreted what the ethnic residential and commercial enclave in San Francisco could be, but also what being Japanese American meant. This book therefore does what historian A. K. Sandoval-Strausz has done for Latinos: centers Asian Americans in postwar urban history.[22] This demographic group is currently the most urbanized in the nation, which should have, but largely has not, forced historians to consider the ways in which, beyond the creation of enclaves, they reshaped US urbanism.[23] The creation of San Francisco's transpacific urbanity is just one example, as we will see; many cities along the West Coast crafted their own contemporaneous variants. The impact of Asian Americans therefore went far beyond what their relatively small numbers might imply; in 1950 San Francisco, for example, Japanese Americans were less than 1 percent of the population, compared to 6 percent for African Americans. Nonetheless, their participation, and their very presence, was the axis for the city's new urbanity that connected San Francisco to a geopolitically critical region and nation.

The story of redevelopment in Japanesetown also adds to the recent work of historians investigating the rapid postwar shift of Asian Americans from, as historian Ellen D. Wu describes it, "unassimilable aliens unfit for membership in the nation" to the United States' "most exceptional and beloved people of color, its 'model minority.'"[24] This literature locates causation for this shift in the pressures of Cold War foreign policy in Asia. Competition with the Soviet Union impelled the United States to appeal to Asian countries and intervene in them in new ways. This motivated white Americans to accept Asian Americans as neighbors and coworkers to an extent previously unthinkable, especially in the US West, a mere decade or two before. Abstract relations with peoples in Asia took on a concrete urgency during the Cold War as ordinary Americans embraced the global struggle in their everyday choices about mundane, yet consequential, aspects of their lives. White Americans, assuming that Asians abroad cared about Asians in the United States, adapted their segregation and discrimination practices accordingly. These reduced constraints still depended on perceptions of Asian American foreignness and conflation with their nations of origin, however; this continuation of "alien citizenship," created by a host of prewar legal and social exclusions, now supported limited social inclusion as Asia became newly significant to Americans during the Cold War.[25] This shift was not solely the work of white Americans. Asian Americans were more than willing to take advantage of the window opened by the Cold War (or even World War II before it) and portray themselves as both assimilative and foreign "model citizens" in ways easily captured by the state, academy, and media.[26]

This dynamic was evident in San Francisco. Certainly, many Japanese Americans in San Francisco assiduously portrayed themselves as cooperative and model citizens, and domestic and foreign-relations milieus were fundamental in reshaping opportunity for the minority. Yet the Cold War facilitated more than just an environment of international pressures and competition. It also fostered a mushrooming set of networks and exchanges, especially between the United States and Japan, within which Japanese Americans could work. An industrious set of Japanese Americans positioned themselves at the nexus of these relations and at the heart of US interest in Japan in order to work toward greater residential and economic inclusion. This hard-forged access depended on conflating Japanese Americans with Japan, the same racial thinking of "a Jap is a Jap . . . whether he is an American citizen or not" that had culminated in their internment during World War II.[27] In the context of friendly US-Japanese relations, however, this conflation reinterpreted Japanese Americans as a useful point of connection to Japan, a nation and culture that many white Americans viewed as fundamentally alien and opaque, if now friendly and alluring. Japanese Americans were understood as naturally suited to interpreting Japanese interests and culture for US audiences, an understanding that augmented their professional or entrepreneurial skills with claims to cultural knowledge. This allowed them to carve out a role for themselves, one that also cast the city as racially liberal, as intermediaries with Japan in the development of San Francisco's transpacific urbanity.

This book situates the development of San Francisco's transpacific urbanism in a history of networks and exchanges, in order to explore the ways that connections with Japan were adapted and transformed in conjunction with the city's built environment, civic life, and racial terrain.

Chapter 1 sets the stage of the story in the landscape and identity of the city through World War II. San Francisco's Gateway to the Pacific identity had a lineage that stretched back to turn-of-the-twentieth-century US imperialism, when conquest and interventions in China, the Philippines, Hawai'i, and elsewhere helped expand the city's national and global significance and its economic fortunes. But while Pacific linkages shaped the city boosters' sense of their metropolis, Japan and its diaspora played a relatively minor role in the city's built environment and identity.

Chapter 2 charts Japan's movement to the heart of San Francisco's Gateway to the Pacific identity and the beginnings of the city's transpacific urbanism. Local, regional, and global conditions reshaped how San Franciscans understood their place vis-à-vis the Pacific Ocean, Asia, and Japan in particular. A cohort of Pacific-oriented local businessmen responded to developments abroad and at home by seeking out new commercial opportunities in Japan's rapidly growing economy. But while some in the business community had a deep investment in Japanese networks, their interest was not shared by city leadership or by many San Franciscans. The Pacific-oriented cohort therefore turned to the sister-city affiliation to rehabilitate connections to Japan rendered dormant or troubled by the war, reshape the popular image of the nation in terms conducive to economic partnership, and foster San Franciscans' investment in Japan and ties to the metropolis. The affiliation helped center the once-maligned enemy in San Francisco's politics and civic life.

Chapter 3 follows the shifting gaze of city officials from the western horizon to San Francisco's streets and buildings. As interest in Japan rose in the city, municipal officials joined their counterparts across the country to develop a renewal program to reshape their city's built environment and, they hoped, future. Using the definitions of blight and urban progress developed by federal programs and academic urbanists, San Francisco officials began to locate their transpacific aims in the built environment. One group of Japanese American residents and business owners responded to both the redevelopment program that threatened their neighborhood and the municipal programs of Japanese exchange with their own Japanese-influenced proposal. Based in their recent experience with incarceration and expressed in the language of urban progress, these merchant-planners proposed a Japanese-themed mall for their neighborhood. Their proposals recast the racialized, segregated enclave and its inhabitants as integral civic participants aligned with the values and goals of city leaders and planners.

The Japanese American merchant-planners assiduously molded their plans to better align with the growth of popular local Japanese institutions, but in doing so they inadvertently exposed their neighborhood to official appropriation. The Japanese American group had sought to preserve their businesses and community, but officials redefined Japanesetown as evidence of a migratory connection to Japan and therefore a hub of the city's Japanese commercial and cultural ties. Chapter 4 examines how the heterogeneous actors involved—local Japanese American merchants, Redevelopment Agency officials, Honolulu financiers, and former territorial elites in Hawai'i—redirected transpacific linkages in the building of the Japanese Center. Successfully doing so, however, required the highly specific set of skills of Masayuki Tokioka, a Honolulu banker and developer. He had built his career at multiple crossroads, between white and Japanese in Hawai'i, between Japanese American and white, and between Japan and the United States, and deftly turned his professional and cultural skills toward his task. San Francisco's Japanese Cultural and Trade Center was built with the city's many ties across the Pacific, redirected toward its Gateway to the Pacific ambition.

Tokioka was just one of the many Japanese Americans building San Francisco's transpacific urbanism. Chapter 5 examines how Japanese Americans redefined themselves as intermediaries, interpreting Japan for white American audiences in the Japanese Center and elsewhere. This position played upon and extended Japan's newfound importance by reinterpreting contemporary popular and academic racial knowledge about Japanese Americans. Prewar conflation of Japanese Americans with Japan had justified their wartime internment; now, in the Cold War, that conflation supported a new understanding of their social and professional value. San Franciscan interest in Japan gave them a new forum to cultivate and demonstrate a cultural fluency that they could leverage toward greater economic and residential opportunity. This was a tenuous inclusion, however, based on continuing perceptions of their foreignness that also made them vulnerable to the shifting winds of foreign relations.

Chapter 6 examines the full scope of San Francisco's consolidating transpacific urbanity in the built environment and racial terrain. The Japanese Cultural and Trade Center was a part of a growing redevelopment project that took an immense if uneven toll on the Western Addition's homes and businesses, especially those of Japanese Americans and African Americans, the two largest nonwhite groups in the district. The process of redevelopment, however, highlighted a largely false division between Japanese Americans and African Americans, as both groups worked toward the same goals of community control, and adopted similar tools of protest and cooperation. Their responses were always in conversation with each other and with other racialized communities in San Francisco. But ultimately, the context and values of the city's transpacific urbanity carved a sharp distinction in public perceptions and redevelopment opportunities for the two long-standing Western Addition enclaves.

San Francisco's transpacific urbanity was developed over the course of decades with the participation—at times unwitting—of a broad range of residents and people overseas. The Takahashis, for instance, participated in the creation of this new urbanity with their imports that lent a Japanese accent to homes all over the city; its ultimate expression in the Japanese Cultural and Trade Center, however, forced the couple to uproot their store from its Japanesetown foundations and resituate it in bustling downtown Union Square, where it added Japanese character to the most recognizable retail space in the Bay Area. The Takahashis' story is a microcosm of the larger story of a city collectively if not cooperatively struggling with regional competition, responding to opportunities near and far, and reimagining its place in a global economy. The story of the Gateway to the Pacific is therefore a story that can be found with local variations across the nation. As San Francisco's example underscores, the imagination and connections of city boosters and local residents roamed much farther than municipal, even metropolitan, bounds. People in the Bay city, as elsewhere, experienced the world at home, and made sense of global shifts from the manageable and familiar places of home, community, and city.

CHAPTER ONE

Japan and Japanese Americans in the Pacific Metropolis through World War II

San Francisco's northern edge sparkled with acres of lights, bejeweled buildings, monumental statuary, and displays from all corners of the globe. Seamlessly integrated into the city's grid, the Panama-Pacific International Exposition of 1915 celebrated the opening of the Panama Canal, the "greatest physical achievement in history," and US transoceanic power. The canal heralded "a new era in commerce," even "a new era in civilization" in which the "circle is now fully circled; the West has met the East."[1] This referred to the west and east coasts of the nation, but it also referred to new US connections with the "nations of the Pacific area." For its San Francisco organizers, the fair showcased their city's "renaissance" after the earthquake and fire of 1906 and its capacity to be the "great gateway opening toward the newly-awakened Orient."[2] City boosters had therefore leapt at the opportunity to host the exposition and show that not only would the United States "dominate the politics and commerce of the Pacific," but that their city was integral to this supremacy.[3] Lawmakers agreed. Senators held that "one of the most cogent reasons" to award San Francisco the exposition was the "growing oriental trade," a "great field for exploitation" in which the United States should "occupy first place." San Francisco, the "greatest port on the Pacific," could facilitate such exchanges with the "greatest ease and facility" and "cement the ties of cordial friendship between America and the nations of the Far East."[4] The exposition flourished the United States' Pacific role, and it proclaimed San Francisco's as well.[5]

San Francisco boosters' transpacific vision of their city, however, crossed troubled waters. San Francisco had by this point sizable Chinese and Japanese migrant populations, enjoyed a rich trade across the Pacific, and was buoyed by military and political interventions in Asia. But transpacific relations could also bring complications. The two Asian nations with the largest diasporas in contemporary San Francisco, China and Japan, demonstrated this. The robust presence of both nations was critical for the city's Oriental bona fides. Yet anti-Asian sentiment hobbled boosters' demonstrations of their city's transpacific fluency, while underscoring Japan's marginal place in the metropolis.[6]

China and the Chinese diaspora in San Francisco confirmed the growing importance of Chinatown for San Francisco. For example, fair organizers created a Chinese Committee to ensure communication with Chinatown's elite. Committee members guided prominent, wealthy Chinese business owners and community leaders through the grounds to assuage any concerns. The majority of the Chinese population, certainly, still drew moral approbation, fears, and racism. But Chinatown's exotic presence was critical to the city's claim to Pacific cosmopolitanism. Promotional materials lauded a visit to Chinatown as "the trip most interesting to the tourist" and conspicuously featured the neighborhood's ornate facades in fair publicity.[7]

The exposition was important to Chinatown boosters, as well. This was an opportunity to revise their neighborhood's much-maligned status as well as that of the newly established Republic of China. They therefore carefully monitored representations of China and the local Chinese diaspora. China's official contributions included exhibits on education, art, and a reproduction of part of the Forbidden City.[8] However, the concession Underground Chinatown in the commercial Zone was far more popular. The tunnel-like exhibit imagined a dangerous, slum-like, and exotic Chinatown populated with opium dens, prostitutes, hatchet men, and gamblers. Chinatown leaders and Chinese officials were outraged and demanded that organizers "suppress the concession" as "a disgrace to the exposition and a slander upon the Chinese people."[9] The exposition organizers needed their participation, and so briefly closed the well-attended attraction in order to remove Chinese actors and rename the enterprise Underground Slumming. The deracinated exhibit remained a clear proxy for Chinatown, something attendees indicated with their continued good business. Yet, as historian Abigail Markwyn has noted, such actions indicated the organizers' willingness to address the concerns of foreign and local Chinese representatives.[10]

Japan's participation was complicated by geopolitics, local politics, and a much smaller and less visible diasporic population. Japanese officials had to grapple with the contemporaneous Alien Land Law, debated in 1913 just as

Japan was committing to the fair. The bill would prevent "aliens ineligible to citizenship"—a euphemism for Asian migrants, the only group prevented by law from naturalizing—from landownership and long-term tenancy; it aimed to halt the Japanese ascendance in agriculture that threatened white industry dominance. In response, Japan's foreign representatives wielded the upcoming exposition as a weapon, threatening to retract their participation if the bill passed. This prompted organizers, joined by President Woodrow Wilson, to vigorously lobby against the California bill, although they lost to proponents who prophesied miscegenation and social upheaval. The passage was a severe blow in Tokyo, where some newspapers even urged war. Nonetheless, the diplomacy of fair organizers helped keep Japan involved.[11]

The exposition organizers thereby ensured Japan's participation, but the episode revealed the troubled position of Japan's diaspora. This was reflected in the fair itself, where Japanese in the United States were absent. The Zone's Underground Slumming concession drew attention to Chinese in San Francisco, albeit in racist ways, and urged visitors to venture into the real Chinatown. In contrast, the Zone's Japanese concession, Japan Beautiful, represented a positive but strictly foreign portrayal.[12] There was, after all, no Japanese analog to Chinatown for the fair to caricature. As the Panama-Pacific International Exposition suggests, the Orient was central to San Francisco's identity as a Pacific city. But it was also unevenly represented.

This chapter explores the changing meanings of San Francisco's Pacific identity and its varying relationship to Japan and Japanese. In the early twentieth century, this remained a meager relationship. Overshadowed by China and Chinatown, both Japan and Japanese migrants had a slight imprint on both city identity and landscape. Japan would move to the center of civic life in the postwar metropolis, but this was little presaged in San Francisco until the end of the Pacific War.

JAPAN IN PREWAR SAN FRANCISCO

The city "on the border line . . . between the Occident and the Orient" had historically been the "Gateway to the Orient," as the Panama-Pacific International Exposition made clear.[13] However, this title had shifting referents. It originally denoted San Francisco's strategic, commercial, and transportation regional importance. Over the course of the twentieth century, it evolved into a domestic story of the city's cosmopolitan landscape and population. And through World War II, Japan held a distinctly marginal position in the city's landscape and civic identity.

San Francisco had long embraced an identity as the "Metropolis of the Pacific," a city with unparalleled reach into the oceanic region's "broad expanse."[14] The Gold Rush turned the Bay city into the first major US metropolis on the West Coast. Well before that, the Spanish empire and the fur trade bound what became San Francisco to ports all over the Pacific Ocean, from Lima to Honolulu to Manila to Canton and more. These links positioned the city so that a 1881 guidebook could reasonably include Japan, China, and the Sandwich Islands as San Francisco's "vicinity" destinations.[15] US imperialism, though, amplified San Francisco's Pacific prominence. The 1898 outbreak of the Spanish-American War in the Philippines was a major boon. The *San Francisco Chronicle* reported, "[M]ost of the supplies in the Quartermaster's line . . . will have to be obtained at San Francisco."[16] The city also had a flood of temporary residents as the troops waited for deployment in San Francisco, their point of embarkation. This was the precedent when foreign powers dispatched troops to quell the Boxer Rebellion in China in 1900. Boosters predicted that "should the war in China continue . . . San Francisco would become an important shipping point for supplies" for all nations involved. California "has become known as a great source of supply the world over of late, and more especially so since the beginning of the war in the Philippines," and San Francisco, as "the nearest point," would benefit the most.[17] By the end of the bloody Philippine-American war, San Francisco had consolidated "an importance formerly undreamed of" as a "vital" staging and supply point for "operations in the Pacific."[18] Furthermore, the four-year-long war had been fought in large part in order to secure US commercial interests in China, opening all of Asia to the "commercial advance of America."[19] As the westernmost major US port, San Francisco was the "natural shipping point" that connected the "commercial centres of America to every corner of the world's greatest ocean."[20]

Over the course of the early twentieth century, this interpretation of the city as a commercial and strategic link between US and Asian markets, investment, and resources gave way to a domestic narrative of the city's history and demographics. No word was more frequently used to describe the city than "cosmopolitan." This shaped accounts of the city's history. Distinctive from the European roots of other US cities, San Francisco began as a "drowsy Spanish hamlet" that also had Russian antecedents.[21] Narratives abounded of the diverse masses of people drawn by the Gold Rush: "Kanaka, Indian, Filipino, Chinaman, Japanese, Lascar, Hindu, Sikh,

Greek, Roumanian, Turk, Italian, Spanish, Portuguese, Swede, Norwegian, Hollander, Frenchman, Mexican, and half a dozen others."[22] Its place as a "strategic point on the international highway" remained, but by the interwar years its commercial and strategic links were subsumed to its status as a "peculiarly cosmopolitan city."[23]

While many contemporary US cities had even larger proportions of immigrants, the "influx of Orientals" made San Francisco distinctive.[24] No single feature was more critical to the city's cosmopolitanism than Chinatown, as the 1915 exposition suggested. The neighborhood came to define San Francisco. As a central attraction for visitors and an integral part of the city's identity, almost every city guidebook and article referenced, and usually pictured, the neighborhood. As one journalist noted as early as 1876, "to visit San Francisco and not spend some time in the Chinese quarter, is equivalent to visiting Rome without seeing the Pope."[25] It was uniquely San Franciscan: there was "nothing like it in any other part of the country," and "there can not be anything like it in China," either.[26]

Chinatown reflected and shaped San Francisco's identity throughout its history. Anti-Chinese animus birthed the neighborhood and defined its exotic allure through the early twentieth century. As historians have richly demonstrated, the neighborhood began as a segregated enclave for a persecuted minority. Morality, disease, gender, sexuality, and race molded perceptions of Chinatown's inhabitants as contagions to be by turns contained, excluded, policed, and exterminated. But the perceived menace also tempted visitors. Early excursions were made "in the company of a detective," a testament to Chinatown's twinned associations of danger and fascination.[27] This changed after the destruction of the 1906 fire. Entrepreneurial residents rebuilt the "exotic city-within-a-city" with tourism and popular acceptance in mind.[28] Architects and property owners crafted storefronts with pagoda-like facades based on the well-liked and colorful "Chinese Villages" of world's fairs. Community leadership combined this "vastly improved" and "inviting" streetscape with anti-crime campaigns to promote a "New Chinatown" suitable for even family visits.[29] As these changes heightened touristic interest in the neighborhood, and after anti-Chinese campaigns ended migration, the city increasingly "took its Chinese colony to its bosom."[30] Journalists began to note a "modern trend in Chinatown" and portray its residents in less foreign terms.[31] The sheer numbers of Chinese Americans in San Francisco lent them some political clout by the 1930s, as local Democrats wooed them into their New Deal coalition. Later, Chinatown leaders cooperated with the war effort and postwar municipal reforms; they also promoted family and gender normativity. This encouraged postwar observers to note the neighborhood's exotic color but also to describe its "new generation of Americans." These included "bobbysoxers," "M.D.'s from Stanford," and "ex-GI's," some of whom lived or worked outside of the enclave.[32] By the 1950s, Chinatown was regularly used as evidence of how San Francisco "squelched" racial prejudice.[33] The neighborhood was a pillar of the city's identity, first of cosmopolitanism and then of tolerance.[34]

There was no Japanese or Japanese American equivalent to Chinatown in either its popularity or its centrality to civic identity, but Japan was not absent from San Francisco's landscape. The main landmark was the "famed and favorite tourist mecca" of the Japanese Tea Garden in Golden Gate Park, the only site marked as Japanese in guidebooks or media.[35] The garden and its buildings had originally been constructed as part of the "charming" Japanese Village in the California Midwinter Fair of 1894.[36] The village's one acre was molded into a "correct representation of Japanese architecture and landscape gardening."[37] After the fair closed, the grounds were turned over to the Parks Commission as one of the first permanent Japanese gardens in the United States. It remained as a garden, gift shop, and teahouse run by the Hagiwaras, a Japanese American family, for the next fifty years. Stuffed with miniature trees, a drum bridge, pavilions, and stone lanterns, it provided visitors with a fantastical vision of Japanese landscapes ten years before Chinatown was remodeled into an "Oriental city [of] veritable fairy palaces"[38] (fig. 1.1). Yet even the tea garden was not without ambivalence. The Midwinter Fair concession had been awarded to an Australian Japanophile, not a Japanese. This provoked Japanese consular officials and the migrant community to call for a boycott of the exhibit.[39] The city's Japanese icon had origins in transpacific contention.

Commerce brightened transpacific tensions. Japan was one of San Francisco's major foreign partners in a trade-dependent economy, although transpacific commerce was not overly significant in prewar years. Through the 1930s, according to a state senate report, coastal and domestic trade with eastern cities was the "base of Bay Area shipping activity," with twice the tonnage of foreign trade.[40] Furthermore, the Pacific city remained well behind New York City, the US commercial capital, in Asian and Japanese trade. Nonetheless, trade with Asia was not insignificant. Between 1900 and 1940, the region accounted for 49 percent of the city's foreign trade. As it did nationally and even globally, the China market received considerable attention from local businessmen. The extent of actual trade between

FIGURE 1.1 The Japanese Tea Garden's fantastical vision of Japan in 1935, cultivated by its stone lanterns, bridge, pagodas, and miniature plants. This was the sole representation of Japan in the San Francisco landscape through the first half of the twentieth century. Photo ID# AAA-7813, San Francisco Historical Photograph Collection, San Francisco History Center, San Francisco Public Library.

San Francisco and China, however, was limited and highly imbalanced. In 1938, the city imported goods worth $4.5 million from China, less than the Philippines' $7.7 million and Japan's $5.4 million. In exports, China ranked only ninth with $2 million. This was far below Japan, whose trade (exports of $24 million) was second only to that of the United Kingdom.[41]

Japan, in fact, was one of San Francisco's preeminent foreign-trading partners for much of the pre–World War II twentieth century. Its trade peaked in 1929, declining slightly after the global depression and Japan's resource-rich colonial conquests. A high volume of exchange nonetheless continued despite the rising animosity between the two empires. In 1939, Japan ranked second in both exports and imports, making up almost one-sixth of the total value of $193 million in the city's foreign trade. This commerce was supported by a host of Japanese trading companies and banks, more than were located in any other West Coast city for much of the century.[42]

This robust economic relationship coexisted with venomous local hostility toward Japan. In the words of historian Roger Daniels, the city was "the mecca of the movement" against Japanese migrants in the United States.[43] Early migrants faced an arsenal of anti-Asiatic organizations and beliefs, informed by preexisting anti-Chinese animus and institutions. From laborers to mayors, San Franciscans turned these weapons on Japanese migrants. The *Morning Call*, followed later by the *San Francisco Chronicle*, led the charge in 1892 with a "crusade against Japanese contract labor," when there were just one thousand such workers in the entire state of California.[44] Mayor James Duval Phelan slandered them in 1900 as "not the stuff of which American citizens [could] be made."[45] Labor unions organized to drive the migrants out of industries where they were seen as competition, and the Anti-Jap Laundry League poured hundreds of dollars a month into boycotts of Japanese laundries. These dynamics consolidated in the Asiatic Exclusion League in San Francisco, a centralized, statewide organization committed to Japanese, and later South Asian and Korean, exclusion.[46]

San Francisco's anti-Japanese racism—locally legislated, socially sanctioned, and supported by violence—shaped US-Japanese relations. In 1906, the Board of Supervisors voted to segregate Japanese students, precipitating an international crisis. Japanese officials and President Theodore Roosevelt both strenuously protested the supervisors' actions. Roosevelt was keen to avoid conflict with a proven military power, demonstrated by Japan's recent victories over China and Russia. The resulting diplomatic negotiations allowed Japanese students to attend integrated schools, unlike Chinese students, whose representatives could call upon no similarly powerful nation as a recourse against discrimination. Negotiations also ended Japanese labor migration. In 1920, former mayor Phelan revitalized the anti-Japanese movement with the formation of the Japanese Exclusion League. Together with other state organizations, league members agitated for their long-cherished goal of complete exclusion. The xenophobic and isolationist post–World War I atmosphere gave them traction. Congress responded with the 1924 Immigration Act, which, among other new restrictions, finally enacted full Japanese exclusion and sparked outrage in Japan. Finally, the Western Defense Command headquartered at the Presidio spearheaded the World War II incarceration of Japanese Americans, lobbying for and eventually organizing the coastal eviction of noncitizens and citizens alike. Local media facilitated this process. The city's Hearst-owned newspaper was one of the policy's most vocal advocates. San Francisco–based anti-Japanese activity was nationally influential.[47]

Japanese migrants and, occasionally, the Japanese government vigorously denounced and opposed all these prejudicial actions, from labor opposition to the World War II incarceration. Japanese workers organized their own protective labor organizations in reaction to boycotts and exclusion from mainstream unions. Migrants filed a lawsuit against the segregation of Japanese students, and held mass meetings against immigration restrictions as well as the Japanese government's handling of the negotiations. Finally, overcoming both fear and panic in the wake of Pearl Harbor, many Japanese Americans in the Bay Area publicly opposed incarceration and the racialized assumptions underlying it. Local publisher James Omura criticized the "punishment" of Nisei for their Japanese "outward features" as "unjust and un-American," comparing it to the actions of the Gestapo.[48] Others pressed the Justice Department to set up hearing boards to verify loyalty on an individual, rather than group, basis. Despite these protests and challenges, anti-Japanese animus remained a defining feature of San Franciscan attitudes toward Japan and Japanese.[49]

Anti-Japanese hostility was an important part of San Francisco's past and the nation's history. However, it was not as central to the city's identity as its anti-Chinese lineage, which in many ways framed subsequent movements and created Chinatown. Japan and Japanese Americans loomed large in San Francisco's politics, but not in its landscape or identity.

A SEGREGATED POPULATION

San Francisco's anti-Japanese animus had national and international consequences, and it also had local effects. Aside from the contentious Japanese Tea Garden, racism shaped the other Japanese feature in San Francisco's landscape, the Japanese American residential and commercial neighborhood in the Western Addition. While known to San Franciscans, the neighborhood was less infamous than Chinatown and nowhere near as central to the city's identity. This was partially because it was smaller. In 1940, there were only 5,280 Japanese, compared to 17,782 Chinese. It was also no attraction. Only the most observant guidebooks mentioned the Japanese American restaurants, shops, churches, and homes near Post and Buchanan. No maps indicated its location, and its very name was changeable. Its Japanese residents called it "Nihonjinmachi" (Japanese People Town) or "Japanesetown," while others used the terms "Japanese district," "Little

Osaka," "Little Tokyo," "Japantown," "Japtown," or "the Japanese streets east of Fillmore."[50]

Despite its lack of notoriety, a Japanese enclave had existed in the Western Addition district since the turn of the twentieth century. Japanese migrants had originally clustered near the ports in the South of Market and Chinatown areas, as well as in the Western Addition. They consolidated in the latter after the 1906 earthquake and fire decimated much of the city but left the Western Addition untouched, joining other displaced San Franciscans who flooded into the district (fig. 1.2). Newcomers rapidly converted middle-class homes into multi-unit housing with added storefronts and new industry. The tide ebbed as other areas rebuilt with new, more spacious accommodations, but exclusions kept Japanese residents from these more desirable locations.[51] Months after the fire, the *San Francisco Chronicle* reported that portions of the district were "now infested by Japanese."[52] By 1910, there was a recognized "Japanese district," and by 1921, it was bounded by municipal policies as well as real-estate practices. The development of the city's first comprehensive zoning policy that year included discussion of an intentionally discriminatory provision to prohibit stores in the Japanese

FIGURE 1.2 Businesses and storefronts on Geary Street attest to the rise of a residential and commercial "Japanese district" in the Western Addition after the 1906 earthquake and fire, circa 1910. BANC PIC 1905.02682A, Photographs of Agricultural Laborers in California, circa 1906–1911, BANC PIC 1905.02634-.02731-PIC, © The Regents of the University of California, The Bancroft Library, University of California, Berkeley.

neighborhood in an attempt to "keep the Japanese population where it is at the present time."[53] Efforts were also made to exclude Japanese from any residential zones, but this was refused by the City Planning Commission and the Board of Supervisors as being in violation of the current US-Japanese treaty.[54] The *Chronicle*'s editors shrugged off the loss: the "least said about the Japanese for a time the better," because "any city will 'zone' itself if let alone."[55] And indeed, what historian Charlotte Brooks has called San Francisco's "racial tradition" of informal social practices, legal tools, and violence was eminently successful in containing Asian residence to a handful of neighborhoods and districts.[56]

San Francisco's segregationist traditions maintained Chinese residence with the most steadfast boundaries up through World War II; Japanese residents found restrictions slightly more permeable. Within the neighborhood's borders, heterogeneity reigned, unlike in Chinatown. As resident Sam Mihara remembered, it was "normal to see various people of various backgrounds."[57] Of the 17,229 people who lived in the census tracts encompassing Japanesetown's twenty blocks, only about a quarter were of Japanese descent. George Matsumoto, who grew up in its confines, remembered some blocks as "almost all Japanese," but his immediate neighbors included "maybe one or two black families; . . . two or three families of Filipinos; the rest . . . white."[58] Nonetheless, its Japanese residents gave the neighborhood its identity. By 1940, the Japanese district included twenty blocks and housed about 4,500 of the 5,280 people of Japanese descent in San Francisco. In fact, some found San Francisco's restrictions so onerous that the Japanese population dropped by one-sixth between 1930 and 1940, as they left for more hospitable environments.[59]

The neighborhood's businesses also defined the neighborhood as Japanese. The most numerous occupation was domestic labor, which engaged almost 40 percent of all employed Japanese Americans and distributed those who "lived in" all over the city. But most Japanese Americans lived, shopped, worshipped, and ate in the enclave. There, a number of small merchants met their fellow ethnics' daily needs. By the outbreak of World War II, there were about 250 of these proprietorships, including groceries, pool halls, barbershops, and restaurants that catered not only to the Japanese community but also to neighboring Filipinos, Chinese Americans, and African Americans. But this cluster of businesses did not represent the entirety of Japanese business and employment. One 1909 account showed about one Japanese-owned establishment for every nine Japanese residents in San Francisco, although labor organizing by whites eventually reduced those numbers. One sector that grew over time was the import stores on Grant Avenue, the main commercial strip of Chinatown where tourists flocked to buy curios and art goods. By the 1930s, Japanese-owned stores in fact dominated the "Chinese" souvenir trade on the street. This Japanese ascendancy, combined with imperialism overseas, added to interethnic tension between Chinese and Japanese but also made the owners wealthy proprietors and the Japanese enclave's largest employers. These stores were in fact the backbone of the prewar Japanese American economy, although they were about two miles from Japanesetown. Another major sector of ethnic commerce was detached and scattered around the city: about 140 dry-cleaning and laundry enterprises, as well as small flower vendors, tailors, and cobblers that catered to non-Japanese customers. Finally, Japanesetown hotels and employment agencies supported an even farther-flung population: the workers who followed crops, manned extractive industries, or labored in other ways throughout the US West but circulated through San Francisco for jobs, information, or rest.[60]

The many businesses of Japanesetown served all residents, whatever their ethnicity. Some of these encounters could cause friction as much as friendship. Japanese and Filipino residents, in particular, had inequitable and sometimes hostile encounters, due in part to their disparate histories in the city. Filipinos did not begin migrating to the mainland United States in large numbers until the mid-1920s, and San Francisco—a city which had played so critical a role in the Philippines' colonization—was the primary point of entry. They therefore arrived well after the enclaves, services, and facilities of Chinatown and Japanesetown were established. They also arrived at a particularly unfavorable juncture for upward mobility: after anti-Asian institutions consolidated and entrepreneurship opportunities had contracted, patterns subsequently heightened by the Great Depression. Additionally, Filipinos were less rooted in San Francisco. Filipinos were more likely to be migrant laborers than Japanese were by the 1930s, circulating through the city and around California and the West. Those in San Francisco tended to stay or live near Chinatown in the ten-block Manilatown enclave that developed during the 1930s; their numbers would not grow substantially in the Western Addition until after World War II. Still, Filipinos at times felt the scorn of Japanese neighbors or shopkeepers, due to their laboring status, national prejudices shaped by Japanese imperialism, and, perhaps, the interethnic tensions that occurred between Japanese and Filipinos in agricultural California.[61]

African Americans were the second-largest minority population in Japanesetown by 1940. The area housed two-thirds of the city's African Americans, but "this was in no sense a Negro area," according to one sociologist, since it lacked the degree of concentration of either Chinese or Japanese neighborhoods.[62] The nine tracts of the Western Addition district had at maximum 14 percent African Americans; most were under 5 percent. While there were blocks almost exclusively occupied by Japanese, there were none exclusively occupied by African Americans. The black enclave centered on Fillmore Street, which bustled with shops, restaurants, groceries, and nightclubs. Most of the city's black businesses and institutions were located here. As prewar resident Franzy Lea Ritchardson recalled, "Very few Negros went downtown because anything they wanted in the business line was on Fillmore Street." This made for a close-knit community: "when walking down Fillmore Street from McAllister to California . . . every Negro person I met I knew."[63]

Although many African Americans remembered a relatively tolerant prewar environment, black San Franciscans certainly faced barriers. Most were domestic or service workers, and black laborers were excluded from dozens of labor unions. There was anti-black institutional and individual discrimination; the enclave was the product of segregation as much as community building. However, San Francisco's African Americans faced far fewer residential restrictions than in other parts of the country, where black urbanites were restricted to fiercely segregated neighborhoods. In a city where white supremacy was built on anti-Asian rather than anti-black racism, and with citizenship that granted protections unavailable to Asian migrants, black San Franciscans could live in every neighborhood. When middle-class African American families bought homes in residential neighborhoods, they found fewer applicable restrictions and less outcry from neighbors than their Asian peers did. Kenneth Finis, a prewar resident who grew up in the Outer Richmond, recalled that his childhood home "wasn't an exception because Blacks lived all over the city."[64] San Francisco's prewar discriminatory apparatus was primarily directed at Chinese and Japanese residents.[65]

WORLD WAR II ARRIVES

World War II was a turning point for Japanesetown. The war placed Japan and Japanese Americans in the center of the city's attention even as it excluded them from its boundaries and reshaped their district. Wartime events also demonstrated the resilience of neighborhood connections during the disruption of war.

The impact of the Japanese attack on Pearl Harbor was immediate. The mayor declared a state of emergency the same day. Japanese Americans huddled together, staying inside to avoid confrontations. As sociologist Charles Kikuchi described, "Everybody [was] in a daze in Jap Town."[66] Other San Franciscans, however, flocked to the neighborhood to gawk at its inhabitants. The curious or hostile crowded in to such an extent that police had to block traffic, adding to the throngs and to resident nervousness. The FBI compounded fears by arresting and detaining hundreds of "suspect" aliens, mostly adult men who left frightened families behind. Aliens—Japanese migrants were prohibited from naturalizing, so this included almost all migrants—had bank accounts frozen, businesses closed, and a long list of contraband items to turn in. These steps threw the Japanese American community into disarray. With just a glimmer of what was to come, people predicted what they thought would be the worst: "all sorts of wild rumors are going around they are going to lock up the Issei."[67] The American-born flocked to health offices for copies of their birth certificates. Many feared mob violence, and stories of assaults spread rapidly. In an exposure of long-simmering tensions, many accounts described Filipinos retaliating for Japan's attack on the Philippines as well as years of discriminatory treatment at the hands of Japanese shopkeepers and neighbors. As Tamotsu Shibutani, a sociologist then an undergraduate at Berkeley, wrote, "Gloom was cast over Nihonmachi."[68]

After the first week or so, however, Shibutani described a "false sense of security" descending for much of December and January as many of the worst fears went unrealized.[69] However, tensions once again ratcheted up as Japan won a series of key Pacific battles. Anti-Japanese organizations renewed their energies, and officials and the public began to debate Japanese removal. First the *San Francisco Examiner* and then Mayor Angelo Rossi, himself the son of Italian immigrants, called for the removal of Japanese aliens and the "detailed and all-encompassing investigation" of their citizen children.[70] Panic returned as the Japanese-language newspapers shut down, Japanese Americans were fired from jobs across the state, the FBI continued to arrest aliens and search homes, and federal restrictions rolled out one after another, creating a swelling band of prohibited areas and restricted zones. Finally, President Franklin Roosevelt acceded to a wave of popular and official demands as well as his own prejudice by signing Executive Order 9066 on February 19, 1942. This order gave control over certain areas to the secretary of war and allowed him to exclude from it any persons deemed necessary. In Karl Yoneda's memory, many "did not realize the significance" of the order at first.[71] But then General John DeWitt, commanding general of the Western Defense Command, designated a restricted military zone. Days later, the army announced plans for "reception centers" for evacuated Japanese Americans.[72] By the end of March, restrictions previously limited to aliens were extended to Japanese American US citizens. All German, Italian, and Japanese noncitizens had been forced to register as enemy aliens and obtain identity certificates in January, but decades of exclusionary laws, practices, and customs culminated in the incarceration only of all those of Japanese descent. Italians were no longer even considered "enemy aliens" within months, and Germans were no longer subject to mass restrictions by the end of the year.[73]

For Japanese Americans, evictions were announced in San Francisco in April. Ironically, this was a relief to some. After months of worry and uncertainty, they "strangely were . . . glad that the long-dreaded ordeal had finally come."[74] Artist Miné Okubo, though, described it as a "real blow": "we had not believed at first that evacuation would affect the Nisei."[75] Psychological reactions quickly gave way to logistical planning. Everyone had to hastily liquidate property, leave businesses and homes in the hands of hopefully reliable friends, put careers on hold, find homes for beloved pets, and generally wrap up an entire life in just a few short weeks; the first group had only six days to prepare. Most Japanese Americans from Japanesetown were carried away by May 10. Ten days later, no Japanese Americans were left in the city except for a hospitalized handful too sick to be moved: "Japanese town was empty."[76]

Most households from San Francisco were bused by armed guards to the euphemistically named "assembly center" at the Tanforan racetrack. Tanforan was a short half hour drive away in San Bruno, but light-years from home. The initial glimpse of the assembly centers was deeply troubling: "the first sight of the barbed wire enclosure with armed soldiers standing guard as our bus slowly turned in through the gate stunned us with the reality of this ordered evacuation."[77] Nothing afterward was reassuring. People were first checked for contraband and then examined for diseases. According to writer Yoshiko Uchida, "Everything was only half-finished" in their new home.[78] The lucky ones received newly if hastily built barrack rooms; the unlucky ones lived in a horse stall with manure, bugs, and horsehair whitewashed to the walls and a lingering, disagreeable smell. Food was served in mess halls, where occupants waited in long lines for unappetizing, sometimes poorly cooked meals very different from the Japanese-derived foods many were used to. In consequence, there was a rush on bathrooms where residents again found long lines and, excruciatingly, rows of unpartitioned toilets. These harsh conditions were ameliorated somewhat by the time they arrived in the permanent detention center. Still, even there the incarcerated found rough barracks, a room of 320 square feet for entire families (larger families had a luxurious 500 square feet; the very largest might have two rooms), and communal bathrooms, laundries, and dining halls.[79]

Familiar faces helped soften the blow of pervasive humiliations. The second sight that greeted arrivals at Tanforan, after the barbed-wire fences, were friends and acquaintances behind the chain links. Those who had arrived earlier strained to find associates on each arriving bus, eager for novelty in the long hours of empty days. For the most part, San Franciscans moved as a group first to Tanforan and then to the Topaz permanent detention center; official policy "move[d] communities together so far as this was possible" so as to maintain a "natural community" and "preserve desirable institutions."[80] Not only could former Japanesetown residents find fellow students from school, friends from clubs, or neighbors from the block, they might also have received some of the same services from the same people. A pregnant woman could, in fact, have her baby with her Japanesetown doctor, Kazue Togasaki, an obstetrician who continued to deliver babies in Tanforan before she was moved to Tule Lake and then Manzanar. They could also buy dry goods from the same man they might have at home: Dave Tatsuno ran the NB Department Store on Buchanan Street before applying his merchandising skills to the cooperative store in Topaz.[81] Many Christians could still attend services with their pastor or priest; Buddhists and Shinto practitioners were exceptions, since many of their religious leaders were interned in Department of Justice camps. Even those who went into the 442nd, the segregated Japanese American battalion, held onto local connections that boosted their camaraderie: "we were like brothers under the skin, so to speak, and I knew all the fellows from San Francisco and from the Bay Area. I guess it was an extension of our life, from before camp."[82]

Not everything provided continuity, of course. Japanese Americans were surrounded by barbed wire and soldiers with guns aiming inward. There were also many absences: friends and family members arrested by the FBI waited out the war in Department of Justice centers, drafted or volunteered service members might be far away or dead, and resettled college students or workers explored midwestern and East Coast cities. Nonetheless, while incarceration was nothing but a life-changing upheaval, the familiar faces and institutions of Japanesetown offered a measure of stability and comfort in the completely alien landscape of internment.[83]

THE WAR IN JAPANESETOWN

Meanwhile, "Japtown," as it was often called by San Franciscans during the war, was in the midst of its own transformation. The streets in April 1942 had indicated their residents' tenuous state. Kikuchi described the neighborhood as "a ghost town": "All the stores are closed and the windows are bare except for a mass of 'evacuation sale' signs." Hordes of scavenging bargain hunters descended on Japanesetown and scoured homes and stores for rock-bottom prices on furniture, cars, merchandise, dishware, and even houses.[84] By July, columnist Herb Caen described the "dismal stretch of yesterday known as Japtown, where the only neon sign alight invites you to the Cherryland Sukiyaki Restaurant."[85]

Vacancies did not last long. Migrants from all over the country flooded into San Francisco to work in the region's thriving war establishments; the Bay Area was the nation's major shipbuilding center and home to scores of military installations. As a result, the city's population increased by 10 percent between 1940 and 1944, from 634,536 to 700,735. The black population grew proportionately faster, from less than 5,000 to 17,395 by 1944 and over 40,000 by 1950. Integrated worker housing was built in Hunters Point and other areas, but not until a few years after the migration began. The Western Addition absorbed many new arrivals as the black Fillmore expanded outward and annexed Japanesetown. By 1943, over 9,000 African Americans lived there, where they found an established black community and relatively welcoming conditions. As poet Maya Angelou recalled, "The Japanese shops which sold products to Nisei customers were taken over by enterprising Negro businessmen, and in less than a year became permanent homes away from home for the newly arrived Southern Blacks. . . . The Japanese area became San Francisco's Harlem in a matter of months."[86]

Black workers eagerly arrived in San Francisco, as elsewhere in the urban north, looking for both jobs and greater freedoms. The Bay attracted migrants from all over, but most black migrants came from either Texas or Louisiana, and 75 percent came from the South more broadly. The booming military industries intensified a rural-to-urban migration that had been ongoing for years in the South. Migrants could have found work closer to home. Mobile, Alabama; Beaumont-Orange, Texas; and Pascagoula, Mississippi, were all defense centers with their own military industries. However, neither military demands nor wartime exhortations of democracy could break the color line in the southern workplace, where racial violence escalated during the war. According to the National Association for the Advancement of Colored People (NAACP) San Francisco branch president Joseph James, himself a wartime migrant, "The newcomers [had] seized a long-looked-for opportunity to escape from the South."[87] Only the most naïve would have expected complete equality in California, but most anticipated reduced oppressions. And California indeed provided respite from the most egregious offenses: the migrant's "children go to the same schools as other children. . . . He can walk down the street without having to move toward the curb when a white man passes. He isn't required on perhaps pain of beating or arrest, to say 'ma'am' to the woman clerks in the stores. . . . He can vote by registering and going to the polls, and no nightshirt Klansmen are going to try and stop him."[88] The new migrants came for both economic and social opportunity.[89]

There was no shortage of jobs in wartime San Francisco. Work was better paying and higher skilled in Bay Area defense industries than in either the South or outside of war work. Good housing, however, was a different story. Heightened anti-black segregation and the tight housing market ensured packed and inadequate conditions. Officials had already considered Japanesetown a "slum" area and debated clearance while it briefly stood empty.[90] Their attention was immediately drawn elsewhere in the militarized city, however, and so nothing was done save a few condemnations. This meant available but undesirable housing for new arrivals. The aging housing stock had suffered from years of neglect from mostly white absentee property owners. The district held over one-third of the city's substandard dwellings, a proportion far outpacing its relative housing stock. Families were forced to pack in: "flats formerly occupied by one family now house a family unit in each room," according to an NAACP survey.[91] This created "crowded and unsanitary conditions in San Francisco's former Japanese district, now largely occupied by Negro defense workers."[92] Furthermore, African Americans' limited options made even these quarters expensive. As one neighborhood resident remembered, "White landlords always charge Negroes higher rents. . . . But this is the worse I have ever seen. We have to stay here, though, because there is no place else we can go."[93]

This "almost impossible situation" aroused "indignation" from some housing officials, but even sympathetic descriptions frequently adopted what one social worker called a "decided racial slant."[94] The district had long been racially marked, and journalists continued to call the area "Jap-town" long after Japanese Americans residents were gone.[95] This historical ignominy colored the new residents, adding another layer of stigma to their burgeoning numbers and segregation, thereby helping to fuel prejudices underdeveloped in San Francisco. The "Japanese problem," observed John-son, "became linked with that of adjusting the new Negro constituency."[96] Anti-black racism built on anti-Asian precedents but took on its own char-acter, especially with the rapidly increasing numbers of black arrivals. Mayor Rossi called for federal intervention to stop the "Negro invasion."[97] Neighborhood groups "viewed this increase in Negroes with alarm," out of prejudice or with the assumption that black neighbors would depreciate property values. They "hurriedly [took] drastic steps to encourage the for-mation of restrictive agreements and other legal and non-legal devices."[98] The rising black population in a long-stigmatized, deteriorating neighbor-hood led to a new awareness of a "Negro problem."[99]

JAPANESE AMERICAN RESETTLEMENT
IN SAN FRANCISCO

Japanese American resettlers returned from incarceration to a very different San Francisco and "Japtown" than they had left. Tight housing conditions, the already crowded state of the Western Addition, and continuing segrega-tion made rebuilding a challenge. Nonetheless, a Japanese American com-mercial and residential enclave reemerged by the mid-1950s.

Japanese Americans slowly returned to San Francisco. The Japanese American Citizens League (JACL) returned immediately after the Western Defense Command lifted its West Coast ban in order to assist resettlers. But as the JACL's representative claimed, "People are afraid that Pacific Coast communities may not accept them."[100] Word traveled quickly about violent incidents on the West Coast. Even though most occurred in rural areas, San Francisco also had incidents that included attacks on a private home and a Buddhist-run hostel. Furthermore, California itself proved unwelcoming. At the end of the war, in an effort to cement their absence and conclusively end their role in agriculture, state officials seized Japa-nese American property bought through loopholes in the Alien Land Laws. While enforcement had previously been slow—the law itself was statement enough—authorities prosecuted more escheat cases in 1944 and 1945 than it had in the past thirty years, actions upheld by the state supreme court.[101] Such conditions led to a slow return. The biggest cluster, 1,200 people, ar-rived in October 1945; these were among those forced back by the detention centers' closure. They were provided $25 and a train ticket to their point of eviction. By April 1946, over a year after the ban was lifted, only 3,001 Japa-nese Americans lived in San Francisco. Another couple of thousand arrived slowly over the next few years from the midwestern or East Coast cities they had relocated to from the detention centers; about half of all those interned lived elsewhere before returning home. And some came from different places entirely. In 1947, federal officials estimated that about 30 percent had lived elsewhere before the war. Whether natives or not, Japanese Americans returned to San Francisco.[102]

Most of these resettlers moved to Japanesetown, where they struggled for space in an intensely crowded housing market. In the years following the war, about one thousand worked as live-in domestic labor for the housing it provided, even those who had never done such work before. Over a thousand moved into public or veterans' housing in Hunters Point and Camp Funston. But the majority settled in the Japanesetown area. When the Buddhist church established the largest returnee hostel to date, a process that took months due to the difficulties of finding adequate facilities, it did so in what had been the commercial heart of Japanesetown.[103] Those who held property generally did so in the enclave. This was the same for those who owned hotels (there were about twenty by 1947). But these fortunate property owners were few: in 1946, the War Relocation Authority (WRA) estimated that Japanese Americans owned only about 150–200 housing units. Furthermore, even property owners were not guaranteed a home. The area "was already bursting at the seams," and their units were likely "being used by negro or other tenants." Many with contracts had leased their property for the "duration of the war," while others had only vague verbal agreements.[104] Many homeless therefore turned to non-Japanese friends: Sumi Honnami moved directly back into her family's prewar rental apartment because of the close relationship between her family and their prewar landlords, an African American couple who had also stored her family's belongings. Most simply rented or occupied whatever and wherever they could. Families doubled or tripled up. In 1946, the WRA estimated that almost two thousand people lived in accommodations built to accommodate a third of that number. Overcrowding also hampered businesses. Every storefront or office in the neighborhood was occupied, some even as housing. Proprietors waited for months before finding a suitable storefront, while other entrepreneurs plied old contacts and every available network for any possible space. By 1947, there were only about 150 Japanese American–owned businesses, down from 450 before the war.[105]

The rebirth of the enclave faced difficulties beyond overcrowding. First, as historians have shown, federal officials discouraged the resumption of enclaves in favor of dispersal and assimilation. Secretary of the Interior

Harold Ickes stated that his department, in charge of the WRA, would "do all in our power . . . to persuade those who formerly lived on the Pacific Coast to locate elsewhere."[106] Indeed, assimilation had become the ideological cornerstone of the internment and resettlement program, reinterpreting an ad hoc and pointedly undemocratic program into, in the words of the WRA director, "an exciting adventure in the democratic method."[107] In this view, wartime relocation gave people in the Midwest and East "their first real chance to see and know the American people of Japanese descent." By prodding Japanese Americans out of their enclaves, relocation diffused the prejudice that had given rise to the camps: "gradually the feeling grew and spread that here were no Emperor-worshipping fanatics but a group of decent, well-behaved, sincere human beings."[108] The applications to leave camps therefore cautioned resettlers against congregation, contrasting it with "American" conduct: "Will you assist in the general resettlement program by staying away from large groups of Japanese?" and "Will you try to develop such American habits which will cause you to be accepted readily into American social groups?"[109] The WRA field offices, other federal institutions, and local agencies that assisted resettling Japanese Americans reinforced this emphasis. One Chicago resettler hostel, for example, advised its residents to avoid "going about town in groups of more than two or three."[110] And the government wielded a big stick in its attempt to shape Japanese American behavior. Internment itself—incarceration on the basis of racial difference—was the consequence of disregarding such assimilationist strictures.[111] The state's interpretation of democratic incarceration moved both blame for the program and the burden of its resolution onto Japanese Americans themselves.

Not all Japanese Americans abided by the assimilationist strictures that had been "pounded for four years into the race-sensitive Nisei," as the *Pacific Citizen* described it.[112] Some found, as one WRA report stated, that "old surroundings continue[d] to exert a strong pull."[113] These resettlers were comforted by neighborhood connections during their incarceration, while others simply found the state's assimilationist pressures onerous or offensive. They therefore returned to the familiarity of Japanesetown because of work, property ownership, family and friend networks, fear, or segregation. The *Pacific Citizen* editors lobbed the blame for this clustering, which they saw as rational and perhaps only temporary, back at the WRA: "the evacuation procedure gave the Nisei new fears—based upon race." Their fears and lingering West Coast hostilities "brought the evacuees together, as only fear can bring together the homeless and unwanted."[114] But other resettlers appeared to take up the WRA's challenge, agreeing that "dispersal was good because we used to live too close together on the West Coast."[115] As historian Scott Kurashige reminds us, the state's perception that Japanese Americans were assimilable was a "paradigm shift in racial discourse" after decades of alienation. As a result, some embraced it as a "present goal and a future certainty."[116]

Japanese Americans did not bear the burden of assimilation by themselves. The state's linkage of dispersal and democracy gave a multiplicity of bodies a stake in their resettlement. Resettlers had considerable support from federal and local, public and private agencies, support not provided to, for example, their new African American neighbors recently migrated from the South. Provisions were not always given with enthusiasm, and assistance varied considerably by locality. As the formerly incarcerated headed home, for instance, Los Angeles mayor Fletcher Bowron at first claimed he could offer no resources for his city's resettlers, while San Francisco mayor Roger Lapham gave their homecoming a tepid yet notable endorsement. All over the West, however, resettlers were preceded by brochures praising their Americanness and contributions to the war effort, while federal public-relations representatives held informational meetings to solicit civic and religious groups into their campaign. Furthermore, when Japanese Americans settled into a number of cities, they could turn to WRA field offices for job and housing notices vetted for nondiscrimination. Officials and volunteers, in fact, swarmed the offices of employers, landlords, and realtors to encourage their acceptance of Japanese American queries. Those in need also had access to other types of support. This included temporary WRA housing—albeit in many places comparable to their detention-center housing—or volunteer-run hostels, some financial assistance, medical aid, facilitated reentry into educational systems, and referrals to local social-service agencies. Multiple federal officials and local representatives were invested in the "success" of the internment program.[117]

This still left a great deal for Japanese Americans to negotiate on their own. They returned to a neighborhood that was already full of people who, like themselves, faced residential discrimination. Observers in San Francisco predicted "racial difficulties."[118] A similar dynamic emerged in Los Angeles, where Little Tokyo had become Bronzeville. There, seventy thousand African American war workers and their families lived where thirty thousand Japanese Americans had been, and observers predicted a race riot when resettlers returned. However, relations were generally cordial. In San Francisco, too, many black community organizations extended "an outstretched hand" to returning Japanese Americans. The San Francisco NAACP president Joseph James sat on the JACL's advisory board, and his group and the JACL "work[ed] together openly and effectively."[119] The Booker T. Washington Community Center sponsored a hostel for resettlers, and its director, James Stratten, hired two Japanese American social workers by 1946. Stratten was aware of the Japanese American history of the neighborhood; the center had moved into Kinmon Gakuen's former building after the Japanese-language school was evicted from the West Coast. He and others at the organizational level were keenly aware of the injustice that had expelled Japanese Americans from their former home. At the same time, they sought to avoid conflict in what had become a new home to thousands of African Americans with few other housing options.[120]

Unsurprisingly, there was tension on the ground. The *Pacific Citizen*, headquartered in San Francisco, described the neighborhood as a "melting pot at slow boil."[121] Some Japanese Americans treated the newcomers with the same disdain that they themselves faced. Alice Setsuko Sekino Hirai remembered with embarrassment "strong racist ideas" against African Americans among some Japanese Americans.[122] This would not have been unusual. Many resettlers held prejudicial views of African Americans. If this racism had not been picked up through mass culture, San Franciscans who grew up in a city where they were at the bottom of the racial hierarchy might have learned it as they passed through cities where anti-black prejudice was systematic and unavoidable. About thirty thousand, or over one-fourth of all the incarcerated, stayed in Chicago for some period while others spent time in other midwestern or East Coast cities. While few found promotion, residence, or socializing to be without barriers, most encountered a definite advantage compared to African Americans. After their own experiences with incarceration, most were willing to accept anti-black racism, or at least not vocally oppose it, on the shop floor or the real-estate office in order to avoid confrontation.[123]

These recent experiences might have added to Japanese Americans' desire to return home and encouraged discriminatory, even predatory, behaviors. One property owner recalled "a lot of pressure" for "Japanese people to push the blacks out as much as possible."[124] Some Japanese landlords might have been "unhappy about the treatment that their properties received at the hands of Negro tenants," and a number of African Americans indeed resented being "pressured by [Japanese] landlords." But, of course, Japanese Americans were not the primary landlords in Japanesetown, before or after the war. Thelma Thurston Gorham, writing for the *Crisis* in

1945, uncovered a "numerical unimportance of Japanese-owned residential property," and instead blamed possible tensions on "the American-style fascists and others who might profit by keeping the two groups at swords' points."[125] Historian Scott Tang has calculated that of the approximately 4,500 Japanese Americans living in Japanesetown in 1940, well under 100 were owner-occupants. By 1950, the census indicated a total of 481 nonwhite (the only racial designation provided in that year's tract-level information) owner-occupied homes among Japanesetown's 6,882 dwelling units.[126] White absentee landlords were the majority, and less than half of the district's property owners actually lived in the district.[127]

Some Japanese renters did have decades-long leases that allowed them to sublet or evict at will, but the neighborhood's racial transition was not solely driven by Japanese American evictions of African Americans. As Hilary Jenks found in Los Angeles, white landlords were either sympathetic to former Japanese American renters or else "preferred the Japanese because they kept the buildings up better, and boosted the land value," as one resident presumed.[128] Racist attitudes in San Francisco had transformed during the war. There was some hostility from African Americans toward resettling Japanese Americans. Resident Willie K. Ito recalled a sense of "this is our territory now" from his black neighbors and a degree of prejudice: "a lot of southerners [had] never experienced living side by side with Japanese Americans" so some were "kind of cautious."[129] However, even the most outraged would have had little recourse if landlords, in a display of San Francisco's emerging new racial environment, favored Japanese Americans as tenants. As one black lawyer noted, "I think that we Negroes do not resent the Japanese so much as we resent the Caucasians who impose restrictive housing covenants on us."[130] Yori Wada, a Nisei social worker with the Booker T. Washington Community Center, mildly described the transition as "a combination of happenstance and bad luck for the black residents" who moved "out of private homes and apartments in order to give way to the Japanese."[131] This description ascribed a kind of benevolence to the many African Americans who were in fact essentially evicted, while downplaying Japanese American complicity. In their desperation for housing and familiarity after their traumatic experience, resettlers inadvertently helped support growing black segregation.

The neighborhood changed dramatically as a result of the war. By 1950, only 60 percent of the 5,579 Japanese Americans in San Francisco lived in and around prewar Japanesetown, a proportional decrease from 1940 that reflected the crowded housing market more than changed discriminatory practices. Japanese Americans were only 16 percent of the neighborhood in 1950, while 34 percent were African Americans. A brisk walk down Post Street would have demonstrated the neighborhood's heightened multiracial status. The African American Cobb's Barber Shop was a block from the Nisei-run Jimmy's Barber Shop. Jimbo's Waffle House (known as Jimbo's Bop City, one of the most famous jazz clubs in the West) was a few doors down from Soko Hardware and less than a block from Nippon Pool Hall. Charles Sullivan, a legendary black music promoter, owned a number of businesses on Post, including a jukebox rental company, hotel, and liquor store. A pedestrian could have heard hymns, at almost the same time, from the black Emanuel Church of God in Christ and the Japanese Presbyterian Church. Also nearby was Wong's Bait Shop and the Luzon Coffee Shop. Resident Benh Nakajo recalled that 1950s Japanesetown had "lots of [Japanese American–owned] stores and businesses, and barber shops and restaurants and dry-good stores and soda fountains, and it looked like the whole Japanese population was living in this area."[132] The Western Addition again held the majority of the city's Japanese American population, but Japanese Americans were no longer the only — or even the primary — racialized minority associated with the Western Addition.[133]

JAPAN IN POSTWAR SAN FRANCISCO

Elsewhere in the city, the Pacific War had transformed other Japanese traces in the landscape. The Japanese Tea Garden had been "dismantled" during the war, and its famed statuary and "nearly 1000 potted shrubs and plants valued at many thousands of dollars" trucked away for the duration. The Hagiwara managers were evicted, and their home was "razed in a misguided burst of patriotism."[134] A lively public debate suggested replacing the site with a number of possibilities, including a hot-dog stand, "a big statue of President Roosevelt" instead of "all that Jap statuary," or a Chinese tea garden. Finally, the Parks Commission simply renamed the park the "Oriental Tea Garden" and hired Chinese employees as staff.[135] After the war, it was renamed the Japanese Tea Garden and, although the Hagiwaras had asked to return, management was turned over to an Australian concessionaire who ran it for another decade.[136]

This marginal place for Japan in San Francisco's landscape mirrored the tiny place imagined for the nation in the city's markets. Japan had been one of the city's most important economic partners in the first half of the twentieth century, despite rising hostilities. As late as 1939, Japan had ranked second and made up almost one-sixth of the city's foreign trade, even in the midst of a global depression, escalating international tensions, and Japan's colonial plunder.[137] But postwar observers saw little to gain from the defeated nation's exhausted capital and decimated industries. The Board of State Harbor Commissioners, the body in charge of the city's ports, virtually ignored Japan when they reassessed the city's potential foreign trade. This was despite the important role they predicted for Asia and the Pacific in the harbor's future: the "Far East and the Pacific Basin have come of age as an economic area."[138] The commissioners therefore outlined myriad possibilities for future economic relations. San Francisco, they believed, should pivot toward that region, but they saw Japan as an irrelevant outlier.

CONCLUSION

Japan held a relatively marginal role in San Francisco's civic identity and landscape from the turn of the twentieth century through World War II, despite the Orient's central place in the city's imagination. Of course, there were explosive moments of tension that culminated in the Pacific War, with clear consequences for local Japanese Americans. But with the exception of the Japanese Tea Garden, there was no acknowledged referent to Japan in the city's landscape. And with the exception of the Japanese district, always far smaller and less notorious than Chinatown, the Japanese diaspora had no large cultural footprint either.

The Pacific War, if anything, further marginalized Japan and Japanese Americans in the landscape. The war whittled down Japan's footprint. The Japanese Tea Garden was replaced during the war to reemerge as "Japanese" once again after it, but no longer managed by Japanese Americans. Postwar San Francisco again had a Japanese American enclave but it had changed, becoming more crowded and decrepit by a wartime population explosion, segregation practices, absentee landlords, and a constricted housing market. And it was marked not only by Japanese Americans, but also by African Americans.

Prior to World War II, San Francisco had developed a long-running identity as a cosmopolitan city linked to Asia and the Pacific world. To a large extent, though, it was shaped by Chinatown and Chinese Americans. In both city identity and social geography, Japanese Americans and their nation of origin were a small part of San Francisco life. This would change only in the mid-1950s, when new opportunities in the Pacific offered Japan a more central place in the civic imagination.

Orienting the Gateway to the Pacific: Reconsidering Japan and Reshaping Civic Identity

San Francisco once again thrummed with the celebration of the Pacific region and the city's place in it in September 1958. The scale and duration of the ten-day Pacific Festival was far smaller than the 1915 Panama-Pacific International Exposition, but projected a much clearer Pacific orientation. Festivities revolved around the nations circling and within the Pacific Ocean, but one stood out: Japan, whose central place in both festivities and municipal attention was little foreshadowed just a few years before.

The Pacific Festival was designed to "focus the attention of the world on the growth and development of cities, States, and nations bordering the Pacific Ocean and thereby foster mutual understanding and cordial relations between the peoples of these areas."[1] The Philippines, Mexico, Indonesia, China, Oregon, and the US territory of Hawai'i were among the celebrants, illuminating the enormous size and rich diversity of the region. Festivities were held from the shoreline to the interior, in the streets and in private venues. Union Square held martial-arts demonstrations and Filipino folk dances. Audiences gathered in the Marina for a Japanese fireworks show, and flocked to major department stores' displays of regional products. A prestigious hotel hosted a grand ball. Participants commemorated Mexican Independence Day on City Hall steps, and walked through exotically decorated Japanese and Chinese American homes in a YMCA-sponsored tour. These combined events brought the vast spectrum of the Pacific region together and emphasized the natural, close relationship between the region's cultures and products with San Francisco civic life and institutions.

The festival drew on the city's "traditional role as the Gateway to the

Pacific." This was an identity city leaders had embraced "since the days of the sailing ships," but it had become "vital" of late.[2] The Pacific War and then the Cold War made the region critical in US interests. Furthermore, the "Far East and the Pacific Basin had come of age as an economic area."[3] As city leaders believed, San Francisco's geography and historic place vis-à-vis the Pacific positioned their city to take advantage of the region's growth. What became an annual festival highlighted the Pacific region's rising importance to San Francisco's identity and future.

Japan was crucial in this regional reevaluation. The recent enemy had more dedicated events and activities than any other. There was a "Japanese Day," Japanese fireworks, visiting dignitaries, and Japanese guests of honor at numerous events.[4] The nation's highlighted place in the festival indicated its special role in city leaders' ambitions. Its economy was, according to *Time*, "bursting at the seams" and "far [outdistanced] its peak wartime production years."[5] For San Francisco business leaders struggling against urban competitors and seeking to establish their city as a US Pacific hub, Japan's growth could spur the city's own.

The Pacific Festival was just one of a number of events and institutions beginning in the mid-1950s that centered San Francisco in the Pacific region and celebrated or cultivated social, economic, and political ties with Japan. Responding to events overseas and urban competition on the West Coast, city leadership and an expanding number of San Franciscans oriented their civic life, built environment, and politics toward Japan. The city had long framed itself as uniquely tied to the Pacific region and had a long history of economic, political, and social ties with the nation. However, these connections were muted in the first decade after World War II. Some had to be rebuilt, others freshly constructed, and some retrieved from coastal competitors before they could be centered in city life.

The sister-city affiliation with Osaka was perhaps the most far-reaching and long-lasting of these efforts to reconstruct San Francisco's transpacific links. Initiated by a small cohort of Pacific-oriented businessmen in 1957, one year before the first Pacific Festival, the affiliation was begun as a way to further their commercial ambitions with a rising economic power. The program provided them with a vibrant set of contacts in the foremost industrial center of Japan and an important trading partner with their city, but it served their interests in more abstract ways as well. The cultural and popular dimensions of the affiliation created a diverse constituency of San Franciscans invested in Japan, from the mayor to young schoolchildren, and provided them broadly public and highly targeted ways for them to express this interest. This expansive participation helped represent San Francisco as a metropolis with deep and extensive ties to Japan, naturalizing its Japanese networks for both San Franciscan and Japanese audiences and encouraging the development of new ones. The affiliation therefore helped redefine San Francisco's long-standing position as the "gateway" between the United States and Asia's industrial anchor.

San Francisco's reorientation was not done easily. The cohort of Pacific-oriented businessmen had a deep investment in Japanese networks, but their interest was not widely shared by city leadership or most San Franciscans at the affiliation's birth. Furthermore, the city's reputation was not especially reflective of Japanese connections, at least not in positive ways.

The affiliation's programs and events addressed these problems and brought links with Japan and its culture, commerce, and people to the center of civic life. First, affiliation events presented an image of Japan conducive to a respectful, equitable relationship, with representations of Japanese maturity and skill. This was very different than the portrayals in film, television, popular magazines, or even law, through the War Brides Act, all of which emphasized women, children, tutelage, and dependency. These feminized and dependent views hindered the local campaign to cultivate support for commercial transpacific partnerships. Proponents had to convince their municipal and business leaders, as well as the general public, that the nation was worthy not just of aid or education but of *partnership*. This is not to say that events did not at times employ stereotypical, nonthreatening representations of Japanese women and children. However, these were judiciously used and carefully balanced with equitable San Franciscan pairings.

Furthermore, the affiliation's popular mandate helped widen its reach. Popular events and lavish spectacles drew thousands of San Franciscans, who experienced their city's commercial, cultural, demographic, and political connections with Japan. Smaller-scale but no less significant exchanges for school-aged youth, dependent on and embedded in a range of social institutions, extended the affiliation to San Franciscans who might otherwise have ignored or even avoided the affiliation. The sister-city program's popular and personal exchanges helped elicit widespread participation in the affiliation and other Japan-related events.

The public spectacles and intimate exchanges of the sister-city affiliation with Osaka helped put Japan at the fore of San Francisco's civic life. Historians have shown how federal officials developed public-diplomacy programs to shore up US foreign-policy aims, but the originators of San Francisco's program never made more than the most token of gestures toward these national aspirations.[6] Instead, they hewed closely to their own local, but by no means provincial, ambitions. The affiliation helped turn the Japanese interest of a cohort of businessmen into one shared by city leadership and a broader swath of the local community. As the affiliation matured into a civic institution with regular events and a dependable constituency, it invested San Francisco as a whole in friendly, productive relations with a formerly hated enemy.

THE "AGING MATRON" OF A TRANSFORMING PACIFIC

When the Pacific Festival opened, *San Francisco Chronicle* editors "heartily salute[d] the occasion." However, the caustic headline precluded any celebratory note: "Queen of the Pacific—or Just an Aging Matron?" The editors saw the festival as part of a pattern of vain and empty self-assurance: "every now and then, it seems, San Francisco requires an excuse to seek reassurance that she is the true and veritable queen of the Pacific basin." However, the city was too satisfied with the "effulgence of four-color tributes" in national media as evidence of this status. "The leadership of San Francisco in the Pacific" instead needed to face hard truths: "painfully reported statistics" revealed that "other ports along the coast—Seattle, Portland, Stockton, Los Angeles—have been overtaking the aging matron of the Pacific basin." The city's "charm and climate" could not alleviate the "stubborn and classic problem of San Francisco's lack of allure to new business and industry."[7] Negligence and competition had eroded San Francisco's Pacific status.

The editorial expressed San Francisco's waning regional importance, a sense that was well founded. For its first half century after the Gold Rush, the city had been the undisputable capital of the US Pacific coast in finance, the arts, manufacturing, population, and domestic and foreign trade, crucial measures of metropolitan ascendency. However, the twentieth-century growth of cities in the US West challenged San Francisco's dominance. As early as 1920, a number of metrics revealed its faltering status. That year's census showed Los Angeles's population gain over its northern counterpart, and San Francisco's population subsequently fell behind other coastal cities as well.[8] For civic leaders who believed that "population is the final measure of activity in all business," this was critical in and of itself.[9] But San Francisco fell behind what had become its southern rival in other important ways. Los Angeles's factories had easily outstripped San Francisco's manufactures, once a sizable part of the Bay city's economy. But the southern city's expanding maritime trade hurt even more. By 1923, the Port of Los Angeles could claim it "achieved Pacific Coast supremacy" in total tonnage, "a leadership which it has maintained each and every year since."[10] By 1937, Los Angeles's trade with Japan, the US West's largest Asian partner, had a higher value than San Francisco's. Military installations and support, another key pillar of San Francisco's historical primacy, were increasingly diverted to other cities through the interwar period. World War II's explosion in military bases, the aircraft industry, and other forms of defense spending aided many coastal cities such as Seattle and San Diego, and it also made Los Angeles's manufacturing second only to Detroit. Even Oakland, a small port in the interwar period, steadily encroached on San Francisco's trade in the 1950s.[11] In the "competitive tournament among world cities," as one Chamber of Commerce president intoned, the "law of selection" mercilessly eliminated the unfit.[12] Geography, an impeccable natural harbor, scenic beauty, and historical connections could not ensure the survival of the "Financial, Commercial and Industrial Metropolis of the Pacific Coast."[13]

These conditions demanded a reevaluation of strategies. The city's economy was quite diverse, so no one single industrial loss was overly burdensome. Nonetheless, the city was uneasily embarking on economic transformations. The Bay's wartime shipyards and naval stations had made manufacturing one of the city's largest industries. This began an erratic decline during the 1950s, which business leaders scrambled to reverse.[14] As late as 1955, the Chamber of Commerce's annual program made its first priority to "attract and enlarge factories." The program mentioned waterborne trade, a historically important part of the city's economy, but the overall outlook was quite provincial. Its scope was the nine-county Bay Area, "the areas for which [San Francisco] is the commercial, financial, industrial, entertainment and cultural hub."[15] This continued a decades-old strategy of regional cooperation that had placed San Francisco as the administrative center of a multifaceted and extensive Bay Area "economic unit" to rival Los Angeles.[16]

The next year's program, however, encompassed a vastly enlarged canvas. In 1956, the chamber proposed a "hard-hitting 'Go-Ahead' program" that made the previous year's plan appear parochial and small thinking. The chamber argued that the city should function "as a service center for one of the world's most dynamic regions": the administrative center not just for the Bay Area, but for the entire US West and the Pacific region. Toward this end, the chamber defined "world commerce" as its foremost priority, replacing the previous year's priorities of factory building and parking and

transit concerns.[17] This altered focus built on the city's historic and continuing strengths in finance, insurance, and exchange but expanded its reach as
the Pacific region gained in new geopolitical importance.

The chamber's enlarged vision corresponded with the shifting goals of
the business community invested in foreign trade. Businessmen associated
with the Port of San Francisco, maritime transportation, and trade were
themselves adapting to new conditions. In the interwar years, domestic
trade was the bulk of the port's traffic. However, changes in transportation technology and federal regulations initiated a decline: by 1949, San
Francisco's trade with East Coast and southern ports had dropped 50 percent, while trade with other Pacific ports dropped almost 80 percent from
prewar levels. Things were no more stable for those engaged in foreign
trade. Prewar Britain had been the city's largest trading partner, but its
postwar commerce never regained its past strength: for example, exports
to the United Kingdom reached $23 million in 1956, the high of the decade,
but much lower than 1938's $32 million. This amplified the importance of
Pacific trade, but that region, too, was changing. The biggest hit was mainland China. The United States placed sanctions on communist China, and
according to San Francisco commentators, trade dropped 15 percent in all
Pacific coast ports by 1950.[18] Actual Sino–San Franciscan trade had always
been more limited than was desired, but the sanctions led to high shortfalls
for individual businesses; Getz Bros. & Co., one of the largest import-export
firms based in San Francisco, claimed to have lost 35 percent of its business
in the aftermath of sanctions. A number of San Francisco–based businessmen, including those who went on to lead the San Francisco–Osaka affiliation, lobbied against the policy, but to no avail.[19] The new economic
environment would require innovation.

Japan's "unprecedented business boom" therefore radiated hope for those
involved in foreign and Asian commerce.[20] The boom would have seemed
fantastical to observers at the very end of World War II. Japan had been devastated: there was not enough food, the economy was in a shambles, and US
officials and Japanese conservatives feared the rise of the Japanese left. But
a few short years brought enormous and consequential changes. Abroad, a
global economic crisis fermented, and the uneasy wartime alliance with the
Soviet Union degenerated into competition. Europe faced its own economic
collapse, and US policy makers clambered to redirect aid from Japan to
Europe to head off Communist expansion. The crisis had implications at
home, too. Officials feared a postwar return of the past decade's depression,
and so sought solvent foreign buyers to keep domestic industry humming.

In response, the United States in 1947 abruptly reversed much of its initial reformist line in occupied Japan. As Joseph Dodge, the policy's major architect, noted, "Only a self-supporting and democratic Japan can stand fast against the Communists."[21] And so officials ended reparations to Asian victims of Japanese aggression, restored conservative elites to political authority, rebuilt the very industrial and financial institutions that had buttressed militarist power, and reconstructed Asian trade networks eerily shadowing Japan's imperialist Greater East Asia Co-prosperity Sphere. This focus on rapid economic recovery turned obsessive after Mao Zedong defeated the nationalist army in China in 1949. For US policy makers, a strong Japan with a network of regional alliances could rebuild and stabilize Asia while reducing the need for China's markets. This strategy also neatly addressed the US need for materials from the resource-rich region and cultivated a perimeter of allies to contain the Soviet threat.[22]

Japan therefore emerged in a "position of strength" in US global strategy as support for "the economic stability of the non-communist area of Asia."[23] In practice, the relationship contained mutuality and conflict, but the alliances granted Japan crucial economic support. When the United States launched its war in Korea, its military-supply needs kick-started Japanese industry while military personnel stationed in the region pumped vital dollars into its economy. This continued through the decade with both outright aid and supportive policies for Japan's industries.[24]

Japan's rapid economic recovery had repercussions for San Francisco. The nation's industrial expansion depended on the importation of basic foodstuffs as well as industrial materials; Japanese manufacturers also increasingly depended on foreign buyers in lieu of robust domestic consumption. In 1951, in the middle of the Korean War, Japan imported $87 million worth of goods, almost one-fourth of all the exports from San Francisco ports.[25] This led the *Chronicle* to report, "The Japanese have filled the gap left by the fading away of mainland China. . . . With Japan buying, the San Francisco Bay Area — the 'gateway to the Orient' — prospers."[26] By 1956, San Francisco observers of Japan's "business boom" were regularly watching the "encouraging" prospects.[27] Japan's postwar resurgence was an entrancing development for those engaged in Pacific commerce. Sufficiently captured, Japanese commerce and relations could help stabilize San Francisco's shifting economic landscape.

Yet even booming conditions could be anxiety producing. There was no guarantee that San Francisco would draw Japan's commerce — or any other business — over and above its coastal rivals. For example, regular transoceanic air service provided myriad opportunities for international connection, but it also required active promotion, political lobbying, and continual research to position San Francisco, rather than Los Angeles, at the coastal hub of such service. Other San Francisco industries, too, kept abreast of federal rulings and industry developments and meticulously challenged any that privileged their urban competitors. Furthermore, the Chamber of Commerce regularly contextualized business activity trends with comparisons to the corresponding sectors in cities such as Seattle, Portland, and Los Angeles. No one in business or municipal circles could remain unaware of the constant drumbeat of rivalry. City leaders, especially those in the transpacific business sectors, were keenly aware of threats from other West Coast cities and creatively innovated to maintain their city's regional preeminence.[28]

THE ORIGINS OF THE SISTER-CITY PROGRAM IN SAN FRANCISCO

In response to these new conditions, business circles devised new strategies to encourage San Francisco's foreign commerce during the 1950s. This included a revival of the San Francisco chapter of the Japan Society, led by some of the same men who would lead the sister-city affiliation. More formally the Port of San Francisco began a program of trade promotion, "the most extensive on the West Coast," that included hiring representatives for important trade locations.[29] This began with a UK representative, but by 1957 focused on the Pacific region. That year, the second representative was hired for Tokyo, and more were sought for other Asian cities. Additionally, new institutions arose to encourage local commerce with Japan. The World Trade Association (WTA) formed the Japanese Affairs Subcommittee in 1956, one of only two subcommittees created that decade to monitor Japan's economic conditions and provide a clearinghouse for businesses.[30]

In May of that year, the San Francisco Chamber of Commerce and the WTA offered another promotional effort, a business tour of Asia. This novel experiment, according to the *Japan Times*, was the first US chamber of commerce contingent to travel to postwar Japan.[31] The twenty-five-day expedition would begin at the Japan International Trade Fair, "the Orient's largest and traditional trade festival," and continue with visits to "business endeavors ranging from cottage industries to shipbuilding."[32] Travelers would have "top level meetings with leaders of government and industry as well as with chambers of commerce in each city visited."[33] The tour would continue with brief stops in Hong Kong and the Philippines, and promised those "contemplating new or increased trade with the Orient a penetrating insight into [its] economic and industrial possibilities."[34] It was costly in both time and fees, but the tour nonetheless attracted forty travelers and became an annual event, reflecting businesses' intense interest in the region.

The tour was designed to promote "San Francisco and its surrounding communities as an expanding market for goods, services and investments with neighboring Pacific Basin nations."[35] This innovation involved fairly straightforward methods of economic development, such as business introductions and economic information. But the tour's boosterism also adopted a fresh tactic: the development of a sister-city affiliation with Osaka, Japan. This was the brainchild of James P. Wilson, a former commercial attaché with the State Department and now manager of the chamber's World Trade Department and president of the WTA. The sister-city affiliation was one of Wilson's many initiatives with Asia that included the business tour itself.[36]

The sister-city program was a new instrument of exchange and in some ways an odd fit with the ambitions of Wilson and other Pacific-oriented San Francisco businessmen. The program was introduced just the year before to share cities' "life, culture, problems, and strengths via classes, exhibits, club meetings, concerts, plays, parades, and city hall affairs."[37] Part of President Eisenhower's People-to-People Program, it was meant to encourage ordinary Americans to "leap governments" and demonstrate their amity directly to people around the world. The program had much higher stakes than just individual enrichment, however. It urged Americans to "communicate with the peoples of foreign countries in the interest of [US] foreign aims."[38] As Eisenhower argued, "Friendship is part of our defense posture," and its bond "must be kept strong and true if our Western way of life is to survive."[39] The program squarely focused on the Cold War aims of overtly engaging Americans toward international peace, while seeking the unpublicized goal of a "favorable atmosphere abroad for U.S. policies."[40]

The sister-city program, and all People-to-People initiatives, was intended to be a "citizens' program" that was "strictly separate from government."[41] However, because such affiliations were too valuable and possibly too fragile to leave to citizens with varying commitments to foreign-policy goals, government officials actively cultivated and directed early relationships. Interested citizens were instructed to create a heterogeneous civic committee and then contact the American Municipal Association, the clearinghouse for advice and possible partners.[42] The US Information Agency and overseas consulates then facilitated communication between the young sisters and kept an eye on their development in the United States and abroad. For instance, when San Diego undertook its affiliation with Yoko-hama, E. Snowden Chambers, program executive with the US Information Agency's Office of Private Cooperation, was "in and out of San Diego sev-eral times," "supplied with suggested programs and materials" to shepherd efforts.[43] In Japan, the public affairs officer of the US consulate acted as a liaison with Yokohama in the apparent absence of direct communication.[44] San Diego was among the first major West Coast cities to affiliate, and this level of federal intervention was higher than for other towns. Still, a range of state bodies carefully guided these seemingly grassroots relations.

San Francisco's route to affiliation was idiosyncratic. First, the formal proposal was rushed to be ready for the tour, bypassing official channels. Chambers did not learn about plans until they were well under way, and he was not particularly pleased with this indigenous initiative. San Fran-cisco's proposed sister was in fact Osaka-Kobe, a conjoined metropolitan area. This directly interfered with a proposed affiliation already under way (along correct procedural lines) between Seattle and Kobe.[45] Second, ex-cept for the formal acquiescence of the mayor and supervisors, the busi-ness community monopolized the program to an unusual degree. Anyone can initiate sister-city "friendships," the program's proponents advised, but it needed "the support of city government, civic and service clubs, school and newspapers."[46] Other cities took this more to heart. Seattle's initiat-ing citizens' committee, for example, represented a broad swath of civic institutions, while the trade-oriented business community had to be con-vinced of its utility before its representatives joined.[47] Not only did Wilson not bother to form a citizens' committee; he and the other San Francisco originators were narrowly focused on commercial possibilities. John Bolles, Mayor Christopher's official representative in Osaka and vice-chair of the chamber's Industrial Development Committee, noted that the affiliation would help the two cities exchange "cultural, economic and civic ideas." But, he added, it really aimed to rectify promotional failings: "we haven't done enough selling San Francisco in this area."[48]

The sister-city affiliation was meant to support boosterism and com-merce, but it did so on different terms than the rest of the tour's itinerary. Unlike business meetings and plant visits, the sister-city proposal evoked the ideals of education, cultural exchange, and kinship. The San Francisco program supporters professed a desire not just for profit but also to "learn about the community life" of Osaka.[49] This layer of goodwill diplomacy earned them meetings and events with the city's municipal and business leadership above and beyond business affairs. In Osaka, Mayor Nakai Mitsuji held a luncheon in their honor and met separately with the group. Members of Osaka social and business organizations "approached [Bolles] with ideas for the exchanges."[50] Japan during its annual trade fair teemed with visiting contingents representing countries and major international corporations. The San Franciscans, an otherwise small and insignificant group, became toasted celebrants with a now-permanent connection to Japan's industrial center. San Francisco would retain this pride of place as the only non-Japanese city represented at Osaka's trade fairs; later, it would be the only city (of three city exhibitors total) with its own pavilion and accorded "nation" status at the 1970 Japan Expo in Osaka.[51]

Osaka was attractive to the San Franciscans as the biennial host of Japan's annual trade fair. The "Manchester of the Orient" had originally been Japan's center of textile manufacturing, and even after the city's heavy manufacturing had overtaken light, Osaka remained the nation's leading industrial city and a top port.[52] Osaka was therefore already well known to San Francisco: among the latter's top exports were supply machinery and raw cotton, both of which helped supply Osaka factories. Osaka, in turn, accepted because its own interest in international relations was even more pressing. Japan's postwar recovery was closely tied to its relationship with the United States. The resource-poor nation had relied on trade and colonialism to supply its prewar industry. Forced to relinquish its colonies after the war and with capital in short supply, the Japanese reliance on US imports led to a looming trade imbalance that policy makers struggled to address. This national problem was magnified in Osaka. As a port and industrial center, Osaka had enjoyed substantial growth through US aid and economic-development policies, but in ways that heightened its dependence on foreign markets. Its textile mills, chemical plants, and developing heavy manufactures relied on imported resources and, increasingly, foreign buyers. The large, Tokyo-based general trading companies, dismantled as part of an overall suppression of zaibatsu (huge financial, industrial, and commercial conglomerates) during the occupation, were now "coming back to life" in strikingly similar forms and drawing a disproportionate amount of Japan's overseas commerce.[53] By the time San Francisco proposed the affiliation, Osaka had already begun its own publicity campaign. The city had originated the Japanese trade fair in 1954, and its Chamber of Commerce had just begun publishing a monthly English-language economic bulletin. The sister-city program aligned with both cities' international boosterism.[54]

It is important to note that selecting one particular foreign city for permanent and privileged sister-city status also had perceived risks. Some contemporary cities hesitated for fear of alienating other foreign towns. The New Orleans mayor pointedly refused to select a sister. He noted that his city was "keenly interested in maintaining contact with foreign countries" and so felt it "wiser not to single out any particular one" in deference to New Orleans's "business and cultural relations with so many different cities."[55] Snowden Chambers encountered this objection "in nearly every town I have visited."[56] This was a significant challenge as the young program's parameters and meaning coalesced, even as it suggested that many cities saw its economic dimensions. San Francisco was certainly not alone in its decision to affiliate; San Diego, Seattle, and Boston—all cities with their own significant international interests—conjoined with sisters at about the same time. Nevertheless, the choice to enter into an affiliation was deliberate.

The commercial ambitions of the affiliation's San Francisco originators shaped the early program. The affiliation's official mission was to "develop this relationship between Osaka and the 'Gateway to the Orient' for the promotion of Japanese-American commerce and mutual understanding in all fields, particularly that of world trade."[57] Furthermore, aside from a handful of municipal employees, the most active participants in San Francisco's affiliation were those with the most at stake in such commerce. They worked for or sat on the boards of the Metropolitan Insurance Company, US Customs, San Francisco Port Authority, State Board of Harbor Commissioners, Fairmont Hotel, and American President Lines; these bodies provided financial services for, oversaw, hosted, and conducted transpacific commerce.[58] There were few exceptions, including the first and quickly replaced chair, Phillips S. Davies, a prominent civic leader. Davies had cultivated youth participation in the affiliation and was committed to the program's ideals: he "sincerely . . . believe[d] it [was] an activity not only to help San Francisco, but to help the whole Country."[59] However, he had no economic interest in Japan. After a year, Davies left to join the finance committee of the national People-to-People Program where his principles were better placed. Lester L. Goodman, the president of Getz Bros. & Co., took over. Not coincidentally, these men were matched by business counterparts in Osaka. The Osaka–San Francisco Town Affiliation Committee had representatives from Japan Air Lines, the Kansai Steamship Company, department stores, and electronics manufacturers, all industries involved in Japan's burgeoning economy and growing export industry.[60]

CHALLENGING VIEWS OF JAPANESE DEPENDENCY

The commercial ambitions of the sister-city committee members shaped the early programming in San Francisco. The initial affiliation events held a common theme: they represented Japan and Japanese in robust, male, and professional ways very different from the feminized, childlike, and dependent portrayals in popular culture. These early affairs were for the most part open to just a few select San Franciscans and Osakans, but they targeted critical audiences. First, the affiliation portrayed Japan in equitable ways for San Francisco business and municipal figures who might not have particularly valued Japanese connections, but who were needed for respectful cooperation. Additionally, events assured Osakan leadership about San Francisco's commitment.

The affiliation began in 1957, just five years after the occupation of Japan ended and within a popular and policy context of inequitable relations. Japan held a distinctly "subordinate position" in US foreign policy, according to Prime Minister Kishi Nobusuke, solidified by treaties, military bases, aid, and international organizations.[61] This was increasingly chafing for Japanese policy makers as their nation rebounded economically and gained General Agreement on Tariffs and Trade and United Nations memberships. Yet the nation's inequitable US stereotype proved stubborn. Historians have shown that throughout this period, the US media emphasized portrayals of Japanese as unthreatening, diminutive, even comic. Stories of war orphans, demure and attractive women, students, and other submissive or immature figures circulated widely. The 1957 film *Sayonara*, which came out the same year that the San Francisco–Osaka affiliation began, is a frequently cited example. The popular award-winning film depicted Major Lloyd Gruver, stationed in Japan during the Korean War, abandoning his previous racism and his white fiancée to marry the beautiful Hana-Ogi over the objections of friends and superiors; Hana-Ogi, too, changes from an independent career woman (an actress) to his submissive, content wife. Historian Naoko Shibusawa has argued that such depictions accommodated Americans to their government's new and costly investment in the recent enemy as well as naturalized US global power. The film used ideologies of gender and age to suggest a postwar United States maturing into a ready global power, with less capable nations such as Japan under its guidance. This drew on decades-old feminine, diminutive images of Japan and helped fuel what writer Mishima Yukio observed as Japan's newfound "fame" for its women, "reputed to make the best wives in the world."[62]

This feminized, distinctly unequal image of Japan had local parallels in San Francisco, especially in the popularity of sukiyaki restaurants and their demure waitresses. During the 1950s, there was a "postwar boom in Japanese sukiyaki and tea houses," according to an enthusiastic critic.[63] These restaurants specialized in "exotic atmosphere" as much as cuisine: "it is not alone the food which pleases so greatly . . . but the entire mood and other-world atmosphere, the gracious glimpses of the Far East."[64] But the most popular feature was the female waitstaff. These "brilliantly kimonoed, beautiful little Japanese girl[s]" were regular elements in reviews.[65] Their work was a performance in submission, a dynamic heightened by the preparation used for sukiyaki in these restaurants. "Geisha-style waitresses" brought out the raw components and then, "kneeling before the electric or gas plate," "cooked [the] main dish right at the table."[66] Thus the women attended to patrons in much more extensive and responsive ways than waitstaff in other restaurants: kneeling by the table, cooking, serving, refilling the tiny sake cups, and replenishing the cooking pan as the customers ate. These women, many of them war brides, and their employers in fact suggest a quite savvy and lucrative use of popular stereotypes.[67] Yet they relied on and furthered the image of feminine, Japanese submission for local patrons and national readers.

San Francisco images of Japanese submission and dependency were not limited to women, as the 1951 peace treaty conference demonstrates. Unequal relations were integral to the conference: it formalized Japan's surrender and sought to create terms for continuing US influence in Japan and Cold War geopolitics more broadly. Not only was Japan largely sidelined by the focus on the "anticipated Russian tempest," but local coverage also diminished Premier Yoshida Shigeru's shrewd diplomacy with belittling references to the "diminutive Prime Minister" with a "high voice." He arrived, moreover, to meet with "leaders of the Nation which beat Japan to her knees."[68] The media highlighted Japan as a "defeated and humbled nation" in ways that echoed long-standing and persistent views of Japan as a "lilliputian Japanese fairyland," amusing, weak, and decidedly inferior.[69]

This view of Japanese inequality was further interpreted in the contemporaneous flood of visitors for reasons other than official state business. During a visit of National Diet members in 1951, an editorial expressed surprise at their embrace of the US occupation: the group "might have been expected to plug for an end" to foreign domination, and so their support was "peculiarly thought provoking."[70] Nonpolitical visitors were similarly shown as seeking US guidance and tutelage. A "great Japanese chef," perhaps unsurprisingly given the food's rising popularity, toured the United States "introducing Americans to the intricate elegance of Japanese cuisine." However, even he was framed as seeking an education: he was seen "prowling around supermarkets," trying to "[learn] something about American fare and its methods of preparation."[71] These trips were often US or worldwide tours, but San Franciscans were especially likely to have a sense of the frequency of Japanese missions, due to San Francisco's status as a hub for transpacific travel. Farmers, judges, sailors, Buddhist leaders, and others came through the Bay city seeking skills, training, or models with which to revise their institutions.[72] Their frequency was enough so that by 1958, Washington political observers disparagingly pronounced them evidence of Japan's "servile democracy."[73]

The national and local perception of Japanese immaturity applied not only to abstracted images of Japan but also to the products that Americans might encounter from Japan. These represented the output that Japan needed to increase and the goods that could fuel San Francisco's expanded transpacific networks. Yet such merchandise struggled to win US respect. Although cameras, transistor radios, and other precision technologies increasingly demonstrated the high quality of some of Japan's exports, many Americans continued to view Japanese goods as "gimcracks" and cheap imitations, the products of an immature economy. San Francisco commentators shared those views, calling Japanese products "always cheap and sometimes shoddy."[74] Even if Japan's economy had experienced a remarkable postwar upswing, San Franciscans and Americans more broadly continued to view its businesses as requiring guidance and development.[75]

The view of an immature and dependent Japan was well established nationally and locally at the beginning of the San Francisco–Osaka affiliation. It is therefore not surprising that the affiliation was first described as an "adoption" of Osaka.[76] Prominent San Franciscans, those needed to support an expanded transpacific relationship, appeared to assume an unequal US-Japanese relationship, a view that clashed with Japanese expectations. In an early exchange, the city donated materials on San Francisco for an Osaka exhibit. The exhibitors and Osaka mayor Nakai Mitsuji also requested a "letter of commendation" to display, which Mayor Christopher duly sent along. In it, he expressed his best wishes for the exhibit and his pleasure in the "cordial relationship which exists between the people of our two cities." However, his letter also betrayed no small degree of condescension: "the citizens of San Francisco" were "very interested in encouraging the economic and cultural progress of Osaka."[77]

Not only did these views comfortably reflect popular assumptions of Japan immaturity; they could lead to dismissive, even rude, interactions. The affiliation had a rocky start after Nakai invited his San Francisco counterpart to his city's affiliation celebration. Christopher, busy campaigning for governor, turned down the invitation and agreed to send a representative only after Nakai offered to cover some of the costs. Moreover, Nakai had to plead for a congratulatory message to be read at the ceremonies. Even worse, ponderous bureaucratic decision-making and untoward economizing kept San Francisco from sending the requisite commemorative gift. All this led a local consular officer to understatedly observe that San Francisco "did not seem to be too well prepared," which resulted in "disappointment" in Osaka. Imbalanced views of Japan could have unfortunate consequences for well-functioning and happy transpacific relations, particularly as they clashed with Japanese perspectives on the affiliation.[78] Unsurprisingly, Osakan officials did not view their position as the grateful recipients of patronage. When the affiliation had first been proposed in March 1957, Nakai was careful to announce the proposed affiliation to his constituents only after the San Francisco Board of Supervisors had formally passed the proposal. Further, this was a proposal up for debate in Osaka: far from jumping at the chance to be "adopted," Nakai and his council deliberated on the decision to accept.[79]

San Francisco's events and exchanges with the sister-city affiliation helped rebalance discordant relations, as evident in San Francisco's inaugural sister-city celebrations of professional Osakan men. To some extent, the professional skew of the festivities simply reflected Japanese travelers: in an effort to keep money at home, international tourism was restricted until 1964, so Japanese travelers were largely limited to those on business, state, or educational missions. However, there were other ways to celebrate the affiliation. San Jose, for example, kicked off its affiliation with Okayama with a week of public events, mostly revolving around local children's contributions and culminating in a public ceremony with an estimated five thousand celebrants. In contrast, San Francisco's first sister-city event was a private reception for Nakai as he passed through the city on a US and European tour. The second event, and the much larger one, honored Sugi Michisuke, president of the Osaka Chamber of Commerce and Industry. The sister-city affiliation, the Chamber of Commerce, and the WTA threw an elaborate, invitational luncheon to celebrate Sugi, a figure of national stature in Japanese economic circles (fig. 2.1). He had been "active in export-import trade in Japan for more than forty years," with executive positions in "nearly

FIGURE 2.1 Sugi Michisuke laughing with E. D. Maloney (*right*), president of the San Francisco Chamber of Commerce, and acting mayor James J. Sullivan at the Japan Trade Center. The affiliation brought leadership from both cities together on equitable terms and countered feminine, stereotypical displays evoked by the mannequin. Here the dummy and its dress is rendered fantastical and artificial by the familiar garb of the three suited men as well as their humorous attitude toward it. The equitable heights of Maloney and Sugi balance Sullivan's much taller stature, suggesting a corrective to such famed, unequal images of a towering, khaki-clad General MacArthur and the slight, tuxedoed Emperor Hirohito. BANC PIC 1959.010-NEG pt. 3 [11-08-57.12] frame 1, San Francisco News-Call Bulletin newspaper photograph archive. © The Regents of the University of California, The Bancroft Library, University of California, Berkeley.

seventy firms and organizations," and was "head" and founder of the Japan External Trade Organization. Sugi had in fact founded the organization in 1951 in order to promote regional trade, but it quickly became a key national institution for international trade promotion.[80] The affiliation thus brought San Francisco leadership into contact with a mature, highly successful Japanese figure and, furthermore, revealed that Sugi too saw the affiliation as an opportunity for "greater cooperation between our two international trading centers" through a "tremendous flow of goods and services to our mutual benefit."[81]

Events such as this were largely private and invitational, but they introduced Osakan figures to the San Francisco leadership most likely to interact with Japanese partners. This included members of the municipal government who might correspond with Osakan counterparts; leaders of civic groups like the International Hospitality Center and the Asia Foundation who might work with Osaka; and members of the Chamber of Commerce and WTA beyond the originators of the affiliation already invested in Japanese trade.[82] These leaders met Japanese men whose status commanded respect and challenged popular stereotypes. Meanwhile, Japanese visitors were assured of their welcome and prestige in San Francisco.

These Japanese visitors were introduced to San Francisco civic life, where they could witness firsthand the municipal support for Japanese partnerships. When visitors arrived under the rubric of the affiliation, the city was responsible for their reception. When an Osaka YMCA group visited San Francisco in 1961, as they did regularly, the San Francisco YMCA director supervised much of their visit. However, John T. Buckley, the president of American President Lines and an affiliation officer, organized additional activities such as a meeting with the mayor and a farewell dinner.[83] What might have been an internal affair of the international YMCA became one associated with a representative organization of the city. Moreover, the municipality at times assumed full responsibility for Japanese visitors. When the Osaka affiliation committee sent their sons on a visit, the boys' schedule—which included tours, housing, games, and transportation—was completely organized by the city and affiliation members.[84] Such exchanges made the city responsible to its Japanese networks in ways that were visible to their Osakan counterparts. The city's hospitality helped reframe San Francisco for these visitors as especially hospitable to their interests.

It is also possible that the sister-city affiliation's exchanges aided participants in more concrete ways. Lester L. Goodman, the affiliation president, noted that San Francisco's finance and trade with Japan had seen a "considerable upward trend of activity since the Affiliation came into be-ing."[85] While a direct causal link is difficult to prove, certainly the program increased and solidified transpacific relationships. Correspondence often passed through the mayoral offices, but the work of organizing, sending out materials for exhibition, and hosting often fell to the men invested in trans-pacific commerce, such as Buckley. Affiliation activities could therefore add to their Japanese business contacts, especially through the sociability and cooperation demanded by the many hosting opportunities. The two cities took "every opportunity to send good will ambassadors, whether they be business men, service club delegates . . . , delegations of students, or just tourists."[86] In fact, they became so numerous that the two municipalities eventually developed a tiered system of hospitality to limit the shared bur-den of hosting.[87] And each of these many visits provided an opportunity for business or political contacts.

FIGURE 2.2 Visiting San Francisco lawyer and goodwill delegate H. G. Serlis was treated to a reception with Deputy Mayor Wani and other officials in Osaka, 1961. Such occasions offered a ready-made network of contacts and opportunities for extra-commercial sociability otherwise difficult for individual travelers to secure on their own. Photo 1 of 3, folder 23, box 43, George Christopher Papers, San Francisco History Center, San Francisco Public Library.

Exchanges could also add another layer to preexisting Japanese ties, allowing participants to demonstrate reliability, commitment, and motivation beyond self-interest. For instance, when Osaka requested materials for a San Francisco exhibit in 1958, Charles von Loewenfeldt organized and shipped them. He was one of the most active affiliation members in San Francisco and a US publicist for Japan Air Lines, one of the exhibit sponsors. This gave him an opportunity to forge a mutual interest in transpacific cultural relations with his client. Such noncommercial bonds could be beneficial. Historian Sven Beckert has shown that in the early years of multinational firms, familial, ethnic, and religious bonds were employed to mitigate the extenuated risk of international capitalist dealings: "reliability . . . came more easily when people . . . had ways of enforcing trust embedded in social connections." Over a century later, international agreements and codified professional practices helped enforce reliability, but "extra-market social relations" could help stabilize trust dislocated by war and generate allegiance from a client with competing choices.[88] This may or may not have helped von Loewenfeldt, but his work with the affiliation paralleled a rise in his Japanese business: by 1965, when he became committee president, he worked for the Japanese consulate and "many of the large Japanese firms"; later, he would work on behalf of the Japanese government itself.[89] In the face of what Americans understood as their enormous differences from "the Oriental mind," as Goodman called it, the sister-city's fictive kinship could create an international understanding far more pointed than the general goodwill conceptualized by People-to-People's founders.[90]

The affiliation helped create a new level of familiarity and respect among Osaka's and San Francisco's business community and municipal leadership. These dynamics relied on and fostered the "goodwill" that the People-to-People Program was meant to engender, but it did so for the specifically local purpose of boosterism, rather than the abstracted Cold War goals of winning hearts and minds abroad.

REIMAGINING SAN FRANCISCO
THROUGH POPULAR SPECTACLES

Small, unpublicized or private affairs were only part of the affiliation's activities. The civic nature of the sister-city program also afforded the opportunity to recast Japan for ordinary San Franciscans. Up through the outbreak of war, a vituperative enmity toward Japan and its migrants had existed alongside thriving commercial relations that had made the nation one of the city's most reliable trading partners. This dissonance had not mattered when San Francisco faced minimal competition from urban neighbors and foreign trade was less significant. In light of the city's lessened position among other West Coast cities by the 1950s, however, San Francisco boosters could not rely on economic precedent or status. Instead, goodwill and civic identity became much more important. Parades and public activities therefore presented Japan as a formidable yet approachable ally, exemplified by a number of friendly and respectful encounters with the Japanese Self-Defense Force. The affiliation's popular events supported amiable connections to match economic ones by centering the city's Japanese connections in civic life, city streets, and urban identity in positive, celebratory ways.

Some of the sister-city events forwarded an innocuous and accessible portrait of Japan through colorful, familiar cultural forms. The affiliation's youth activities provide examples. Before he resigned as San Francisco's affiliation president, Phillips S. Davies had invited the San Francisco Youth Association, an umbrella organization for high-school student groups, to join the affiliation. Thomas Rowe, the association's director, made his organization an integral part of the program and quickly became a rare non-economic committee member and an affiliation vice president.[91] Rowe incorporated the sister city into the association's events, designating a "San Francisco–Osaka Day" in the group's annual Youth Week programming. Students produced a local television special and held a public program at a local department store, in which they sang, performed "authentic Japanese dances," and recorded a greeting of "hundreds of students" for Osaka's schoolchildren.[92] The group also arranged for the store's display of Japanese dolls from Osaka. The dolls had some contextual specificity—they portrayed a centuries-old Osakan festival—but the program emphasized a colorful and familiar view of Japanese culture.[93]

Clearly, Osakans helped promote this accessible, attractive, even stereotypical view of Japan. The dolls were characteristic tokens. Among Osaka's other gifts were multivolume histories of their city and a gold medal with the Osaka coat of arms, but mementos such as lanterns, carp streamers, and artwork were far more numerous. Such gifts were accessible, if exotic, to Americans audiences, and drew on tropes such as delicateness, aestheticism, and timelessness integral to US views of prewar Japan.[94] They were also chosen for that reason. When the Osaka committee sent a group of watercolor paintings, for instance, members explained that they were "typical" to Japan and "rich in Oriental taste," but had been selected in accordance with "American taste."[95] Seeking popular reception for their goods and relations, Osakans consciously used US stereotypes about Japan in the service of their own interests.

Such gifts, however, were interpreted through and contextualized by San Franciscans' local goals. The dolls, for instance, were displayed alongside Japanese cultural performances, but Japan itself was represented by an adult male figure, the Japanese vice-consul, who was interviewed for the Youth Association's television program. Once again, the affiliation balanced stereotypical Japanese images with a male figure of maturity and professionalism.

Japanese formidability and professionals were especially on display at the affiliation's first large-scale events. They featured the Japanese Self-Defense Force, as the reconstituted armed forces were known, raising the specter of the recent enemy in its most direct and immediate form. As historian John Dower has argued, wartime demonic images had not been limited to the Japanese leadership or military during World War II, as they were with Germans and Italians. But no figure had so directly embodied Japan's savage threat as the combatant, the "Japanese superman."[96] Through parades, visits, and celebrations, San Francisco affiliation events helped reinterpret that most frightening of Japanese personages into a friendly, approachable ally.

San Francisco's first postwar public encounter with the Japanese armed forces was during the city's inaugural Pacific Festival in the fall of 1958. A visiting naval squadron was the most prominent feature of the festival, crowding a ten-day schedule full of activities celebrating Japan and sister city Osaka. Vice Admiral Yoshida Hidemi was a special guest at the opening ceremony, and his officers occupied seats of honor. The mayor hosted a private reception for them that same day, and Yoshida gave a featured speech at the "patriotic observance" later that week. His crew offered a martial-arts demonstration in Union Square and attended nightly USO dances while the ships docked prominently off San Francisco piers.[97] Indeed, the squadron was at the heart of the Pacific Festival.

The squadron's visit could have been a potent reminder of Japan's former enmity, but the activities of both Japanese servicemen and San Franciscans painted a friendly picture of Japan. The war had been over for more than ten years, and representatives of the Japanese military had since been positively received in the United States.[98] Yet the careful framing of events suggested a lingering local unease with the Japanese military, and the cooperative efforts of Japanese and San Franciscans endeavored to defuse any fears. For instance, local media underscored the contingent's lack of threat. The visitors came on warships, but they were "former American naval vessels" in Japan's "'self-defense' force mobilized while the nation was still under American occupation." In addition, while the fifty-four-year-old vice admiral had decades of service, his "World War II career was spent entirely in shore commands," not in direct combat with Americans.[99] Finally, San Franciscans were given the opportunity to experience the postwar Japanese force's "naval hospitality" for themselves. The crew warmly welcomed visitors at an open house aboard the "gaily decorated" destroyers and frigates, where "eager" crew members acted as guides. Enthusiastic San Franciscans took up the invitation. "More than 5000 visitors"—"hundreds of families"—took advantage of the opportunity to clamber aboard the ships and meet the crew.[100] The squadron could have been a looming reminder of war in a city that had served as an embarkation point for the Pacific theater. Instead, the Japanese Self-Defense Force was rendered friendly and familiar by their own activities and those of participating San Franciscans, who exhibited an acceptance of and friendliness toward their old enemy.

The squadron's visit was integral to the festival's activities, but its participation had been a fortuitous coincidence that demonstrated how the sister-city affiliation amplified underused transpacific networks. The festival had been "somewhat hastily organized," as a critic noted, and the mayor's announcement the preceding month included few details aside from a list of nations "invited" to participate.[101] It was only after that skeletal statement was issued that organizers learned of the Japanese squadron's concurrent stay. Clearly, the links of city leadership to Japanese institutions were not without weaknesses. However, other networks rectified them. The local Japanese American community, which had almost no role in the affiliation up to this point, had known about the squadron's visit and pointed out its overlap with the festival dates.[102] Their need to interject themselves into the affiliation was a contrast with other city's sister-city programs, in which Japanese Americans might have much larger roles. But San Francisco affiliation proponents' singular focus on transpacific commerce meant they had to be educated in Japanese American utility.

Japanese Americans began to take on more of a role as a result of events such as the squadron's visit, and in ways that also furthered the integration of the Japanese navy into San Francisco life. Local Japanese Americans had known of the visit through an established tradition of hosting Japanese sailors. Japanese ships had long docked in the city on training missions, goodwill visits, and ship deliveries. In fact, two separate squadrons had stopped in the Bay three years earlier; the first returning from their delivery in South Carolina and the second simply on a goodwill visit. But these were neither officially received by the city nor granted anything but perfunctory media coverage. This limited reception grew, through the affiliation and through Japanese American activity, into a public and citywide series of events. The Japanese American community still held its picnic for the crew, but it was incorporated into a much longer list of civic activities. The sister-city affiliation, and the larger scope of Pacific-oriented interest it reflected, gave the city a way to assimilate preexisting Japanese connections that had apparently lain outside the realm of the city's municipal and business leadership. The affiliation brought these links into the public eye, connected them with city institutions, and contextualized them in a broader set of interests and networks with Japan, all while highlighting the contributions of the city's long-marginalized Japanese diaspora. The affiliation regularized this and subsequent visits into affairs for the entire city to enjoy.[103]

Two years later, another cohort of the Self-Defense Force traveled to San Francisco. This group underscored not just the strength of the city's transpacific ties, but also their unique historical specificity and expansive popular support. It was the centennial of the first Japanese mission to the United States and the first commercial treaty between the two nations. President Eisenhower declared 1960 the "United States–Japan Centennial Year," and there were observances in Washington, DC, Boston, and other cities.[104] San Francisco, however, commemorated the centennial of the *Kanrin Maru*, the first Japanese ship to arrive in the United States. This centennial was linked to the national one, as the ship had been the vanguard for the diplomatic mission traveling separately on a US ship. But the *Kanrin Maru* was unique to San Francisco, the only US port it visited.

The *Kanrin Maru*'s centennial was equally, if not more, important for San Francisco's counterparts across the Pacific, but for slightly different reasons. The Osaka newspaper *Osaka Shinbun* marked the centennial of the commercial treaty as historic, but the embassy's ship was even more so; the *Kanrin*'s journey marked the first time the Japanese people had crossed the Pacific Ocean alone.[105] As one of the *Kanrin*'s distinguished passengers noted, the venture was "epoch-making . . . for our nation" and marked Japan's rapid mastery of Euro-American knowledge and technology: the sail- and steam-powered ship had been piloted by a Japanese crew, none of whom had prior oceangoing experience and who in fact had seen their first steamship just seven years prior.[106] The emphasis on the *Kanrin* was a gratifying commemoration, far more so than the US-Japanese treaty itself. That had been the unequal result of US gunboat diplomacy and led to considerable domestic outcry and dissent. To commemorate the centennial, therefore, Osaka sent both its assembly president and the striking gift of a polished black granite monument to its sister city. Furthermore, this was an anniversary the Japanese government took seriously. The *Kaiwo Maru*, a merchant-marine training ship, carried the monument to San Francisco. In addition, the former premier Yoshida Shigeru attended the festivities in San Francisco, on his way to celebrations along the East Coast.[107]

The final parade for the *Kanrin Maru*'s centennial gave San Franciscans the opportunity to publicly celebrate their city's historic links with Japan, including diasporic ones. In order to appeal to a maximal audience, the parade included an array of images of Japan. These included female, decorative, and nonthreatening figures. The grand procession of almost 2,000 participants included a legion of "600 Japanese girls" from "schools and churches throughout the Bay Area." The dancers performed a "classical folk dance" in "multicolored kimonos," evoking a colorful and accessible Japanese culture.[108] They were part of a contingent of about "1,000 Japanese from throughout the Bay Area," representing the Nisei Foreign Legion, churches, and other organizations.[109] This participation was a token gesture for a group still facing employment and residential discrimination. Yet it broadcast an embrace by their city and illuminated another bond between the city and Japan.

The Japanese American girls' performance was unusual, but the truly uncommon spectacle came from the Japanese cadets of the *Kaiwo Maru* (fig. 2.3). Forty paraded in the procession, "dressed in green and blue robes of feudal samurai . . . similar to those worn by men of the first Japanese sailing vessel that docked [in San Francisco] in 1860."[110] Descriptions of the "black tortoiseshell helmets or saucer-shaped straw hats" and "white handled swords" suggested the novelty for viewers. Their striking performance brought to life a storied, martial figure evocative of "the splendor of old Japan," before militarism and hostilities with the United States, and "delight[ed]" the ten thousand San Franciscans in attendance.[111] Such samurai figures had reentered US consciousness with Akira Kurosawa's *Seven Samurai* (1956), his next hugely popular film after *Rashomon* (1950), although were still largely overshadowed by geisha and female stereotypes. San Franciscans were given a close-up view of these stalwart figures. Alongside their fellows in contemporary uniforms, the cadets offered a starkly martial view of Japan with serious-faced young men in "elaborate robes."[112] Their activities communicated a long history of Japanese strength and formidability, a contrast to the dancing girls in kimonos.[113]

The centennial events helped make the longevity and breadth of San Francisco's ties to Japan visible and public. Certainly, this was a self-presentation of the Japanese merchant marines and their government. Yet

FIGURE 2.3 Japanese cadets dressed in robes of feudal samurai march past a reviewing stand with Mayor George Christopher (the tallest front-row figure) and other municipal officers during the 1960 *Kanrin Maru* centennial. Their striking costumes and solemn faces drew popular attention and portrayed a martial and robust Japanese past, one with unique historic ties to San Francisco. Negative 7292, subject file: Parades—Japanese, San Francisco Historical Photograph Collection, San Francisco History Center, San Francisco Public Library.

the display was framed by San Francisco organizers who coordinated the parade and alerted the press and the public of the cadets' plans. Additionally, the display of Japan's martial (not militarized) history was contextualized in the city's own past and migrant links with Japan. Finally, thousands of San Franciscans demonstrated their own interest in these Japanese connections and their city's Pacific bonds by flocking to the parade. Together, the organizers and the attendees showcased a diverse set of transpacific connections for their city.[114]

FOSTERING MASS AND ROUTINE LOCAL ENGAGEMENT WITH JAPAN

Parades and festivals were spectacles that made San Francisco's Japanese ties public and popular, but they were unusual occasions that also drew those already amenable to Japan. The sister-city affiliation integrated Japanese connections into San Franciscan life in more mundane and expansive ways, too. Exchanges between schoolchildren, especially those with young women, reached large numbers of even reluctant San Franciscans through the everyday and intimate spheres of families and schools. This appeared paradoxical, given that such images of Japanese femininity and immaturity upended the affiliation's portrayal of Japanese maturity and equity. But the symbolism of youth and women was unthreatening. Their cultural associations granted the affiliation meaning beyond mere commerce and offered the sister-city affiliation entry into otherwise inaccessible spaces.

Although businessmen steered the affiliation, youth had an important symbolic role. On his first trip to Japan in 1959, taken to solicit participation in the second Pacific Festival, San Francisco mayor George Christopher couched the affiliation's significance in terms of youth. He told his Japanese audience that he hoped to deepen the two cities' relationship through the "cultural interchange and exchange of children visitors" so that "young school children in both cities" could "know each other."[115] These future "good citizens" would be steeped in a global perspective and could ensure that the "civilized world . . . [would] be composed of civilized people."[116] Christopher drew on a romantic view of youthful innocence as well as contemporary views about the crucial role of childhood in an individual's development. One 1965 US survey of sister-city programs found "unanimous agreement" that adult exchanges were "secondary to the main purpose of fostering understanding among the young leaders of tomorrow."[117] Children were, in fact, secondary in the San Francisco program, but they potently symbolized the program's ideals and elevated the relationship above the solely commercial.[118]

Youth were also important participants. The San Francisco Youth Association and the school district were the first institutions outside the business community to join the sister-city affiliation, fulfilling the People-to-People Program's popular mandate. Their participation meant that most children in the city engaged the affiliation in some way. Students in high schools across the city partnered with Osakan pen pals. Youth of all ages traded artwork and scrapbooks across the Pacific. Schoolchildren were a useful constituency for the affiliation because they were involuntarily entangled with geographically dispersed and functionally disparate institutions.[119]

Children's participation also brought the sister-city affiliation to San Franciscans of all ages. The Osakan schoolchildren's creations were often publicly displayed in department stores in both cities, as many such gifts were. In San Francisco, this put the exchanges before casual pedestrians in the busy downtown shopping district, where they could seek out or bump into these exhibits. These public exhibits provided even the most provincial or resistant residents some degree of engagement with the affiliation through the most innocuous and nonthreatening means possible: the unchallenging, universally recognizable creations of children.[120]

As dependent members of households, children also ensured a degree of participation by parents and family members. This might only mean an audience for events such as the "Youth Week" affiliation programs. But the demands of children could also lead to more extensive contributions. For example, when a group of high-school boys from Osaka visited San Francisco, the San Francisco Youth Association arranged a party, tickets to a Giants game, a sightseeing tour, a visit to Stanford University, and families to billet each member of the tour. The latter was a generous display of hospitality that required three nights of hosting and attendance at some activities. There were many ways that adults could join the affiliation's exchanges on their own; the Lions Clubs of both cities affiliated in 1960 and beauticians in 1961. However, adults could easily avoid or remain ignorant of the affiliation simply because they had no equivalent to mandatory school enrollment. Primary and secondary students extended the reach of the affiliation to a much larger swath of the city populace.[121]

Student exchanges relied on childlike portrayals of Japan, demonstrating the utility of such tropes for widespread appeal. But the "Miss Osaka" exchange, begun in 1961, also showed the ways that San Francisco leaders reworked those stereotypes into a semblance of parity. That year, the San Francisco committee invited a "Miss Osaka" and chaperone for a free, weeklong visit in order to "show San Francisco to our Sister City through the eyes of an Osaka school-girl."[122] The chosen ambassador was young and pretty—the San Francisco committee had expressly requested a "photogenic" delegate—and in many ways resonant of the most familiar media portrayals of Japanese women.[123] Keiko Tsuji was a "real Japanese-type girl," in the words of the Osaka committee.[124] She was a quiet, accomplished student with considerable skill in Japanese dance, music, embroidery, and singing. Tsuji's "demure" and "captivating young charms" were warmly received in San Francisco, where she was consistently described in diminutive if affectionate terms.[125] When "little Miss Osaka" was treated to a shopping spree, for instance, the store's fashion coordinator had "trouble finding things small enough for her."[126] Tsuji was also often photographed in the kimono specially provided by a Japanese department store—the young woman wore dresses at home—and presented a figure that Americans across the country could find recognizable and sympathetic.

Tsuji's apparent adherence to stereotypes was a choice on both sides of the Pacific. While instigated by the San Francisco affiliation committee, the "Miss Sister City" program was developed in partnership with Osaka committee members, who accepted the invitation and chose the delegate. Members' choice in gifts had already demonstrated a preference for accessible, attractive representations of Japan, something Tsuji certainly fulfilled. However, this was a departure for the San Francisco committee. Partly it demonstrated their national ambitions. Lester L. Goodman, who conceived of the program, saw it as an "opportunity to dramatize" San Francisco to a US audience and predicted that "newsreels, television, the press—Washington . . . would love it."[127] But it was also perhaps dictated by the private spaces that Tsuji would enter: the homes of San Franciscan host families, the dinners with locals, and attendance at San Francisco schools and activities. The demure and unthreatening Tsuji was better positioned to enter those vulnerable spheres than, for example, the martial young Japanese cadets. In fact, it had been the specter of Japanese young men in San Francisco schools that had instigated the Board of Supervisors' attempt to segregate Japanese students in 1906; it was true that those young men tended to be older, of an age when they could work to support their studies, but their sex as much as their age had sparked San Franciscans' fears for their daughters' safety.[128]

However, the Japanese trope of femininity and immaturity was not allowed to stand alone. The next year, upon invitation from their Japanese counterparts, the San Francisco committee sent their own representative: "Miss Sister City." The program thus paired two young high-school women, equitably representing their respective cities in the nonthreatening ways perhaps necessary for the extended and intimate exchange. The San Francisco committee chose a student, Ellen McGinty, with very different attributes from the demure Tsuji. Her biography had entire paragraphs listing her academic and extracurricular achievements, leadership, and service. There was no discussion of rectitude or demureness; instead, she came across as a self-possessed and confident young woman. Nonetheless, as an attractive and unthreatening young woman, "Miss Sister City" was very much the equal of "Miss Osaka."[129] Like the youth programs of art exchange and pen pals, the affiliation paired youths from both cities in parallel relationships. The affiliation's San Francisco leaders were not entirely willing to forgo the program's goals of equitable transpacific relationships.

Children's participation shaped the affiliation in crucial ways. Their activities embedded San Francisco's program in ideals of goodwill and world peace. The particular status of children also extended the affiliation's reach among the San Francisco populace. Finally, even though the program's youth events drew most heavily on representations of Japanese immaturity and feminization, the program's exchanges balanced these images with parallel portrayals of young and female San Franciscans.

SAN FRANCISCO'S RISING COMMITMENT TO JAPAN

The development of the sister-city affiliation with Osaka signaled a shift in politics and civic life in San Francisco. The affiliation's public and celebratory exchanges were just one component of Japan's newfound importance in the city. Municipal and business leaders broadly incorporated Japan into their programs and institutions in ways that supported new and preexisting transpacific links.

Rising interest in Japan and a willingness to work equitably grew in the years after the affiliation began. After the first few years of the affiliation, and after years of clear, well-publicized evidence of Japan's economic "boom," a growing number of civic leaders became interested in transpacific ties. For example, the San Francisco Chamber of Commerce developed new means to cultivate Japanese commerce. In 1959, the Chamber of Commerce published its first "Directory of Importers and Exporters" in the Japanese language, the only foreign-language edition aside from a Spanish version produced in this period.[130] The following year, the chamber's World Trade Department collaborated with Japan Air Lines to publish a "tremendously popular" Japanese-language guide to San Francisco, "one of many successful efforts the World Trade Department is making to expand trade."[131] Such efforts were critical because, by the early 1960s, Japan had become, according to James P. Wilson, "our biggest single overseas trading partner."[132]

The municipal government, too, grew more attentive to San Francisco's transpacific ties. The dismissive attitude apparent in the early days of the affiliation evaporated, replaced with a new level of commitment. In 1958, Mayor Christopher sent a "State of the City" letter to the Board of Supervisors that departed from his past tallies of municipal departments' activities. Earlier statements had reflected his reformist vision and maintained his predecessors' concentration on internal administrative functioning. The 1958 version, however, echoed the Chamber of Commerce's goal to make San Francisco an administrative, commercial, financial, and visitors' center for the Bay Area, the United States, and the Pacific region. The latter was especially significant for a "cosmopolitan" city that was "the gateway . . . with the Pacific nations," as the area's "rapid economic growth" meant more Pacific-area travel and commerce.[133] These were not just words. Christopher acknowledged the importance of Japanese relations the following year with a goodwill visit to Japan in 1959, where he met with local and national officials in Osaka and across the country. This was the San Francisco mayor's first visit, but trips from both sides of the Pacific increased. In just the first decade of the affiliation, there were mayoral missions from Osaka to San Francisco in 1957, 1959, 1961, and 1965 and from San Francisco to Osaka in 1959, 1963, and 1964; visits by other "ambassadors" occurred with even more frequency. These trips involved a considerable local investment of funding and time, taking the mayors away from the city for weeks and involving the scheduling of a host of administrators. Furthermore, the mayor's office, the police department, the fire department, and other divisions hosted transpacific guests in order to give them broad exposure to the city and to open opportunities for mutual education. Municipal officers thus invested in the transpacific sharing of policy, institution building, and programming.[134]

These visits, communications, and exchanges with Osakan counterparts sustained a new level of familiarity with Japanese partners and a willingness to support transpacific partnerships in San Francisco. In 1959, at the mayor's urging, the Board of Supervisors granted the sister-city program a small annual budget to supplement privately raised funds. Municipal staff also supported the affiliation, applying the fluency they gained through regular contact with Osakan officials and institutions. One longtime staffer, for example, patiently explained the opaque titles of upcoming Japanese visitors to Christopher's successor. Years of exchanges had granted the administrator a working knowledge of Japanese bureaucracy, social hierarchies, and some understanding of how to address them. Such knowledge smoothed the reception of travelers and created respectful transpacific municipal relations across mayoral administrations.[135]

CONCLUSION

This work of transpacific connection building was not unique to San Francisco. Similar exchanges occurred along the West Coast and around Japan, through institutions such as the annual Japan-American Conference of Mayors and Chamber of Commerce Presidents, begun in 1955. San Francisco boosters attended such events. In 1959, Goodman went to represent his mayor, and the mayor himself attended in 1963; the latter trip was coordinated by the San Francisco chamber's annual business tour of Asia. But because these programs were expressly intended to connect many US cities with Japanese counterparts, it had less utility for San Francisco boosters who desired unique and exclusive networks. Such activities made the sister-city program decreasingly useful for their needs by the early 1960s. Other cities also employed the sister-city affiliation to solidify their own transpacific exchanges, and in fact, Japan had the most numerous relationships of any participating non-US country. By 1960, the nation had nineteen affiliations in ten states. San Francisco was not the only city to see the value of overseas promotion and exchange.[136]

Still, the sister-city affiliation had been key to remaking San Franciscans' relationship with Japan during the late 1950s. The affiliation proved a useful instrument to challenge popular views of Japan and rebalance a local incarnation of the US-Japanese relationship. From intimate private introductions to large-scale public events, many of the early San Francisco events highlighted Japan as an equal, ready for and capable of partnership. The affiliation thus encouraged municipal and business leadership, outside of those already invested in Japanese networks, to reimagine Japan as a viable partner. Furthermore, the large-scale events and festivals provided a means to revise the city's civic identity into one more conducive to a friendly partnership with Japan. Through the accessible, approachable, yet still formidable presentations of Japan, the affiliation gave locals a way to rethink their own possibly still hostile views and begin to see Japanese partners as useful allies. Diverse San Franciscans from city supervisors to schoolchildren, Chamber of Commerce members to diversion seekers, were provided low-stakes and highly leveraged ways of engaging Osaka and Japan with events such as craft exhibits, parades, pen-pal programs, Japanese tours, and the "Miss Sister City" program. This resulted in an expansive and far-reaching sister-city affiliation that told a story of San Francisco's long-standing and valued connections with Japan, replacing their city's historic transpacific conflicts with a story of celebrated and long-lived connections.

The affiliation also had unintended consequences for Japanese Americans. They had been overlooked in the early stages of the affiliation, but they used their social and ancestral connections with Japan to make themselves public and valued participants. This was a lesson that could be applied toward greater inclusion in their home city. Their new knowledge would be put to work in other forums, particularly in the urban redevelopment program that San Francisco used to mold its landscape to forward its emerging transpacific ambitions.

CHAPTER THREE

Redeveloping Citizens:
Planning a New Japanesetown

As San Francisco's sister-city program developed, another municipal program took shape, this one far more divisive and controversial. On a Tuesday afternoon in October 1952, three hundred San Franciscans gathered in the downtown Commerce High School auditorium as San Francisco Redevelopment Agency officials unveiled plans for their "Project Number One" in the Western Addition district. The agency had already alerted the project area's community organizations, but this was the first public hearing on plans, procedures, and projected time lines for the extensive, twenty-eight-block project. Most of the area would be completely cleared and resold to private developers for housing, commerce, and institutional use. About 9,600 people lived within the project's boundaries, many of whom could not afford to live elsewhere or were prevented by racial restrictions from doing so. Fears of displacement and exclusion from the redeveloped district had, in fact, prompted a raucous and lively debate at a public hearing in 1948, where three thousand gathered and many publicly expressed their opposition to San Francisco's nascent redevelopment program.[1]

The public hearing in 1952 was quite different. For the most part, while some residents still raised "individual problems," the meetings in general were "marked by lack of opposition."[2] The Redevelopment Agency enjoyed the support of many of the attendees, even though the proposed upheaval was far more immediate. In part, the agency had taken steps to address past concerns. For instance, the Council for Civic Unity, a local interracial and interdenominational civil rights group, had conditionally supported the redevelopment program before but now unanimously and enthusiastically endorsed the plan, based on newly added provisions for residents and assurances of fair treatment. Others, though, may have simply found less incentive for public protest. The first public hearing had generated heated criticisms but had done little to change the Board of Supervisors' actions.[3]

One speaker, Victor S. Abe, stood out. He endorsed the program as others did, but requested some modifications. Abe represented the second-largest nonwhite group in the Western Addition, known in prewar years for its "Japanese district." The Japanese American population had now shrunk—it was much smaller than the city's Chinese American population and was even rivaled by growing numbers of Filipinos—but it still occupied about 20 percent of the redevelopment project area. As a resident and lawyer with offices in the district, Abe recommended that the agency set aside one block for "Japanese neighborhood shops." Past demands based on justice and protection had been largely disregarded by city officials, so Abe did not bother to reference the rights of merchants or residents. Instead, he emphasized the benefits to consumers and tourists of a proposed shopping center with "Japanese architecture . . . featuring Oriental art goods and restaurants."[4] Abe did more than join other speakers in praising the project. He and the groups that he represented, perhaps skeptical of their abilities to change the project in any fundamental way, attempted to carve out a place for themselves within it.[5]

Abe responded to San Francisco's sweeping new redevelopment program. Well before the city redefined itself through its Pacific location and Japanese connections, San Francisco began to envision its future as, in the words of one urban planner, a "new city."[6] This process, like the sister-city relationship, sought to redefine the city, but through its infrastructure, layout, buildings, and neighborhoods. In a period when the suburb was a national ideal of community and family life, urban planners looked to redefine what a city could be and to remake its built environment to attract new residents and businesses.[7] The postwar redevelopment program would dramatically reshape the physical landscape in the decades after World War II.

Ideas about urbanism and data on city conditions guided the massive redevelopment program, and both tended to abstract the people who lived or worked in the targeted areas. This was particularly true of those who were identified with the infectious "blighted" districts and who were understood to hinder the vitality of the "new city."[8] Abe attended the public hearing as a representative of some of those people, the Western Addition's Japanese Americans. He and a group of Japanese American merchant-planners therefore sought to extract their neighborhood from these descriptions of blight and protect their recently reestablished homes and businesses. In the process, they learned important lessons about civic participation and city values, which they then applied to Abe's proposals for their neighborhood and to their relations with city officials. The designation of the neighborhood as part of a redevelopment area allowed for a public, if highly selective, redefinition of Japanesetown, its residents, and its place in San Francisco's urban identity. Abe and his group of merchant-planners offered a participatory portrayal of their neighborhood aligned with the values and goals of city leadership and planners.

The merchant-planners' proposal for a modest, one-block commercial development was ultimately unsuccessful, but it had significant unintended consequences. The redevelopment program gave the Japanese American would-be redevelopers the impetus and the framework to argue for their community's alignment with city values and goals. Along the way they cultivated a working relationship with city officials that helped redefine their community and its place in city plans and civic life. Their Japanesetown ideas also inspired a municipal reconsideration of their neighborhood. Their proposal would dovetail with the city's rising Japanese interests, leading toward its capture by the city's Gateway to the Pacific ambitions. But getting to that point took the work of a cohort of Japanese Americans who were spurred by their purely local and immediate fears of redevelopment's consequences.

PLANNING FOR A "NEW CITY"

San Francisco planners were swept up in the national tide of urban planning that began during the New Deal. The city was among the first in the country to advance a redevelopment program, in a state that held the first US redevelopment agencies.[9] San Francisco's urban planning had begun during the war, in part in response to real deterioration in the city's built environment brought on by wartime overcrowding and decades of delayed new building. Planning was also motivated by new forms of knowledge about the city environment, government abilities, and possibilities for intervention. These tools created a postwar redevelopment program that would inform official and community definitions of urbanism and civic participation over the course of the next decade.

New institutions and laws at the federal, state, and local levels articulated an emerging urban thinking animated by systematized conceptions of slums and blight. The Depression had, indeed, halted new building or repairs, compounding overcrowding and adding to problems of outdated infrastructure and environmental health problems. But such conditions were by no means new. Perhaps more significantly, the New Deal brought a heightened social consciousness while officials implemented Progressive ideals that brought government into activities once considered outside state purview. Furthermore, real property surveys and the census of housing, newly inaugurated in 1940, made decay visible and quantifiable in new ways for planners and policy makers. In San Francisco toward the end of the Depression, the San Francisco Housing Authority and other bodies began collecting data on the city's infrastructure in ways that called attention to deterioration and overcrowding. One means of addressing those conditions was provided by the Wagner-Steagall Housing Act of 1937, which provided funds for the city's first public-housing construction in 1940. This built housing "out of the slums" for poor families and ostensibly smoothed the path for the remaking of the deteriorated places they left.[10] The war halted actual work, but the municipality hired staff and prepared a master plan to guide city planning. Officials also lobbied for favorable state redevelopment legislation in Sacramento. In 1945, California legislators passed the Community Redevelopment Act to enable cities to designate redevelopment areas, create redevelopment agencies, claim property for public use, clear land, and sell it for private development. California's 1951 Community Redevelopment Law and the federal Housing Act of 1949 provided funding mechanisms. These new laws allowed for novel, expansive, and expensive state intervention into private property. Alert to signs of decay and newly energized by the potential for government mediation, bodies such as the San Francisco Housing Authority and the new City Planning Department laid the groundwork for postwar programs.[11]

Contemporary views of cities rendered the "benevolent intervention," as historian Samuel Zipp termed it, of redevelopment necessary.[12] The San Francisco master plan was characteristic: "a city is like a gigantic machine. It should run smoothly, like a ball bearing, with all its parts fitting and working together in harmony."[13] This imagery carried its own logic. All "parts"— homes, transportation, infrastructure, services, and so forth—were tightly bound and required the smooth operation of the rest. This joined the condition of one small neighborhood to the functioning of the whole. It also portrayed a city as a man-made, inorganic thing, something that could be rightfully tinkered with and repaired. These explanatory metaphors worked in conjunction with diagnostic metaphors. Vocabularies of contamination and sickness—usually descriptions of "blight"—supported the need for intervention, since the "injurious effects" of one blighted neighborhood could "spread to and pull down surrounding neighborhoods." And because "single owners are hesitant about making repairs on buildings," only local governments could reliably undertake the "reorganization and redevelopment" of entire neighborhoods.[14] These views paved the way for the extensive redevelopment of acres upon acres of US cities.[15]

San Francisco city planners immediately targeted the Western Addition for redevelopment (fig. 3.1). Depression-era research had highlighted problems of blight in the district: families in "murky cubicles, damp basements," crowding, tuberculosis, "blocks of firetraps," delinquency, high police activity, low tax contributions, and disproportionate use of social services.[16] For instance, reports noted the age of the buildings. As one of the few neighborhoods to escape the fire of 1906, it had some of the oldest housing stock in the city. But its escape, paradoxically, marked the "beginning of the decline" for the district, according to planners.[17] The once middle-class and stolid buildings were subdivided and intermixed with retail and industry to accommodate burned-out residents. The Depression and war only amplified the district's classic indications of blight.

Crowded conditions, segregated residents, a tight housing market, and absentee landlords had indeed taken their toll on the Western Addition in ways that inequitably reflected on its residents. As in other cities, pre-existing poor conditions made the district available to segregated residents. The resulting segregated overcrowding was profitable for some property owners, who were able to charge captive tenants comparably high rents with little incentive to maintain their buildings. Yet city officials tended to cast blame for the district's conditions on the residents themselves. As one public-health official noted, while many of the district's units were "previously considered substandard . . . the individuals now occupying them have certainly not added to the desirability of the dwellings." He also added, with astonishing blindness to the barriers of low wages and segregation, residents "seem perfectly content" and "apparently are not interested in obtaining more desirable quarters."[18] In contrast, experts removed responsibility from the actual property owners with empirical yet passive and subjectless assessments. "Blight" was a resultant condition in areas that were no longer economically feasible to rehabilitate or remodel; it described a substandard and stagnating or deteriorating area "of some size," not a single building.[19] This conception absolved individual property owners from responsibility. They were simply "discouraged" from expensive rehabilitation, given that a newly remodeled building would still sit among "dingy surroundings."[20]

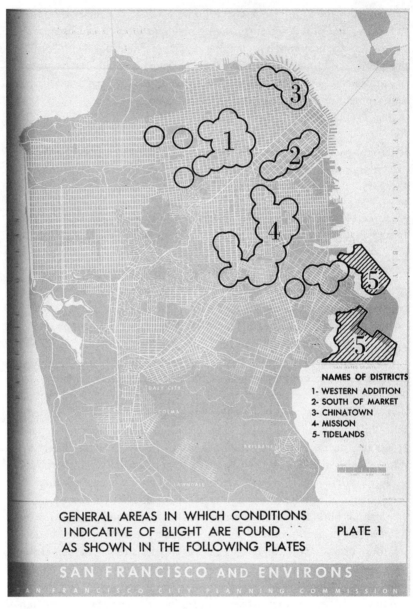

NAMES OF DISTRICTS
1- WESTERN ADDITION
2- SOUTH OF MARKET
3- CHINATOWN
4- MISSION
5- TIDELANDS

GENERAL AREAS IN WHICH CONDITIONS
INDICATIVE OF BLIGHT ARE FOUND
AS SHOWN IN THE FOLLOWING PLATES

PLATE 1

SAN FRANCISCO AND ENVIRONS

FIGURE 3.1 The Western Addition was pinpointed as the most urgent problem of blight in the city. Map from San Francisco City Planning Commission, *The Master Plan of San Francisco: The Redevelopment of Blighted Areas: Report on Conditions Indicative of Blight and Redevelopment Policies* (San Francisco: San Francisco City Planning Commission, 1945), plate 1.

Planners found and explained the district's poor conditions and elaborated on their consequences. The public-health official above was more blatant in his bias than most planners. But while the planners who drew up the official reports flipped the causation, they arrived at approximately the same characterization of the residents: impoverished conditions made for economically and socially impoverished people. These planners allowed the empirical language of data to "reveal" the "complete story of [blighted conditions'] impact upon people, and their drain upon the resources of the community."[21] The master plan and subsequent documents carefully depicted the location of condemned dwellings, housing in need of repair, building fires, criminal offenses, and juvenile delinquency (fig. 3.2). This aggregated

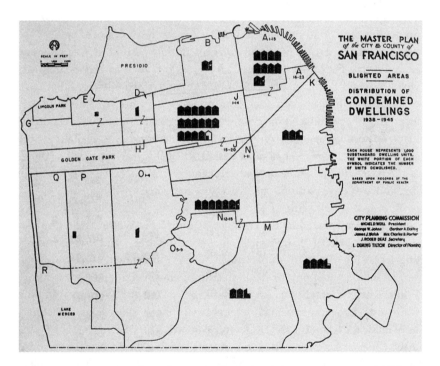

FIGURE 3.2 Maps visually and empirically depicted the concentration of blight in order to present the need for drastic intervention, and placed the most aggravated problems within the Western Addition district. Here the district is shown to have the most condemned dwellings, more than twice that of the next highest area. Map from San Francisco City Planning Commission, *The Master Plan of San Francisco: The Redevelopment of Blighted Areas* (San Francisco: San Francisco City Planning Commission, 1945), plate 4.

information became a call to action because poor conditions were not only harmful but, as planners understood them, produced all manner of social ills. Overcrowding, proximity to toxic industrial materials, and poor sanitation could lead to "ill-health, transmission of disease, [and] infant mortality." But social problems such as juvenile delinquency and crime, for example, also "result directly and indirectly from the substandard and slum housing conditions, overcrowding, lack of recreational space and intermixture of deleterious influences." Thus, an area's physical conditions affected its social conditions and, indeed, constituted "a menace and threat to the health, safety, and welfare" of the city as a whole.[22] By their very definition, these blighted areas were a drain on public and private treasuries. For instance, planning documents frequently noted the district's crime rates: although it housed only 1.2 percent of the city's population, it accounted for 2.8 percent of all property crimes, 4.6 percent of offences against the person, and 17 percent of arrests for prostitution.[23] And this was only one example in a list of overburdened city programs such as fire, health, and welfare services. In fact, contemporaneous studies demonstrated that most residential neighborhoods, including middle-class ones, cost more in services than they produced in revenues. San Francisco municipal documents made no such comparisons, however. Their narrow view gave planners the confidence that sweeping actions would "literally clear away the mistakes of the past" and make the district "an economically healthy, well functioning, and attractive part of San Francisco's cherished environment."[24]

Yet while it had signs of blight, which in its technical definition was a "stage of decay preceding the category of slum," the Western Addition was not deemed a slum. "San Francisco's main slum area," and indeed its only one according to a 1939 property survey, was Chinatown, a neighborhood consistently singled out for its dilapidation, congestion, and poor health.[25] While wartime conditions had definitely battered the Western Addition, Chinatown had simultaneously experienced a surge in acceptance that may have made it a less attractive target for planners. The US alliance with China made Chinese Americans the "good Asians" during World War II and ended decades of exclusion, allowing migration from China (albeit for a token 105 migrants a year) and naturalization in 1943. Furthermore, the need for workers at home and combatants on the battlefront gave Chinese Americans opportunities to rally around the war effort. Men and women of Chinese descent took jobs in the war industry, often in types of work and firms once closed to them; bought war bonds; and enlisted in the military. One writer in *People's World*, a communist labor newspaper, stated, "I don't think any other section of the American people is doing more per capita toward victory [than] San Francisco's Chinatown. . . . They are giving their sons, their money and blood donations from their bodies to aid the Red Cross. Their patriotism is beyond question."[26] This was entirely different from the neighborhood's earlier image as a marginal and fundamentally exotic place. Just as important to city officials was the neighborhood's long-standing status as San Francisco's foremost tourist attraction. Crowding and mixed usage in this context translated into excitement and allure. This energy would be lost if Chinatown were redeveloped as sanitary and modernist. The neighborhood was therefore still considered blighted, but planners singled out the Western Addition as the residential area most urgently in need of redevelopment. This rank was not simply an objective conclusion drawn from data, but a choice that weighed more factors than could be mapped in reports.[27]

The perception of the Western Addition's "social and economic liabilities" drew the attention of planners, but so did its possibilities.[28] In a small, hilly city bounded by water on three sides, land was at a premium. The Western Addition had particularly attractive land for many reasons. San Francisco is a city of microclimates; a neighborhood's temperature and clarity is calibrated by a precise triangulation with the many hills and forms of water. The district's "sunny climate" brightly contrasted with some of the more fogbound neighborhoods. In addition, its perch on the side of a hill gave it "sweeping" views of the city and the bay.[29] Furthermore, the district's central location made it convenient to the excitement and features of city life. To the north was the posh Pacific Heights neighborhood, where the city's elite presided in stately homes. A straight mile west—"a pleasant walk to your office downtown"—lay the central business district. The shops of the adjacent Union Square were "only five minutes away by trolley coach." The classical architecture of City Hall and other government buildings were even closer, to the south. Also in Civic Center, theaters, restaurants, the Opera House, the Herbst Theater, and the San Francisco Museum of Art offered the upscale, cosmopolitan culture that an urbanite might desire. Yet its slight distance from these amenities also offered a respite from urban busyness. To planners, the redeveloped Western Addition would be "urban living at its best—all the beauty and restfulness of the suburbs combined with all the advantages of 'the City.' "[30]

City planners hoped these attractions of the renewed district could draw some of the "thousands of families, of all income groups, now living in distant suburbs," at least those with a cultured bent.[31] Realistically, planners did not seek to entice the large nuclear families flocking to those new bedroom communities. The renewed Western Addition would be urban multi-unit living. Planners built for those they assumed would be interested in such a lifestyle, such as young professionals, couples, and small families. These new urbanites would not only occupy the newly built homes, but also enlarge the city's coffers with their property and consumer taxes.

This desired in-migration was an implied, at times explicit, replacement of current residents with new ones. The potential residents were necessarily distinct from the district's many segregated residents: most of the new suburban areas maintained racial boundaries. In addition, the future denizens would be wealthier. One estimate noted, "Perhaps the lower middle income group will find it difficult to meet" proposed post-redevelopment, market-rate prices, but those with "good incomes" would find them reasonable. By the city's own estimates, more than a third of the households in the district had incomes less than $2,100; the city's average family income three years later was $3,000, almost 50 percent higher. Sixty percent of the district's residents had incomes at or below the city average.[32] In 1947, the city's very lowest estimate for future rents was $25 per room; the same report estimated that only about 15 to 20 percent of the district's surveyed families would be able to afford adequate space at that cost. The redeveloped district would require new residents as well as new infrastructure.[33]

The many racially segregated or poor, or both, people living in the district had at best a tangential place in the city's plans. Which is not to say that planners ignored their prospects. State and federal laws required some accountability for their rehousing. One early proposal suggested that the new housing built in redeveloped areas would have a trickle-down effect, easing the market in other areas; ultimately, public housing was built in adjacent areas, although in insufficient numbers. Nonetheless, planners squarely focused on the city as a whole, understood to require the replacement of ill-fitting components of the urban machine when necessary. As the director of city planning mildly stated, "Improvements of this magnitude cannot be achieved without inconveniencing some."[34] This was justified through pocketbook reasoning. City planners estimated that the "optimum use of land" could bring in more than three times the tax revenues and cut the cost of social services. This made it cheaper to "completely eradicate" the Western Addition and rebuild it than to allow it to continue.[35] While they were concerned about decrepit conditions and their injurious effects on people, contemporary planners dealt with the built environment. They aimed to eliminate bad physical conditions for the city's overall good, not necessarily to better the life of all those residing in it.[36]

City officials were blandly confident that the Western Addition's residents would find adequate, if not better, housing. Residents and their advocates did not share their optimism. They were only too aware of the dire
housing shortage, and many had also experienced the difficulties of finding
housing in a racially restricted market. On the other hand, they were also
familiar with the district itself and had no illusions about its condition. As
one resident noted, "The Western Addition is a blighted area and should be
rebuilt."[37] People saw redevelopment as a useful and even necessary use of
state power to improve homes and communities. Current city plans, however, appeared to them to neglect those already living or working in the
district.

Many observers weighed two factors in evaluating the proposed redevelopment project: the need to better the conditions of the Western Addition and the need to protect residents' interests. The rights of residents, of
course, were not the only basis for apprehensions about redevelopment.
Property owners envisioned huge losses from investments, taxpayers
dreaded increased taxes to fund the program, lawyers challenged the use
of eminent domain, others worried about the continuation of a district that
was a "menace" to the rest of the city.[38] Still, the district's residents and
business were the primary issues for observers.

This emphasis was evident at the first public hearing on the district's
redevelopment held by the Board of Supervisors in 1948. The hearing was
held at the Civic Auditorium and drew an enormous crowd of three thousand. The narrow question under consideration was whether to designate
the district as a redevelopment area; this was the first step under state law
toward a local redevelopment authority, project plans, and eventual land
acquisition and private development. Nonetheless, many of the twenty-six
speakers took the opportunity to weigh in on redevelopment as a whole.
For example, Edward Howden, the representative for the Council for Civic
Unity, argued that redevelopment was "very necessary." But the overwhelming majority of Howden's testimony underscored the need for "safe-
guarding, not only the property rights, but also . . . the human rights of the
people living in the Western Addition," views echoed by organizations as
divergent as the Veterans of Foreign Wars and the San Francisco Building
Trades Council.[39] Howden therefore laid out a number of provisions necessary to get his council's support: antidiscrimination measures for resultant
developments, satisfactory housing for displaced residents, facilities for displaced businesses, and neighborhood and nonwhite participation in future
developments.[40] The Council for Civic Unity subsequently worked hard for

these anti-discriminatory policies. Members crafted such ordinances themselves, prodding the Redevelopment Agency into reluctant support, and ultimately winning unanimous approval from the Board of Supervisors. Such assurances unified the council's support and demonstrated their belief in the efficacy of such measures.[41]

Other speakers, though, were more dubious. As Dr. Carlton B. Goodlett, representing the San Francisco chapter of the National Association for the Advancement of Colored People (NAACP), stated, "Experience has taught minority peoples that if we don't start out right we might not end right." He opposed the current redevelopment program because it placed "the burden upon the minority people to . . . see that their rights are protected."[42] African American lawyer Ulma A. Abels lived and worked in the district and claimed absolute support for the concept of redevelopment.[43] Yet even Abels assumed that residents would be disadvantaged, citing the example of property prices. He knew that the district's property had inflated prices; satirically, he wondered if "real estate agents were somewhat surprised to hear that the property that they had been selling at such exorbitant prices was so dilapidated." This inflation, he argued, meant that assessed values would almost certainly be far lower than the owners' purchase prices. Renters would be even harder hit once land and development costs made rents too high, and this was an estimated 90 percent of the district's population.[44] As a result, Abels argued that "if the Government is going to subsidize private enterprise, they should also subsidize the parties to be displaced."[45] Experience suggested that such support would not be forthcoming. Abels, a savvy observer, undoubtedly understood that the district's bloated property prices and rents were caused by segregation as much as a tight housing market; this was an inequity unremedied by the state as of 1948.[46] His statements were popularly received by the audience, who responded with outbursts of laughter and applause fruitlessly shushed by the hearing chair. The proposed redevelopment program's supporters had far less vocal backing from attendees.[47]

Despite the highly charged hearing, the Board of Supervisors easily passed the ordinance in their next meeting without addressing any of the safeguards advocates demanded. Their vote set the district's future. Shortly thereafter, the city formed the Redevelopment Agency and formally accepted the Western Addition redevelopment project. Residents and their advocates had challenged the city's plans with their own definitions of municipal responsibility and city betterment. But the supervisors swept past their concerns. The district's present residents and businesses were, in many ways, understood as part of the problem.[48]

CHALLENGING BLIGHT IN JAPANESETOWN

Many of the hearings' speakers were African American, but Japanese Americans also opposed the redevelopment measure. The San Francisco chapter of the Japanese American Citizens League (JACL) also argued that state law did not "contain provisions which we believe are necessary for the protection of the people now residing in the designated area."[49] But they and other Japanese Americans expressed their dissent in less public ways. The JACL registered their dissent by letter, so while it entered the public record, there was no public face of Japanese American opposition. Others also expressed their ire about the city's plans for their neighborhood in community forums. Dr. Kazue Togasaki, who both lived and owned property in the district, contended that it "would benefit only a few of the persons now living in the area. Most of them would suffer from it." The JACL's organ, the *Pacific Citizen*, compared it to the recent incarceration as "the final evacuation," viewing it as "a double hardship" of racial restrictions and resettlement challenges.[50]

The most public and persistent response by Japanese Americans came from the lawyer Victor Abe and his small group of Japanesetown property owners, professionals, and merchants. They proposed a commercial development for the redeveloped district that would retain their businesses and their claim to their neighborhood. They did not mount a challenge based on social justice and protections for minorities. Instead, the group sought cooperation based on shared goals and values with the city. Their proposals were motivated by personal interest, their recent internment, and a genuine community concern. They sought to extricate what they argued was a thriving economic community from the municipal portrait of blight.

Just a few thousand Japanese Americans had returned to the Western Addition by the time of the first public hearing in 1948, and the community was still rebuilding homes and businesses when the Redevelopment Agency publicly announced its plans in 1952. The city now had a schedule and a master plan for the redevelopment area, about twenty-eight blocks of the much larger Western Addition district. Japanesetown's inhabitants quickly construed the consequences. The *Nichi Bei Times*, a Japanese- and English-language newspaper located in the district, spelled out the impending destruction. It would cut Japanesetown in half, threatening blocks of homes and half of the commercial center that stretched down Post Street, the northern boundary of the redevelopment area (fig. 3.3). All the businesses on the south side of the street would be demolished, while those on the northern side would be weakened from the fracturing of the commer-

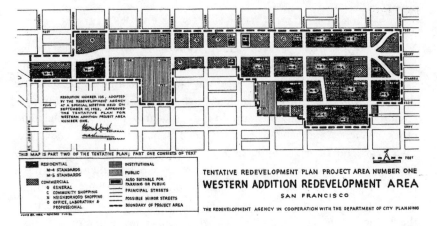

TENTATIVE REDEVELOPMENT PLAN PROJECT AREA NUMBER ONE
WESTERN ADDITION REDEVELOPMENT AREA
SAN FRANCISCO

THE REDEVELOPMENT AGENCY IN COOPERATION WITH THE DEPARTMENT OF CITY PLANNING

FIGURE 3.3 Although there were businesses and residents in the surrounding blocks, the most thickly Japanese American were within the six between Post to O'Farrell and Webster to Octavia. The Japanesetown commercial strip lay primarily along Post Street, the northernmost boundary of the project area. Map from the Redevelopment Agency of the City and County of San Francisco in cooperation with the Department of City Planning, *The Tentative Plan for the Redevelopment of Western Addition Project Area Number One and Related Documents* (San Francisco: San Francisco Redevelopment Agency, 1952).

cial and residential cluster.[51] Demolitions had been looming for years, but the new time line had an urgency and specificity lacking in earlier, indeterminate projections.

After speaking out at the 1952 public hearing, Abe requested a meeting with representatives of the agency to clarify the ideas he had proposed. Abe was an active member of the local chapter of the JACL as well as the Japanese Chamber of Commerce of Northern California. Representing the chamber, he "recommend[ed] that Japanese American merchants be given assurance that they will have the first opportunity to purchase adequate land fronting on Post Street between Bush and Laguna." If land could not be "set aside for Japanese neighborhood shops," he "recommend[ed] that the Plan be amended so as not to include this particular area."[52] He wanted to shield his cohort's businesses, but he also made it clear that San Francisco could gain as well. His proposal was a "great opportunity" to create a shopping center "designed in Japanese architecture and featuring Oriental art goods and restaurants" that would attract local visitors and tourists. Abe's was a basis of preference grounded both in race and in geography, based on "Japanese participation" and on the contemporary boundaries of Japanesetown's commercial heart.[53]

As in any redevelopment scheme, San Francisco planners depended on the participation of private investment to carry out the actual development. This was a critical basis for redevelopment policy and funding. Therefore the agency director, James Lash, and the attending staff were noncommittal but willing to listen to Abe's ideas. While Abe was told the agency was "interested" in his proposal, Lash expressed doubt about the merchants' capabilities. He did not speak directly to the needs or context of the Japanese American merchants and professionals and was apparently unaware of the longevity, even with the wartime disruption, of some of their businesses. Nor did he seem to explain the possibilities of financing or the steps for would-be developers. Instead, Lash generalized about the Western Addition's businesses, explaining that the considerable lag time between eviction and project completion would be difficult for small businesses to endure. Temporary relocation space might become available to bridge this gap, Lash conceded, because the district had "many small businesses . . . considered marginal, which start up and do not last too long, thus creating vacancies."[54] His comments aligned with city planners' characterization of the district's businesses as "cursed by frequent failure and turnover," symptoms of the area's "conditions of blight."[55] It was up to Abe and his colleagues to challenge this characterization. Lash asserted that "much more information was needed about the business people" of the Japanese American commercial area—such as property ownership, types of businesses, and extent of businesses—before "any determination of their future needs could be made."[56] Otherwise, they would be presumed to be among the district's "marginal" enterprises.

Abe and his group needed to extricate their businesses and property from the "blight" understood to encompass the district if they were to preserve them. Their subsequent activity addressed both goals. The group's proposals illustrated a community defined by race but familiar with municipal standards and assimilated to civic goals. This was accentuated by their cooperative advocacy, moving in an acceptable fashion through institutional channels. Abe and others set out to demonstrate that redevelopment was not simply speeding up inevitable "failure and turnover" but disrupting viable businesses with much to offer the city as a whole. They did this using the education they received, beginning with the first meeting with Lash, where Abe learned the kinds of data and values that resonated with city officials. This learning process continued throughout and shaped the merchant-planners' campaign of community self-fashioning.

The Japanese American merchant-planners had a smaller hill to climb than they would have before the war, or even directly after it. In the years since their return, a number of changes had occurred at the state and federal level that suggested shifts in their status. Federal and local bodies had supported their resettlement in various ways, connecting them to social-service and civil organizations previously cut off to them by Japanese migrants' alienage. Of course, the reevaluation of Japan and the growing local and national interest in the country provided new if limited avenues of inclusion. Just as important, the state began addressing long-standing legal concerns. In 1948, the US Congress passed the Evacuation Claims Act, allowing formerly incarcerated Japanese Americans to file for reimbursement for their wartime losses. The bill paid minimal sums on tremendous losses, leaving the need for real redress down the road. But it acknowledged their hardships and the need for rectification. Additionally, California's Alien Land Laws, so vigorously enforced less than ten years before, were overturned. Voters in the state overwhelmingly rejected Proposition 15, which would have fixed the laws' authority in the state constitution; court decisions in 1948 and 1952 fully eliminated them.[57] Most significantly, the controversial 1952 McCarran-Walter Immigration Act ended the exclusion of Japanese migrants and the prohibition on their naturalization. Japanese migrants were no longer "aliens ineligible to citizenship," with all the resultant exclusions that status enabled. In terms of the law, Japanese Americans had moved with astounding rapidity from suspect subjects to full citizens.[58] This change was not necessarily reflected in social practice, though. Japanese Americans still faced numerous discriminations in the housing, employment, and social spheres. In San Francisco, they still had to convince the Redevelopment Agency that they were not marginal.

After Abe's meeting with the agency, a small group of "Japanese American business and professional men" from the Japanese Chamber of Commerce formed a planning committee.[59] It was composed of Abe and five others, who all lived or worked in the neighborhood and so had a great deal at stake in their district's redevelopment. Kikoroku Honnami chaired the committee and owned a small general-merchandise store, Richard M. Seiki owned a hardware store with his brothers, Carl Hirota was a dentist with a practice from the interwar years, Yoshiaki Moriwaki owned an insurance and real-estate business, and James Hirano owned an import business. They had varying degrees of property ownership in Japanesetown: Hirano owned the building that held his offices and home; the Seiki family owned two flats and later their store property; Hirota rented an office, although he owned a home in the residential Richmond district a mile west; and Hon-nami rented both his family's apartment and storefront.[60] The differences in property ownership suggest a difference in financial resources, but they had all built their businesses in the interwar years and depended on nearby clients and customers. Dislocation from a local customer base could be costly for any small business. This was emphasized by the Blyth-Zellerbach committee, a redevelopment booster group that included some of the city's biggest businesses and was not especially known for sympathy toward dis-placed residents and businesses. Yet the committee suggested compensation for lost customer "goodwill," the loyalty that came "after devoting many years to establishing a purely local trade" and that may not transfer to a new location.[61] This loss was especially detrimental to businesses that depended on not just local, but also segregated, racialized clientele. Dr. Hirota, for example, could not simply turn to non-Japanese patients to replace his local ones. In fact, ten years earlier, he had taken even more drastic measures to retain his patient base. He had avoided internment by moving his family inland, but he nonetheless remained tied to the camps because he relied on detainees as clients. As a result, he established a clinic in Utah to serve those held in the Topaz detention center for the duration.[62] As Abe stressed to a city supervisor, the "business establishments of Japanese Americans" depended "upon their location in the same area," with access to the same customers.[63]

The group depended on the Japanese American community for their live-lihoods, but they were in no way representative of it. In 1950, 51 percent of Japanese American men in the San Francisco–Oakland area were laborers, domestic workers, or in some other form of nonprofessional service. Only 15 percent were professionals, managers, officials, or proprietors. In 1949, the median income in San Francisco was $3,009, while the median income for Japanese Americans was fully one-third lower, $1,937. This low-income rate was not surprising. Although the War Relocation Authority (WRA) and contemporary scholars found that Japanese Americans faced somewhat less employment discrimination after the war—and were, in fact, invested in such positive findings—they had long been excluded from work com-mensurate with their education, were newly relocated back to the city, and still discriminated against by employers. Given these facts, the merchant-planners had an unusual level of occupation as well as of property owner-ship. A 1956 survey of three hundred Nisei in San Francisco found that just a quarter of them owned homes, and this was in a sample dispropor-tionately skewed toward professional, managerial, and clerical occupations.

In contrast, at least three of the six merchant-planners owned at least one property. They were thus a privileged group, whose plans did not attempt to represent the community as a whole. Instead, their proposal painted a picture of Japanese Americans reflective of themselves, a selective portrait that served their interests but was also the most likely to find acceptance and support from city and federal officials. They had tried to have the Japanesetown blocks excluded from the plan. But clearly aware of its unlikelihood, they sought to preserve what they could.[64]

The group's redevelopment proposal for a "Japanese Garden Center" therefore refined a vision of community that allowed for the continuation of their businesses. Their ideas built on the rising local interest in Japan. They proposed a one-block "tourist shopping center" that was carefully oriented to outside visitors, unlike the prewar Japanesetown. Their two-story shopping center was conceived as an attraction in itself, with a central courtyard and "Japanese garden." Intimate Japanese restaurants and "Oriental gift shops similar to the prewar stores on Grant Avenue [in Chinatown]" would be featured. The proposal thereby combined Chinatown's tourism with the new shopping malls so popular with postwar developers. At the same time, local Japanese Americans could obtain the familiar goods and services that were part of life in Japanesetown. The proposal included space for branch banks, drugstores, and professional offices: many of the businesses Japanese Americans and other nearby residents used every day.[65]

Of course, such a modest development would leave many out. Japanesetown had pool halls, soda foundations, and several auto repair shops owned by Japanese Americans. There were also a multitude of laundries and dry cleaners scattered around the city, still the most numerous business among Japanese American proprietors. These, however, did not figure into the merchant-planners' edited version of Japanesetown. Businesses like an auto shop would require too much space, of course, and be noisy and pungent to boot, but the absence of such enterprises was telling. Included were businesses that were appealing, like décor and souvenir shops, and those that were professional. The Garden Center was not a cross section of the Japanese American commercial world, but rather an orderly, middle-class, professional, and attractive outsiders' view of it.

Although it was selective, the merchant-planners envisioned their project as a community endeavor. Funds would come from an investment corporation, the only way that small entrepreneurs could raise sufficient capital for a "million-dollar venture."[66] This would attract investment from a much wider swath of the community. The merchant-planners and their investors could buy stock in the corporation; the balance would be covered by loans. Therefore, while the project was clearly a commercial enterprise, it was also a way for residents as well as proprietors and investors to "join together" as a "Japanese community" toward their future in the redeveloped district.[67] This recreated and revised the ethnic economy that Abe and his group were a part of and depended upon for patronage. The Japanese Garden Center plans framed the still low-income community as organized and economically viable.

To do this, the Japanese Garden Center ethnically homogenized a neighborhood that included black, Chinese, Filipino, and even Mexican residents and businesses, a process that resettlement had not been able to do. It also relied on unifying and redefining two transforming elements of the Japanese American economy, long geographically separate: the ethnic staples and services marketed to Japanese American consumers in Japanesetown and the prewar "Oriental gift shops" in Chinatown that had catered to white tourists. Although the latter had been central to the prewar Japanese American economy, only a small number had reopened. Many proprietors found long-term leases occupying their prewar locations, and long-time hostility from Chinese American merchants hindering their postwar return. Meanwhile, proprietors rebuilt the specialty groceries, hotels, restaurants, pool halls, and services around the prewar commercial heart of Japanesetown at the Post and Buchanan Street intersection. These establishments relied on nearby Japanese American patrons seeking Japanese-speaking staff or goods such as pickles, fish, and utensils unavailable elsewhere.[68] But these had also changed. They were now neighborhood stores as much as ethnic businesses, convenient to diverse locals seeking hardware supplies, groceries, and car repair close to home.[69]

These postwar Japanesetown businesses gained new customers, but they also struggled with the loss of ethnic ones. They had been born in the segregated confines of the enclave, but those conditions were changing. Aside from the district's impending redevelopment program, Japanese Americans had begun to move to other parts of the city. Some lived elsewhere because the severe housing shortage prevented their return. Other middle-class Japanese Americans began to find footholds in previously white residential neighborhoods in the 1950s.[70] For instance, Hirota, one of the merchants' planning group, had moved to the Richmond district in 1948, when it was still only about 1 percent nonwhite. He encountered some hostility, but by 1960, enough people of Asian descent had moved in to make up 10 percent of the district's population. This pattern indicated a degree of declining housing discrimination for Asian Americans—the Richmond, with its decades-old housing and location adjacent to the Western Addition, did not have the same racial restrictions as many of the East Bay suburbs did—but it also posed a challenge to the businesses that had relied on proximate Japanese American customers. Still, a dispersed community might continue to sustain ethnic businesses. For example, Uta Hirota, Dr. Hirota's daughter, continued to visit Japanese American stores on Post Street after her family's move: "You had to have o-sashimi and things like that. . . . I don't know if my [Richmond] butcher knew how to cut that. So we always came to Japantown for Japanese foods."[71] Those businesses that specialized in Japanese products could anchor the businesses with non-ethnic specializations.

Supported by this continuing Japanese American patronage, a redeveloped Japanese American shopping center was therefore feasible but far from assured. The acceleration of residential dispersal might in turn accelerate integration into the wider San Francisco economy, at least for basic consumer needs. Historian Lon Kurashige has shown that Los Angeles Little Tokyo merchants worried about declining consumer loyalty from the Nisei even before the war, when most were still living in the neighborhood.[72] Likewise, postwar San Francisco Japanesetown merchants needed to find an entirely new group of patrons. The burgeoning sukiyaki restaurants spoke to a larger and growing interest in all things Japan. Among those who flocked to those restaurants, some were interested in bringing either the cuisine or the décor home. The proposed relocation of art-goods stores from Chinatown's Grant Avenue to Japanesetown's Post Street was therefore intended not only to revive an important sector of the Japanese American economy, but also to lure in white visitors. Once hooked by the curios in souvenir shops, they might forgo the city's major sukiyaki restaurants and try one of the smaller, neighborhood-serving Japanese-food eateries, browse specialty food items, or even pick up a few things at the hardware store. These patrons were an untapped market for future expansion.

In uniting two major sectors of the Japanese American economy—the businesses that served white sightseers and the businesses that served local Japanese Americans—the proposed Japanese Garden Center integrated two disparate groups of customers. This process downplayed the businesses' other nonwhite customers, and emphasized the Japanese Americans who could add authenticity to the garden atmosphere. After all, one of the great attractions of Chinatown was that it was a "city within a city," a place where a visitor could buy exotic souvenirs but also see "Chinatown people on the street" and get a glimpse of their life full of unfamiliar customs and sights.[73] This side-by-side browsing, buying, and eating might be seen as a complement to the contemporaneous limited but increasingly tolerated residential integration of Asian Americans in the Bay Area and elsewhere. The proposed center intended to integrate customers, but within a commercial center segregated by race in ownership, management, and tenancy.

Chinatown could serve as a model of this selective use of racialized and segregated difference, too. The increasing acceptance of Chinatown's inhabitants demonstrated that a neighborhood could be a segregated, racially distinctive place yet still enjoy a rising level of familiarity and tolerance. Many journalists playfully described the "American" workplaces, habits, language, and clothes of the younger Chinatown inhabitants; clearly amused by the juxtaposition of "American" and "Chinese," they nonetheless noted their intersection. Such an integration was effected officially in the city in 1955 with the termination of the "Chinatown squad," a separate police force whose existence, independent of the regular police beat, had accentuated Chinatown's image of criminality and foreignness; for national observers as well as local advocates, this elimination signaled Chinatown emergence "out of the shadows" and into mainstream politics and civic life.[74] While the neighborhood's assimilation into the citywide system of policing was largely celebrated in Chinatown, both city officials and Chinese Americans viewed full residential integration with some hesitation. When the city dedicated Ping Yuen, its first public housing project in Chinatown, in 1951, Chinese American activists made sure that the project was open only to those of Chinese descent, even though the San Francisco Housing Authority had adopted a nondiscrimination policy for new projects the year before. Chinese Americans had advocated for the project as recognition of their wartime loyalty and worthiness as Americans, as well as in the clear-eyed awareness that real desegregation would be slow in coming. The Japanese American merchant-planners reflected a similarly cautious balance of integration and assimilation in their proposal.[75]

Unlike in Chinatown, however, the Japanese American entrepreneurs forwarded a proposal based on a commercial enterprise and not a residential one. A housing development was certainly a possibility and, in light of the continuing restricted housing market, in fact desirable. Moreover, available federal financing would have made a residential project more feasible than a commercial one. The Nichi Bei Kai (Japanese Association) made the community aware of new federal loans through the vernacular newspapers in 1956, well before any concrete plans for the Garden Center was under way. Two members of the association, including Shichisaburo Hideshima, its president and also a member of the chamber planning committee, met with a Redevelopment Agency official to gather information about available aid for residents and businesses. As they were told, "very attractive financing" was available to any group interested in building multiple-unit housing.[76] Creating "a decent home and a suitable living environment for every American family," after all, was the stated goal of the 1949 Housing Act that supported San Francisco's program.[77]

The favorable possibilities of such a residential project was demonstrated by another racialized Western Addition community institution. Jones Memorial Methodist Church, a black congregation, was located on Post Street just a few blocks from where the merchant-planners proposed their Japanese Garden Center. Like the merchant-planners, the Reverend Hamilton T. Boswell saw the redevelopment program as an opportunity to "participate instead of evacuate."[78] But his church chose to build housing. His congregants raised over $40,000 themselves and took out $466,000 in federal loans to build the Jones Memorial Homes, thirty-two units of low-income senior housing. Boswell similarly wanted to contribute to city redevelopment efforts, but he chose to focus on residents rather than businesses.

Boswell paralleled Abe and the merchant-planners in his relative privilege, although his came from civic familiarity rather than entrepreneurism. Boswell was known in San Francisco's political circles through civil rights activism, which involved him with municipal offices and local interracial organizations. Additionally, he served on the city's Juvenile Justice Commission and later became chair of the Housing Authority Commission. So while Jones Memorial did not, as he claimed, "have any rich people in our church," his church did hold a significant and unusual degree of social capital. No other black group appeared able to propose a similar development; Boswell claimed that the Jones Memorial Homes were the "first in San Francisco to be financed by a Negro corporation."[79] This was not surprising, since the scale and funding mechanisms of early redevelopment favored large developers.[80] Jones Memorial Methodist Church's congregation was therefore comparable to the merchant-planners in their unusual attempt to participate in the redevelopment project as a nonwhite group, and to do so for community betterment.[81]

Despite these similarities, the two cohorts diverged considerably in their proposed or realized projects. Jones Memorial not only built housing, but did so for vulnerable elderly and low-income people. Moreover, while the church built the homes "for the community," it was not necessarily the black community. Certainly, congregants wanted to offer housing for black seniors, a demographic overlapping the two that the Redevelopment Agency most struggled to rehouse: minority residents and low-income seniors. But no project built through the redevelopment process could explicitly favor any one group, so, as Boswell laughingly said, "we're going to integrate with a vengeance."[82] By serving those largely underserved by the Redevelopment Agency, Boswell joined his development scheme with his social activism and produced an integrated, low-income residential development. This required highlighting marginal members of the black community, something Abe and his group were reluctant to do. The Japanese American merchant-planners, with concerns shaped by incarceration and business, focused on presenting a coherently ethnic commercial development that favored the most economically stable among Japanese Americans.

The merchant-planners did briefly consider the possibility of a residential project, but a number of factors made it undesirable. Practically, local Japanese Americans had little experience in large-scale apartment management. While a handful were landlords, their properties were the Victorian houses and small multi-unit buildings that housed themselves and a handful of mostly Japanese American families. Furthermore, they believed costs of land and building would put any for-profit residential development into a "high-rent class" when "it is questionable whether Japanese residents would be interested in high-rent apartment buildings."[83] This suggested Abe and his cohort's assessment of the economic means of most other Japanese Americans, who either could not afford expensive units or were more interested in saving for purchase. But it also demonstrated the merchant-planners' inclination to build for the Japanese American community. They did not envision themselves, as the Redevelopment Agency and Jones Memorial Methodist Church did, as building for those who were not their co-ethnics.

Uncomfortable with an integrated residential project, Japanese Americans, unlike African Americans, may also have been uncomfortable with a Japanese American residential development because of their recent experience with the WRA's assimilationist strictures. Although most resettling Japanese Americans had not abided by them, the rebirth of enclaves had been an unplanned, informal return to familiar people, landmarks, languages, and services in a neighborhood where they met little resistance. A planned residential project, even if accepted by the Redevelopment Agency, might have raised residual hesitations about creating a permanent, racially homogeneous, and intentional residential community. Moreover, assimilation might have been a concept that some of the merchant-planners were invested in. Abe, for example, had been active in camp politics while he was interned, part of a vocal group who supported limiting internee office-holding to US citizens. He had argued for these measures because, he believed, "all energies and efforts should be put into demonstrating our loyalty."[84] He had been conscious about images of Japanese Americans and the way in which portrayals of assimilation with US ideals and institutions could shape them. A lingering pressure for assimilation, however unenthusiastically embraced, might have colored the merchant-planners' choice for a commercial, not residential, development for their neighborhood.

The merchant-planners were not alone in their discomfort with residential segregation. The San Francisco bilingual newspaper, the *Nichi Bei Times*, indicated a similar apprehension in their coverage of the neighborhood's impending redevelopment. While the newspaper regularly reported on the program's progress, it did so less frequently in terms of the eviction of families—cognizant of the era's concern with social isolation or maladjustment and valorization of the nuclear unit, editors usually spoke of families, not individuals—and more frequently in terms of business displacement.[85] This was consistent with the newspaper's coverage of the Japanese American community since its postwar reestablishment in May 1947. The Japanese- and English-language newspaper reported the opening of almost every store, hotel, and restaurant under Japanese American ownership. In contrast, the rebuilding of residential Japanesetown remained understated. Residential patterns might have been apparent: while the district-based newspaper's readership extended through Northern California, those living in the neighborhood might not have needed journalistic reminders of what they saw every day. Yet the way in which writers referred to the neighborhood demonstrated a reluctance to even name the residential community. Ethnic journalists consistently used the "Uptown" moniker in reference to Japanesetown. In a departure from prewar patterns, this name lacked any reference to race but instead called attention to the neighborhood's geographic location in the wider city.[86]

The Japanese Garden Center addressed the concerns of displacement within the Japanese American community by demonstrating its assimilation with contemporary values and municipal programs. Abe had approached the Redevelopment Agency with a purely commercial project, seeking opportunities for the participation of local merchants rather than the retention of local residents. Their plans redefined the racially mixed neighborhood as Japanese American and then framed the Japanese American community as an organized, ethnic economy that cohered an increasingly integrating population. As entrepreneurs rooted in the clientele of the Japanesetown community, they had interests at stake as well as the resources with which to act upon them. While ostensibly representatives of a racialized community, Abe and his group proposed the alternative to an enclave. Their development would be a contribution to the city's wider social and economic life, "an asset to the tourist attractions of San Francisco."[87]

Throughout the process, the merchant-planners adopted the language of urban development. Their proposal would "improve" a select parcel with a project that would be an "asset" to the city.[88] Their businesses were not the cause or a symptom of blight, they argued. They were instead active partners with the city in alleviating it, and their proposed Japanese Garden Center signified cooperation with civic goals. Although still defined by race, theirs was a relationship with the city that was participatory, contributing, and integrated with the surrounding concerns.

REDEVELOPING THE COMMUNITY

Abe and his group of Japanese American merchant-planners framed their Japanese Garden Center to municipal officers as a project of the "Japanese community." The group was small, however. The original cohort therefore reached out to other Japanese American merchants and property owners in Japanesetown, connecting in various ways with others who were similarly situated and like-minded. Redevelopment's slow crawl gave them ample time; although plans and time lines were announced in 1952, few results were seen until 1958, when property acquisition began. This gave the merchant-planners years to develop ideas and advance a selective community identity that reflected the one in their proposal. They portrayed the entire ethnic neighborhood as an institutional and economic community rather than a residential one, aligned with both post-incarceration pressures and municipal values.[89]

The merchant-planners formulated their ideas as San Francisco officers slowly developed their redevelopment program. The Redevelopment Agency was established in 1947, and as early as 1952 was criticized for excessive delays by the press and public. This was partly due to the instability of the sweeping and changeable federal and state policies that bodies at all levels struggled to implement. The California legislature had quickly passed the Community Redevelopment Act in 1945 to establish local redevelopment agencies and give them power to acquire land; the 1951 Community Redevelopment Law added a tax-increment financing mechanism. The federal Housing Act of 1949 provided generous funding—giving local authorities two-thirds of the purchase and clearance costs of blighted properties—but what historian Jon C. Teaford has called "bureaucratic roadblocks" hindered its implementation. Furthermore, the constitutionality of these laws was not tested in the courts until the 1954 US Supreme Court decision of *Berman v. Parker* confirmed the use of eminent domain for public-use projects. Until that point, local redevelopment agencies had held uncertain authority and thus struggled to secure loans.[90]

There were also obstructions specific to San Francisco. Over the course of the decade, the Redevelopment Agency cycled through four different directors and multiple commission chairs. Frequently, these officials were appointed by the mayor as part of political bargains or for their personal ties rather than planning expertise. These inexperienced leaders were hampered by the agency's internal conflicts, charges of nepotistic corruption related to land-acquisition pricing, and inexperienced staff. Legal challenges also held up the process. Diamond Heights was the city's first implemented project, although it was conceived after the Western Addition one. Diamond Heights' 325 acres of hilly topography were largely undeveloped, and so the Redevelopment Agency had hoped that its plan for thousands of new housing units would generate little protest and loosen the housing market. However, its legality as a "blighted area" was a weakness, and the resulting lawsuit held up all the city's projects until the California Supreme Court decided in the agency's favor in 1954. Political entanglements also slowed city activity. The Redevelopment Agency had limited autonomy in some essential matters and depended on other city bodies to carry out some of its activities. These departments, particularly the city attorney and the Planning Department, refused to cede authority to the fledgling agency. As a result, bitter turf battles hamstrung the redevelopment process for most of the 1950s.[91]

All these delays gave Abe and his merchant-planners plenty of time to craft their ideas and organize the business community along complementary lines. In response to agency staff's 1953 request for more information, and in an effort to gather data on the Japanesetown business community and enlarge their numbers, the group made a general call to other Japanese American merchants in the area. The chair of the merchant-planners, Kikoroku Honnami, announced the start of regular public meetings to discuss plans for a Japanesetown shopping mall.[92] Out of these meetings came a group of merchants, entrepreneurs, and property owners, the Japanese American Merchants and Property Owners Association. The very name provided a veneer of organization to the community and furthered its identity as an ethnic economy. This significance was not lost on Abe, who communicated its formation to the Redevelopment Agency.[93]

The Merchants and Property Owners Association seems to have lasted for only a year or two, but its issues and its advocacy continued in other forms. Japanese American merchants and professionals met regularly for years in forums such as a weekly luncheon club.[94] The merchant-planners were also part of many Japanesetown-based organizations whose members they drafted. For example, Abe was an officer in both the Japanese Chamber of Commerce and the JACL. Shichisaburo Hideshima, another original merchant-planner, was also a member of the chamber and president of the Nichi Bei Kai. These organizations all had differing primary interests but held common class and ideological biases with each other and with the original cohort of merchant-planners. The Nichi Bei Kai, the oldest of these groups, began in the early twentieth century with, in the words of historian Yuji Ichioka, a "special relationship" with the Japanese government.[95] The group took on more of a social and cultural cast after the war, but it had originated as an elite institution with extraordinary powers to police the behaviors of what officers worried were the unruly laboring migrant masses. The JACL was also an elite group of professionals and strivers; it began as a way to prove the second generation's Americanism, with crucial support from migrant leaders, during the rising international tensions of the 1930s. The group had then skyrocketed to national renown as Japanese American spokespeople who maintained resolute support for the US government and internment policy. This had made the group suspect among many Japanese Americans during the war, but it continued to hold a great deal of mainstream influence afterward, as the community's main advocacy organization. The chamber, composed of Japanese-national businessmen and local merchants, proprietors, and professionals, had backed the merchant-planners' proposal from the start; the group had formed as a chamber subcommittee. These prominent community institutions had much in common with the merchant-planners.[96]

The broad group of Japanese American merchants, property owners, and entrepreneurs embraced the merchant-planners' values and goals and similarly demonstrated a concerned and participatory civic engagement. For instance, Japanesetown merchants proposed an annual festival, the Kiku-Matsuri, or Chrysanthemum Festival.[97] Like the Japanese Garden Center proposal, and just two years before sister-city activities, the festival used Japanese culture to attract sightseers. This was a departure from earlier neighborhood traditions. The Japanesetown Buddhist church had regularly hosted an Obon festival, honoring the dead. This had been a spectacle, usually involving a parade down the commercial strip, choreographed dances, a fair at the church, and colorful kimonos. Its participants and audience had generally been local Japanese Americans, although by the 1950s it had begun to attract an interested non-Japanese audience.[98] The Kiku-Matsuri built on that interest. The festival was not among the most prominent Japanese festivals, yet it appealed to an unfamiliar audience. It occurred in early September, one of the Bay city's most pleasant months. It obliquely referenced the Japanese imperial symbol and was visually engaging, with dolls costumed in fresh chrysanthemums and other displays of the blossoms. The festival does not seem to have become a regular event, but it expressed goals very much aligned with the Japanese Garden Center: draw in non-Japanese customers with exotic cultural displays and boost the neighborhood's cachet as a racially distinct but integrated part of city life.

As attractive as the festival might have been, Japanesetown merchants were ultimately animated by matters that were more mundane. Small targeted requests to city hall were a regular part of their activities. In 1954, for instance, the Merchants and Property Owners Association lobbied the Board of Supervisors for revised parking regulations for the commercial strip on Post Street. In 1955 and 1956, the Japanese Chamber of Commerce requested more parking meters and traffic lights. In 1957, an informal group requested more police protection against a rash of petty crimes in the neighborhood. These appeals would benefit them as businesses, of course, since they facilitated customer access. But they did so in ways that expressed a concern with order and acceptable public behaviors. Their requests also pitted their desires against those of others in the neighborhood. Parking meters, for example, claimed the street for business at the expense of residents' long-term parking needs and helped render the streetscape commercial. Additionally, the group defined themselves against disorganized and unlawful behaviors, those that were associated with the blight in the district and that were increasingly associated with "Negro crime."[99] In contrast, the black newspaper the *Sun-Reporter* was much more skeptical of anti-crime initiatives. The editors saw "the whole problem of crime" as too easily blamed on "the Negro populous [*sic*]." Additional policing, in this view, was less a solution than another problem: the department needed to "improve its techniques and methods of dealing with racial minorities."[100] These routine requests, then, defined a heterogeneous neighborhood as Japanese American, and homogenized an otherwise diverse ethnic community around common sentiments and goals. This reinforced the image of Japanesetown as a viable, assimilated marketplace in order to make claims on the municipality.[101]

Redevelopment planning was only one part of the much larger institutional life of postwar Japanese America. As historians have shown, its many groups and organizations worked in diverse ways toward a culturally assimilated and "American" image of Japanese Americans at local and national levels. Recreational or volunteer groups, like the Nisei Fishing Club or Nisei Lions, provided organization and familiarity to Japanese American leisure. Occupational organizations demonstrated their work ethic and suitability for the mainstream economic sector. The JACL continued their vigorous campaign for the rehabilitation of Japanese American citizenship that they had begun after the Pearl Harbor bombing. As historian Ellen D. Wu has shown, Nisei veterans composed a particularly resonant cohort, given the widespread recognition of the Japanese American 442nd Regimental Combat Team.[102] These groups provided an everyday foundation to Japanese Americans' new legal status, no longer "aliens ineligible to citizenship." The Japanese American merchant-planners intersected in many ways with the people and ideas of these other organizations. But they also depended on their resonance in San Franciscan and US civic life in order to make their claims for an orderly, commercial, ethnic community.

Like the original Japanese Garden Center proposal of Abe and his group, the broader activities of Japanese American merchants and professionals echoed the values and goals of city government and demonstrated their concerned and participatory civic engagement. These small demands gave Japanese American leaders experience with articulating their needs to city government and presented an organized and assimilated community to city officials.

NEGOTIATING OPPOSITION

While the broad swath of merchants, professionals, and property owners developed lines of communication with the city, a core group remained interested in a redevelopment project for Japanesetown. Under the leadership of Honnami and Abe, the merchant-planners retained many of the original members and gained some new ones by 1958. They made informal surveys and held occasional meetings, reiterating their commitment to other Japanese Americans and to city officers. They maintained frequent contact with the Redevelopment Agency, hosting meetings and even dinners with officials to keep themselves informed of the progress. Until 1959, they had little in the way of concrete plans or proposals, although a name change from the "Japanese Garden Center" to the "Japanese Village" suggested that discussions about the project continued. Throughout, they kept officials aware of their developments, maintaining a cooperative and supportive relationship with municipal officers.[103]

The Japanese American residents, business owners, and property owners of the merchant-planner group employed cooperation toward their businesses' future. But beyond that, like their neighbors, a number also fought their neighborhood's redevelopment through lawsuits or by simply ignoring the city's orders to move. They challenged the agency's right to take their property, the prices they were offered, and the municipality's very right to deny an individual owner's choice to redevelop their buildings independently. These property owners cooperated with *and* challenged the city: they proposed cooperation as merchant-planners or through community organizations, but they also contested municipal actions as individuals. Such conflicting activities represented their desperation to keep their newly rebuilt homes and businesses, even as it suggested their increasing fluency in the language of urban development.

Once the city began acquiring property in the Western Addition project area in 1958, a number of Japanese American property owners, along with other property owners, openly challenged the process. This situation was complex for the members of the merchant-planners. For instance, Masao Ashizawa, the second-generation owner of a hardware store on Post Street, attended agency meetings with other community members for updates on the redevelopment process throughout the decade. Sometime during the late 1950s, he joined the merchant-planners and eventually became an officer of the group. But this membership did not impede his fight to retain his home and his income property on Post Street. He remembered being one of the last holdouts, his home "close to the last one standing" when "everything else was bulldozed around me."[104] While his opposition and his cooperation appeared to be at odds—ironically, his property stood on the very same block that the merchant-planners wanted for their project— they worked in tandem. Participation with the merchant-planners gave him access to more information about redevelopment's progress than he would have gleaned through the agency's notices or media coverage.[105]

Masateru and Dave Tatsuno, the father and son owners of NB Department Store, also split their energies between cooperation and challenge. On one hand, they tried to fight the city's acquisition process. The elder Tatsuno established the store in 1902 and had already rebuilt it twice: once after the 1906 earthquake and fire and again after his wartime incarceration. The family was reluctant to disrupt their business once more and so was among the six Japanese American property owners in one block sued by the city in condemnation proceedings. This was either hard-knuckle business sense—wanting more for their property than the Redevelopment Agency had offered—or a simple challenge to the city's claim on their property. Either way, these property holders chose not to comply with city actions. The Tatsunos took the further step of countersuing the city based on the assessor's valuation on their property. A month later, they were charter investors in the incorporated merchant-planners. Despite their oppositional stance over their property, the Tatsunos also cooperated with the redevelopment project itself. As for Ashizawa, the merchant-planners' project was their "Plan B," their last-ditch effort to maintain their building.[106]

The legal representative of both the Tatsunos and Ashizawa was Victor Abe, who also had a complicated relationship with the redevelopment process. Not only was he the original spokesperson and legal representative for the Japanese American merchant-planners; he also represented a number of Japanese American individuals engaged in legal battles with the city's rights of eminent domain or property transference. These conflicting roles spoke, in part, to the dearth of Japanese American lawyers in practice at the time and to Abe's prominence through extensive work with community organizations. Regardless of why he represented so many seemingly conflicting interests, his private practice gave him room to advocate for residents in a way that was proscribed among the merchant-planners, who as a group chose to advocate for their community as an ethnic economy rather than to exercise their claims as residents.[107]

Japanese Americans such as Ashizawa and the Tatsunos challenged the city's proceedings as individuals. They certainly benefited from being a part of a community. Their shared lawyer had a wealth of experience to offer them. They also pooled information. Representatives of the merchant-planners, community organizations, and bilingual newspapers attended meetings with the Redevelopment Agency and received press releases, sharing their information through organizational channels or in the ethnic press. While these conditions belied individual contestation, however, opponents chose not to protest as a community through, for example, a collective lawsuit or organized resistance; these were measures some of their African American neighbors would adopt in later years. Instead, Japanese American property and business owners articulated their opposition in the familiar postwar language of individual property rights and ownership.

They emphasized their community's economic foundations and mainstream values rather than their racial clustering or persecuted marginality. Their approach was somewhat ironic, as they shared the language of property rights with white homeowners seeking ostensibly race-neutral justifications for segregation. As historian Charlotte Brooks has shown, however, by this point white San Franciscans had begun to see a shared set of values with Japanese and other Asian Americans, as they sought to rationalize limited Asian American residential integration in the context of Cold War interventions in Asia. Japanese Americans struggling with their neighborhood's redevelopment were more than willing to urge this view further, as they demonstrated a similar concern with property and social values through their willingness to work within the redevelopment program.[108]

Japanese American merchants, property owners, and residents found that their homes and businesses, so tenuously and tenaciously rebuilt after internment, were again under threat. So they used all the weapons available to them, whether challenge or cooperation. They fought, though, in terms established by the municipality. They did not challenge the principle of redevelopment itself, but instead the city's valuation of their property or their rights to their own individual parcel. Their opposition was expressed as fundamentally consistent with the city's values. Redevelopment, after all, was meant to increase property value in the city and raise tax revenue for its coffers. They argued their property too had value, not only sentimental or cultural but explicitly monetary value. Even the opposition of many Japanese Americans to redevelopment was carefully crafted using the language of city government and mainstream businesses.

CONCLUSION

San Francisco's postwar urban planning laid out a "new city" of model neighborhoods, commerce, and housing. It also established a framework of civic thinking. A city, in this view, was an intricately connected machine in which each neighborhood or even block contributed to the health of the whole. In addition to age, infrastructure, and zoning compliance, a calculus of costs, services, social conditions, and revenue helped determine an area's status. These factors determined the Western Addition district's blight, targeting it for the city's first wholesale demolition and rebuilding program. But a group of Japanese American merchants and property owners turned that civic thinking toward the protection of their ethnic economy. They, the merchant-planners argued, were distinct from the rest of the district as contributing, valuable members of San Francisco life who could add even more to its cosmopolitan assets if given the opportunity.

Through both their plans and their actions, the merchant-planners offered a public and participatory representation of Japanese American citizenship. The project was one part of the struggle to rebuild after the years of internment, to reestablish an economic, ideological, and physical space for Japanese Americans in San Francisco life. This process did not follow a natural or predefined route, but was learned in years of working with local government and institutions. Along the way, they gained valuable knowledge and experience in presenting themselves and their community to local government. From Victor Abe's first meeting with Redevelopment Agency staff, he learned the kind of information about the Japanese American community that he needed to provide in order to be taken seriously: property ownership, business volume, capital, and customer relations. This information was evidence of the feasibility of the Japanese Garden Center (later the "Japanese Village") as a commercial space, but it was also shorthand for Japanese American economic citizenship. The information that Abe and the Japanesetown planning group provided was constructive of this image, but so were their actions in and of themselves: their cooperative approach with city officials, their contributions toward official goals, and their public assimilation with overall definitions of urban improvement and progress.

This community portrayal was telling. The Japanese American merchant-planners were ambivalent about ethnic community, at least in public discourse. They were nonetheless very invested in redefining Japanese Americans' public image. To forward this, as we shall see again in later chapters, their community portrayals rested on difference rather than on common circumstances with their African American and low-income neighbors. Departing from their original concordance with the NAACP, even the JACL chose to abandon their oppositional stance based upon civil justice. At the same time, the merchant-planners increasingly drew connections with Japan. Their proposed mall highlighted the Japanese culture that was rising in popular interest, even as it provided space for continuing Japanese American services and businesses. This presaged the turn that redevelopment would take in the district, as their proposals showed the city how the Gateway to the Pacific framework could be used to reshape Japanesetown.

CHAPTER FOUR

Pacific Crossings: Japan, Hawai'i, and the Redefinition of Japanesetown

The Western Addition reached a turning point in 1958, with palpable consequences for those in the district. Property owners and residents had been contacted in January; by February, the Redevelopment Agency had snatched up one hundred parcels. This process, already swift compared to years of stasis, was accelerated for those in Japanesetown in the northern project area. The commercial strip along Post Street bordered Geary Boulevard, an artery marked for expansion into an expressway and slated for some of the earliest construction. The agency hired eight negotiators to quicken acquisitions in the area. By autumn, demolitions began, scattering dust and leaving flat, emptied lots behind. In the bulldozer's wake were crushed homes and a catalog of livelihoods: the Fuji Hotel, an establishment that had housed Japanese laborers from the prewar years; Yamato Auto Repair, another prewar, multiple-generation enterprise whose owners, after selling their building to the Redevelopment Agency, were never able to afford that kind of economic security for their business again; Takahashi Trading Company; Post Pool Hall; Nakagawa Apartments; Capitol Laundry; Sakamoto Grocery; Kik's Smoke Shop; and many others. Officials aimed to clear all the project's Japanesetown blocks by the end of 1959. Furthermore, amid the debris, the agency announced tentative plans for its second project in the district. As the *Nichi Bei Times* mournfully reported, "San Francisco's Japanese business and resident district, already sliced in half . . . may be doomed for complete destruction."[1]

The Japanese American merchant-planners watched these developments and returned to their plans for a shopping center with renewed vigor. They did so as assiduous students, applying the values and ideas around them to maintenance of their businesses and livelihoods. In addition to the official redevelopment program, the merchant-planners drew inspiration from local and national popular culture. Their proposals for a "Japanese Garden Center" responded to San Franciscans' increasing interest in Japan over the past few years, evident in the sister-city affiliation, restaurants, and events. Using demonstrated popular interest in Japanese food, designs, and other elements, the merchant-planners conceived a blueprint for their neighborhood that was commercial, tourist-oriented, and increasingly Japanese, not Japanese American.

The Japanese American merchant-planners' ideas fell on fertile ground as city boosters began to link their Gateway to the Pacific goals with the built environment. City leaders had pushed for a revitalized redevelopment program and private downtown construction, but they were also increasingly attuned to the salience and popularity of Japanese cultural activities. They saw such activities not simply as colorful spectacles and recreation, as the merchant-planners did, but as vital components of San Francisco's Gateway to the Pacific identity. City boosters—supported by influential and wide-ranging groups—sought to make San Francisco's landscape evidence of its Pacific prominence.

Mayor Christopher and city planners reassessed the concept of a Japanese-town development in this light. While they enthusiastically supported the idea of a Japanese center in the Western Addition, they had no interest in the one proposed by the Japanese American merchant-planners: a shopping center run by locals putting new facades on preexisting businesses. Instead, city officials envisioned a more lucrative Japanese center, with businesses, goods, and capital crossing the Pacific to institutionalize San Francisco's preeminent place in the Pacific region. The merchant-planners had astutely gauged local interest in Japan. But city officials had ambitions for a different type of Japanese project, one that far exceeded the merchant-planners' abilities and resources.

Building such a Japanese center was a challenge. It required a developer proficient in a specific set of skills, over and beyond the necessary financial capabilities: language skills, working relationships with Japanese businesses, and real-estate development experience. This was decades before the well-publicized, oft-maligned Japanese rush on large-scale real-estate purchases and investment in the United States, and so suitable prospects were few.[2]

The Redevelopment Agency found a developer in Hawai'i, Masayuki

Tokioka. Tokioka, a banker and community leader from Honolulu, had the requisite skills: he was experienced, capitalized, and able to bring Japanese businesses and investment to San Francisco. He had gathered these resources through education and hard work, but his home place was also crucial. The same year that the agency solidified plans for a "Japanese Cultural and Commercial Center" project, Congress made Hawai'i a state. The statehood debates had highlighted the former territory's historical identity as a hub for transpacific travel, investment, and ideas, a position that complemented San Francisco's own Gateway to the Pacific ambitions. Both the territory and its inhabitants were framed as having unique, valuable connections with Asia, a launching point for mainland relations westward. While in many ways unique, Tokioka epitomized these intermediary ideas. He had built a career bridging Japanese and US interests as well as Japanese American and white elite communities, while employing opportunities possible only in the "Crossroads of the Pacific." He arrived in San Francisco via the city's historical networks with the islands, which fostered the transpacific links that he would use for the Japanese Center.

Western Addition redevelopment quickly progressed from 1958 through early 1960. Redevelopment Agency officials not only cleared project land, but also solidified the residential, institutional, and commercial plans for the finished district and, in February 1960, put out calls for developers. Over the course of those two years, city officials adopted the Japanese American merchant-planners' proposed "Japanese Garden Center," but revised it into a center for Japanese goods, investment, and culture in ways that would come to exclude the merchant-planners. Tokioka, the selected developer, better fit the needs of agency officials because of his transpacific connections built through Hawai'i's "Crossroads of the Pacific" position vis-à-vis Asia and the mainland United States. San Francisco's Japanese Center drew upon the city's preexisting transpacific connections, extending and redirecting them toward its Gateway to the Pacific ambition.

FROM A JAPANESE AMERICAN
TO A JAPANESE PROJECT

The accelerating process of redevelopment in the Western Addition lent new urgency to Victor Abe and his group of Japanese American merchant-planners. Prodded into action by the grim reality of redevelopment in their neighborhood, they solidified their own plans for Japanesetown. The merchant-planners built on local interest in Japan to promote a project that was increasingly Japanese, rather than Japanese American.

By the end of 1958, Abe announced renewed planning for a Japanese-town development, reclaiming the Japanese American merchants' stake in the neighborhood. His modified group proposed to incorporate in order to build a "Japanese American shopping center."[3] Abe steered the group as its lawyer and president. He had amassed years of interactions with the Redevelopment Agency and so continued as the group's primary city liaison. Newcomers joined the old veterans, continuing the original group's variety and relative privilege as established business owners and professionals: they included Tamotsu Sakai, the owner of the Uoki K. Sakai grocery store, begun by his father in the early twentieth century; Masao Ashizawa, owner of Soko Hardware; Dr. Tokuji Hedani, an optometrist; and Mike Inouye, owner of Mike's Richfield Service Station. About half were Japanesetown residents. Whether they worked or lived in the neighborhood, they had an investment in its future: they themselves received notices of demolitions, conferred with neighbors, and worried over their businesses as they watched buildings fall.[4]

The renewed group of merchant-planners revisited the original plans. They maintained its investment corporation form, a "million-dollar venture."[5] Their proposed shopping center also retained the same mix of businesses, with Japanese souvenirs and imports as well as "a community shopping section with professional offices and Nihonjin-machi stores."[6] This also kept its two distinct clienteles: the "tourists" and the "issei and nisei trade."[7] The merchant-planners therefore continued to advocate for a shopping center to maintain and expand their businesses. Their small establishments depended on the shared circumstances and ethnic loyalties—as well as the limited options—of a segregated enclave. Their plans sought to continue these economic ties, even in the face of declining residence, by stamping the neighborhood with ethnicity in order to attract new customers.

Japanese elements were more firmly centered in the merchant-planners' revised project. The original plans had featured a landscaped Japanese garden but made few other gestures toward Japanese design. In contrast, the new plans featured "Japanese-style architecture."[8] The shopping mall would include a "Japanese garden and possibly a Japanese teahouse" as components in an entirely Japanese-influenced edifice (fig. 4.1).[9] These ideas appear to have borrowed from Chinatown's touristic pagoda-like rooflines and exotic motifs as a successful commercialized enclave.[10] Abe and his group reconceived their mall as a similar cultural attraction, where markers of Japan were not limited to the merchandise for sale but shaped the visitor's entire experience. The Japanese goods, food, architecture, and landscaping would be augmented with featured cultural demonstrations, including "dancing,

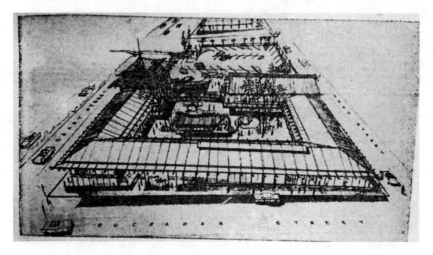

FIGURE 4.1 A 1959 drawing of the Japanese Garden Center by Arthur Iwata, Berkeley architect. This iteration of the merchant-planners' proposal conscientiously evoked Japan and Japanese style. Note the exaggerated roofline, pagoda-like structure in the far left corner, and landscaped garden. "Preliminary Plans," *Nichi Bei Times*, November 26, 1959. Courtesy of the Japanese American National Library, San Francisco, CA.

music exhibits and art."[11] These elements would create a "Japanese festival atmosphere" that would "show the contribution of Japanese culture and the local Japanese community to the city of San Francisco."[12] By mid-1959, the group had formalized their plans around this reconceptualization, incorporated, hired an architect, and produced drawings.[13]

These adaptations drew inspiration from the rising popular interest in Japanese culture. Nationally, the 1950s had seen a "Japanese boom," fueled by Japanese movies, foods, depictions of women, home décor, and other symbols of Japan.[14] Locally, sukiyaki restaurants, the San Francisco sister-city program, and other activity had increased the visibility and popularity of Japanese culture and people across the city. The Japanese American merchant-planners had experienced this interest firsthand. Groups like the Japanese Chamber of Commerce, Japanese American Citizens League, and Nichi Bei Kai organized much of the cultural labor behind these Japanese-themed events. For example, the chamber had hosted the Japan Day Fête in Golden Gate Park in 1957, featuring Japanese music and dancing, traditional forms attractive to a general audience. The fête was organized in conjunction with a Japanese-art exhibition at the de Young Memorial Museum that itself contained many items donated by local Japanese Americans. Such popular and well-regarded events demonstrated that Japanese American cultural representation was well received by San Franciscans and easily integrated into mainstream institutions.[15]

The merchant-planners' new ideas emphasized the Japanese culture they and others had been called upon to present in so many civic activities. Indeed, their new embrace of Japan was well calibrated. The *San Francisco Chronicle* lauded the "natural merit" of a "Japanese Village in the Western Addition Redevelopment Project" in the "predominantly Japanese neighborhood." Such a project would, the editors argued, "contribute substantially to the city's cosmopolitan charm in addition to the local economy."[16] Abe and his group reshaped their plans in the last years of the 1950s around rising local interest in Japan and Japanese ties. They also encountered a public far more receptive to development projects that could place these transpacific ties in the built environment.

REBUILDING DOWNTOWN FOR REGIONAL PROMINENCE

While the merchant-planners developed their ideas, San Francisco boosters and downtown interests were increasingly linking their goals to its physical infrastructure. If the city was to be, as they intended, an administrative hub for the Pacific region, its built environment had to accommodate the necessary expansion of businesses and transactions.

From the late 1950s, San Francisco Chamber of Commerce leaders envisioned their city as the future "Headquarters of the West." Their peninsular city could certainly never be the biggest, but its "strategic position on the Coast and on the Pacific Basin" could make it the "financial nerve center of Western America and of the great Pacific world beyond."[17] San Francisco had been built by financing and managing the extractive industries of its hinterlands, which extended not only to its Los Angeles rival in its early years, but also to neighboring states. This imprint remained. In 1957, there were already thirty national corporations headquartered in San Francisco, with combined assets of $32.6 billion. Plans to attract more business included personalized invitations to corporate executives, but boosters also looked to reshape the city as welcoming to new transplants.[18]

The chamber backed "a number of redevelopment projects" intended to "strengthen the city's position as the 'Gateway to the Pacific.'"[19] These included the project in the Western Addition, where new housing could shelter the planned upsurge in white-collar, professional employees. But corporate employers would also need homes, so boosters looked to expand the central business district. Leaders of some of the city's most prominent businesses organized to support downtown redevelopment, including the heads of the Crown-Zellerbach Corporation, Fairmont Hotel, and Crocker-Citizens bank. Eventually, they formalized in 1959 as the San Francisco Planning and Urban Renewal Association in order to mobilize the necessary "citizen interest and support," especially from the "top business leadership of the community," who can "influence the climate of opinion at both the political level and at the citizen level" in favor of a robust renewal program.[20] This group pushed for the restructuring of the city's slow-moving redevelopment process. It became the official Citizens' Advisory Committee required under federal law and provided a clearinghouse for research for and information about the city's planning.

Once downtown businesses had marshaled their support for the redevelopment program, it gained a new level of momentum. At the state level, court decisions and legislation eased the process of eminent domain and supported renewal programs. Locally, Mayor George Christopher ushered in a new era of redevelopment with downtown's support, overhauling the Redevelopment Agency for speed and professionalism. Christopher, a pro-growth Republican, had been elected on a reform, clean-government platform and brought with him a respect for expertise and close ties with downtown. His first move was a series of fresh appointments. After the resignation of the San Francisco Redevelopment Agency Commission chair in 1958, he appointed as replacement Everett Griffin, a prominent local executive and former director of the Chamber of Commerce; Griffin had criticized the agency's sluggishness in the past and supported a far-reaching redevelopment program.[21]

The next appointment was even more significant. Previous directors of the Redevelopment Agency were selected for politics rather than experience, but the new head, M. Justin Herman, brought extensive knowledge of federal funding and local implementation to the job. Herman had been a "career civil service employee" who had led the local regional office of the Housing and Home Finance Agency from its inception in 1951.[22] He was thus at the local helm as the housing agency expanded from its Washington, DC, headquarters to branches across the country, and oversaw redevelopment projects in a four-state region. He was an advocate of aggressive, extensive redevelopment and was all too familiar with San Francisco's "cumbersome and costly" program.[23] Herman came to the Redevelopment Agency charged by the mayor to remake it, on Christopher's promise of minimal interference and amenable commission nominations. Thus empowered, Herman installed professional planners at the agency, developed working alliances with other city agencies, and oversaw a huge growth in the city's redevelopment program, eventually making it one of the most active and federally well-funded programs in the country.[24]

Supported by downtown business interests, Herman pushed for the redevelopment of vast swaths of downtown. His first project in the area was Golden Gateway. Approved in 1959 and begun in the early 1960s, the project bordered the central business district and hugged the shoreline. It cut through the city's longtime produce market and warehouse district and replaced it with commercial, office, and hotel complexes; largely expensive and modern high-rise apartments; and public plazas. The "exciting new prospect for San Francisco's growth" was joined in 1966 with a South of Market project, which uprooted thousands of poor tenants from the area's residential hotels for a sprawling convention and commercial center.[25]

Municipal activity joined an explosion of privately financed high-rise building. The final years of the 1950s saw the beginning of what the chamber called the "Big Build," a record-breaking "building boom" that would reshape the city's skyline. Together, public and private activity reflected the "widespread faith in San Francisco and its destiny as the headquarters city of the Pacific Coast."[26] City and business leaders eagerly adopted public and private urban renewal as a way to meet their perceived needs.

"THE CULTURAL AND BUSINESS GATEWAY TO ASIA"

City boosters increasingly began to see demolition and construction as a way to build more than just skyscrapers and high-rise apartment buildings. They also believed that interventions into the built environment could create material expressions of their city's Gateway to the Pacific identity.

The *San Francisco Chronicle*'s critique of the city's first Pacific Festival in 1958 included an example of a potential project to connect the city's Gateway ambitions with the built environment. Not only did city leaders need to rethink their "unconcern toward major projects and redevelopments," but, editors urged, they also needed to embrace cultural initiatives that would highlight San Francisco as the West Coast center for all things related to the Pacific region. In fact, the editors' example of "planning for the good of the city and [its] healthy growth" was not a redevelopment project at all, but the "remarkable collection of Oriental art" offered to the city by Chicago industrialist and International Olympic Committee president Avery Brundage.[27] The editors insisted, "As the Western gateway to the Orient, San Francisco would be a most suitable home for it."[28] Because of the offer's contingencies, the city would have had to fund a new wing for the de Young Memorial Museum, the city's landmark art institution, to house the collection and do so, for Brundage's tax purposes, on a verbal offer. Some officials balked at the necessary bond and the burden on voters' trust, and funding stalled for years. In 1958, Brundage withdrew his offer, complaining that "there has been a great deal of talk, but nothing has been done."[29] Finally, the mayor, the museum, and city representatives hurried to take steps to acquire the collection.

Two years later, the city ran a bond to build the necessary wing next to the Japanese Tea Garden. The bond elicited "the widest range of enthusiastic endorsements ever accorded a municipal proposition."[30] In an election that also saw propositions for school loans and state indebtedness, the chamber backed what it called "the most important municipal ballot measure" with "all possible support"; groups from the San Francisco Real Estate Board to a Cooks Union local issued affirmative statements, while every major municipal, state, and federal elected official in the city endorsed the measure.[31] Art teachers added Oriental-art lessons to their curriculum, and local experts gave lectures on "The Brundage Collection and Its Significance for San Francisco."[32] "Record crowds" flocked to a promotional exhibit of Japanese art at the de Young in advance of the election.[33] With "no formal opposition," voters approved the measure by a forty-point margin. The passage of the city bond measure, and its massive voter support, transformed San Francisco "overnight" into "one of the great centers in the world for scholars and connoisseurs of Oriental art and, through that art, of Asian civilizations."[34]

The Brundage collection demonstrated the utility of culture, construction, and popular support for San Francisco's role as not only "the top United States center of Oriental art" but also "the cultural and business gateway to Asia."[35] The new Brundage Wing provided ballast for the city's economic and political networks in the context of the city's "Asian civilization," as the sister-city's cultural exchanges and institutions did. Furthermore, the collection added its own small but significant numbers to the people building networks across the Pacific: the "scholars and connoisseurs," the patrons, those San Franciscans whose Asian interest might be sparked by the new trove of art. Although the pieces in the collection were old, the addition to the city's cultural life was not static: its content would be kept "alive and moving" by "the latest scholarship and latest discoveries," and "attract the best minds in the various fields of Oriental scholarship." Moreover, "this community and the world" were kept aware.[36] The Brundage Wing's grand opening included the five-day "International Symposium on Asian Art," in which scholars from all over the world gathered to discuss the collection and "tell the outside world again what San Francisco has acquired in the Oriental arts."[37] The publicly financed development—which eventually grew into the city's famed Asian Art Museum—was viewed not only as a recognition of the city's Pacific status, but also as a way to reinforce San Francisco's function as a Pacific hub.[38]

BRINGING JAPAN TO JAPANESETOWN

The reception of the Brundage collection demonstrated both the potential for similar, Asian-oriented building projects and the voracious public appetite for such institutions. This environment provided Japanese American merchant-planners a similarly amenable official audience, if on a smaller scale. However, as Western Addition redevelopment went from clearance to calls for developers, city officials redefined the merchant-planners' project. While officials borrowed their basic idea of a Japanese-themed tourist attraction, they also extracted it from the ethnic economy.

M. Justin Herman, a self-described "Japanophile," made a Japanese center project one of his first priorities upon becoming director of the Redevelopment Agency, although he also foreshadowed the merchant-planners' exclusion from it.[39] At his initial meeting with the Redevelopment Agency Commission in 1959, he called for the formal designation of a "Japanese Trade and Cultural Center" in the Western Addition project. His ideas traced back to the Japanesetown merchant-planners' mall with souvenirs, professional offices, and sukiyaki hosted by small, local proprietors. However, Herman wanted the center to contribute to "our excellent trade relations with Japan" and the "fortunate status San Francisco enjoys as a major American gateway to and from Japan."[40] This would be a center for the most significant Japanese corporations and manufacturers. The unintended consequence of the merchant-planners' appeal to popularity was a disparity between their proposed development and the ethnic economy. By making their project more "Japanese," the merchant-planners inadvertently weakened the logic that they—local Japanese American merchants and professionals—were the best group to carry out their plans. City officials saw themselves as better served by actual, well-capitalized Japanese developers and investors.

Mayor Christopher took these ideas to the public at the Pacific Festival's "Japan Day" two days later. The mayor borrowed Herman's descriptions of a Japanesetown development almost verbatim and placed the idea in the larger context of the city's transpacific ambitions:

> In furtherance with our excellent cultural and trade relations with Japan, I am asking the members of the redevelopment agency to work with the citizens of San Francisco, Japanese interests of this city, our friends in Japan, particularly those in our sister-city Osaka, to create in the Western Addition a Japanese trade and cultural center. . . . San Franciscans have a deep affection for the Japanese traditions of that area and would be proud to see that recreated in the form of a trade and cultural center. It will further strengthen the fortunate status San Francisco enjoys as a major American gateway to and from Japan.[41]

The center would support and increase relations with Japan by giving them a new, consolidated home and by demonstrating their longevity and popularity in the city. The mayor was revealingly ambiguous about the idea's original proponents. His call to "Japanese interests" could be read as either local Japanese Americans or businesses with financial or commercial interests in Japan. Likewise, the Japanese American enclave was euphemized as the neighborhood's "Japanese traditions," alienated from actual Japanese American people. He promoted a Japanese center that would not only raise the city's Pacific stature but also bring "other international groups" to "adjoining renewal areas."[42] A month later, when the sale of "cleared slum lands" was announced for the following spring, one block was specified for a "Japanese cultural or commercial center."[43] Herman and Christopher had laid the groundwork for the adoption of the project into the Western Addition's redevelopment.

Official enthusiasm was supported by evidence of Japanese retailer enthusiasm. Department stores were among the earliest businesses to thrive in postwar Japan, evidence of rising industry and consumer strength; they were also among the first to turn to overseas markets. This growth, however, threatened small retailers, who pressured Japanese lawmakers to pass a 1956 law limiting their domestic expansion. Large Japanese retailers therefore looked for overseas locations, a move supported by industry and government. These sectors hoped that overseas stores could promote Japanese goods by demonstrating their rising sophistication and, ideally, help balance US-Japanese trade by returning profits to Japan. Furthermore, overseas stores were a "useful market research tool" to gauge the "ever-changing American market which is of so great importance as the major export destination of export dependent Japan."[44] Before the marketing of Japanese goods directly to US consumers took off in a big way, Japanese retailers were important early pioneers in getting Japanese-made goods into US hands.[45]

By the late 1950s, Japanese department stores were "raring" to open their own overseas branches, according to one Japanese evaluation.[46] At the same time as Christopher's speech, Takeshimaya, a major Japanese department store, established its first overseas branch in New York City on Fifth Avenue, while Seibu, another large department store, made plans for a Los Angeles location. These stores sold antiques, lacquerware, silks, pearls, and other high-end, high-quality goods from Japan, not, as an executive emphasized, "cheap, souvenir items."[47] Still, as souvenirs were intended to do, the stores and their contents aimed to evoke Japan in the shoppers' minds. Because not all of these items were necessarily unique to Japanese producers, the imported décor and kimono-clad, female Japanese clerks reinforced this function. The stores also sold themselves as cultural institutions, promising exhibitions of traditional dances, flower arranging, and tea ceremonies.[48]

These department stores were not quite the kind of trade center that Herman or Christopher had in mind, but they suggested intriguing possibilities. The stores brought together commerce and culture and stood as public and concrete testimonials to the host city's transpacific networks. However, they posed challenges for San Francisco. Seibu and Takeshimaya had instigated the New York and Los Angeles stores. If Herman and Christopher wanted their own Japanese development—and one more complicated than a single retailer—they would need a developer capable of building it.

SEEKING A DEVELOPER

Herman was enthusiastic and invested in the Japanese center project, claiming a "fond[ness] for Japanese culture." His support reflected his belief that redevelopment could do more "than just replace slums" with new buildings or infrastructure.[49] His career had not been entirely devoted to urban planning—before working at the Housing and Home Finance Agency, he had been an industrial relations officer with the San Francisco Naval Shipyard—but he had fully embraced its potential to revitalize cities. The Japanese center project seemed to him a possible "nucleus" for a "new international center for the city."[50] He therefore used the tremendous leeway granted him by the mayor to usher it through what turned out to be a long process filled with roadblocks. The project had staunch backing from a strong-willed and quite powerful official.[51] This turned out to be bad for Abe and the merchant-planners. As Herman solidified potential developers, his behind-the-scenes activity made it clear that he had little interest in the merchant-planners. He focused his attention on Japanese possibilities.

Herman's zeal for a Japanese-backed project may have reflected a confidence based in past experience. During his tenure with the housing agency, he had hosted the occasional visiting Japanese city planner or official. The educational value of these tours appears to have been mutual. Japanese visitors examined US urban problems and solutions, of course, but Herman observed US-influenced Japanese growth and development firsthand. This left a favorable impression. He contacted at least one of these visitors even before he assumed office in order to solicit his participation in the Japanese Center: Matsui Kakuhei, with the National Construction Industry Association of Japan. Such a center would, he argued, be a "good business development and a great cultural exchange opportunity between our two countries." By offering "a foretaste of the delights to be experienced in your country," US visitors might be tempted to visit Japan and spend their valued tourist dollars. At the same time, San Francisco could flourish its status as "the major U.S. gateway to the Orient."[52] Herman's experiences in transpacific exchange had given him contacts and precedent for Japanese partnerships.

Matsui's silence in return indicated a troubling imbalance in interest. Herman made repeated appeals to him and even met with the San Francisco consul general of Japan for assistance.[53] But the Japanese builder never responded. San Francisco officials saw the redevelopment process as an effective and logical way to build a Japanese center, but their Japanese counterparts might have disagreed. For Herman and the mayor, the project's location made sense: it was the "heart of the Japanese community of the West."[54] But historically, Japanese government officials and other elites had viewed emigrants, especially laborers, as an unruly lower class needing control and regulation; the allegiances and identification that Herman and other city officials assumed between Japan and its diaspora might not have been particularly compelling. Furthermore, the proposed project's location might have given Matsui pause. He had toured the "increasingly blighted" Western Addition during Herman's introduction of the city's redevelopment program.[55] Herman had intended to reveal the challenges and miracles of US urban renewal, but it is also perfectly likely that the visitors were more attuned to the challenges of the location, less so to the miracles not yet in evidence. Even executives from one of the eventual investor companies were "particularly concerned with the blighted area" around the project's location, and, in an open expression of bias, worried about "the low social and economic level of some of the surrounding population."[56] At a time

when Japanese capital was tightly controlled and its international investment closely scrutinized by the Ministry of Finance, Japanese developers
avoided as many risks as possible; both Seibu and Takeshimaya had built in
stable, established commercial locations.[57] In the face of apparent disinterest, and having run out of personal Japanese contacts to call upon, Herman
gave up soliciting a Japanese developer directly.

The project was by no means lost, however, because there was by this
point a growing crowd of interested US developers. Foremost among these,
of course, was Abe and his group of Japanese American merchant-planners.
Happily ignorant of Herman's activities, they were heartened by the city's
commitment to a Japanese center project. Having already appointed an architect to study the area, they now developed preliminary drawings. In the
midst of talking to the press, coordinating with the architect, and meeting
with his group, Abe was undoubtedly aware of his growing competition.
When the agency announced plans for the center, it predicated details on
"negotiations now under way with prospective builders," and the *Nichi Bei
Times* had reported on other interested prospects.[58] These included a prosperous South Bay Nisei restaurateur who had begun to branch out into
housing developments, and a group that included Jay D. McEvoy, a San
Francisco banker. Celebrated Southern California architect Richard J. Neu-
tra also queried Herman about the project. When Herman called a meeting
about the project in November 1959, he presided over a crowded gathering
of dozens of interested architects, investors, and builders.[59]

The proposal for a Japanese center had come a long way since Abe first
raised the idea to a skeptical agency director in 1953. Abe nevertheless maintained a stout confidence rooted in the fact that his group was the only
one with actual plans and drawings, printed in the *San Francisco Chronicle*
as "under study" by the Redevelopment Agency.[60] Unbeknownst to him,
however, the proposed project was shifting in ways far beyond his capacity. Two developers, both from Hawai'i, helped move the project out of the
merchant-planners' reach. The first, Hung Wo Ching, reshaped the project,
and the second, Masayuki Tokioka, would ultimately be awarded it as the
last man standing—the sole developer who could meet its new terms.

EXPANDING AND REDEFINING THE
JAPANESE CENTER PROJECT

Hung Wo Ching was a Chinese American businessman from Hawai'i, already well versed in California real estate and known to the San Francisco
Chamber of Commerce. He therefore came to the project with interest,

experience, and a critical eye. He emerged from the November meeting as Herman's most promising lead, and the two engaged in an extensive correspondence for months. Eventually he dropped out, but he left his mark. As the first outsider to carefully critique the project, his serious assessments expanded and clarified it.

With advanced degrees from Harvard Business School and Cornell, and work at Beijing's Yenching University, Ching was a well-trained and successful man of business. Among his many accomplishments, he had rescued the struggling Aloha Airlines from near bankruptcy and built eight subdivisions in Honolulu. He also had interests in Southern California and a residential and commercial development in the booming Bay Area suburbs. Indeed, by this point his feats had earned him recognition in San Francisco business circles. Ching was one of a handful of notable businessmen, and the only nonwhite speaker, invited from Hawai'i for a San Francisco Chamber of Commerce event.[61]

His experience convinced Ching that the Japanese center project was too limited in scope and space. First, he and the meeting's other attendees argued that the designated one block was inadequate: "one block does not make a trade center."[62] Furthermore, the proposed limited parking could ill compete with that of the new, rapacious suburban malls. Herman was open to the addition of another block, possibly two, to the center, but initially pushed back against suggestions of underground parking: "the City would shrink from making a capital investment."[63] Given pressure from Ching and other interests, however, Herman worked out the addition of project land and the development of financial mechanisms to feasibly construct an expensive, underground parking lot. Herman came to embrace such measures as "essential to the success of the Center," a "project which people from all over the world would regard as an unique tourist attraction."[64]

Ching pushed unsuccessfully for other enlargements. He felt that the Redevelopment Agency's ideas were "too confined and limited."[65] He wanted a "market place, a show place, and a place where Californians and their visitors may become directly acquainted to the cultures of the Orient."[66] Unsatisfied by the idea of a solely Japanese development, he proposed a "Far East Cultural-Commercial Center" that would "include all countries of the Pacific Basin and not be limited strictly to Japanese culture."[67] Using the familiar language of goodwill diplomacy, he argued that this widened scope would better address international goodwill, mutual trade, and understanding.[68] In fact, popular culture supported Ching's ideas. Best-selling writers such as James Michener and films such as *The King and I* suggested a widespread interest in all parts of the region.[69]

Herman, however, never took this "Far East" amendment seriously. He firmly responded to Ching's ideas for a "Far East Cultural-Commercial Center" with specifications and details for a "Japan Cultural and Trade Center."[70] A Japanese center fit with the neighborhood's "traditions," as Herman euphemized the Japanese American enclave. The development was not to be a vague gesture of goodwill but a feature connecting the city with modern Japan's "cultural contributions" and "her contemporary and traditional architecture, her products and services."[71] A cultural center like the Brundage museum wing could encompass the Far East, but this commercial center looked only to the Asian nation with mature investment and commerce. Certainly, businesses in San Francisco were interested in the developing markets of Asia, but redevelopment officials sought "high-priced merchandise" for the center's predicted "three and a quarter million dollars of retail and international trade," a connection to Japan's booming "new future."[72] These disparities in thinking between Ching's and Herman's visions proved insurmountable. But the conversations clarified the Japanese Center's function: as a link to the trade and businesses of Japan, the industrialized hub of the Pacific.

The interest and scrutiny of serious developers demonstrated the viability and potential of a Japanese commercial and cultural center for those at the Redevelopment Agency. The project became a "top priority."[73] But newly implemented ideas raised the stakes for whoever would develop the center, reimagined by this point as an expensive development filled with Japanese tenants and investors. Creating such a complex was outside the scope of most of the other interested developers. The chosen developer would have to be heavily capitalized and would need to be familiar with the Japanese language, the US redevelopment process, and modern Japanese business circles. On the eve of large-scale Japanese direct investment in the United States, outside of import/exporters, few people were up to the task. Ching, for all his building experience, was not. He had repeatedly followed up with Herman's contacts in Japan to no avail and seemed to have no leads of his own.[74] The proposed developer would have to be an intermediary, a bridge to Japan and Japanese interests, in a way that none of the other interested developers were.

HAWAI'I, THE US AMBASSADOR TO ASIA

Hawai'i, the home of both Ching and the eventual developer Tokioka, was framed by those very transpacific ideas and networks. A rich history connected San Francisco and Hawai'i and situated them in a Pacific world, connections forged in the transpacific commerce and imperialism that had made Hawai'i the fiftieth state and the United States a Pacific power. Popular perceptions of these links were heightened in 1959, the year that the Redevelopment Agency formalized its plans for a Japanese center and the year that Congress made Hawai'i a state. The debate over statehood portrayed the territory in ways that resonated with Herman's goals for a developer and with San Francisco's foreign-relations strategies toward Japan. These same links brought Ching and Tokioka to San Francisco and shaped the capabilities, networks, and background that Tokioka would use to develop the city's Japanese Cultural and Trade Center.

San Francisco had a historic, if unbalanced, relationship with Hawai'i. From the mid-nineteenth century, the city was the primary mainland port for Hawaiian trade, as it was for most of the traffic crossing the Pacific to the West Coast. The sugar industry tightened this bond. San Francisco banks capitalized early sugar planters, and San Francisco sugar brokers extended them credit. This financial relationship increased with the Reciprocity Treaty of 1876, which allowed duty-free US importation of Hawaiian sugar and, to secure the support of California refiners, stipulated a minimally refined grade for sugar exports. This provision ensured that the islands' primary export was for almost a century fodder for Bay Area refineries. Claus Spreckels was one San Francisco refiner who sought other avenues of profit. He arrived on the same boat as the treaty itself to snatch up Hawaiian land for sugar production. Within a decade, he had grown so influential in the islands' economy and politics that his rivals unceremoniously ousted him from the islands. Yet his family retained a monopoly on West Coast sugar refining until the formation of the C&H Sugar Refining Company, owned by a group of Hawaiian sugar plantations, just off San Francisco Bay.[75]

Sugar lost its economic centrality in Hawai'i after World War II; by statehood, defense expenditures had dwarfed sugar's annual value in the islands, and the rising tourism and pineapple industries were encroaching.[76] Nevertheless, San Francisco's position as regional hub of corporate headquarters and finance continued its close relationship with Hawai'i. For example, Matson Navigation, headquartered in San Francisco, was the main carrier for Hawaiian sugar and remained the most significant cargo company between the mainland and Hawai'i even after sugar's decline. By the 1980s, three of the Big Five sugar factors—the companies who made up the pre-statehood economic and political oligarchy—had relocated their corporate headquarters to San Francisco.[77]

This historical relationship took on a new meaning as San Francisco boosters remade their city as the Gateway to the Pacific and the debates over Hawai'i's statehood amplified its "Crossroads" identity. The mid-Pacific archipelago had long acted as a "meeting ground for the Occident and the Orient" and had historically been defined by the United States through its proximity to Asia.[78] US and European traders involved in Asian markets brought the islands into the global economy as the "stopping-place for almost all vessels passing from continent to continent," making Honolulu the busiest Pacific port for much of the nineteenth century.[79] Its mid-Pacific location was also crucial to US maritime expansion. After a small but powerful group of planters descended from US and British missionaries and traders overthrew Queen Lili'uokalani in 1893, annexationists argued that a Hawaiian colony could establish the United States as a Pacific empire. As the navel strategist Alfred Thayer Mahan claimed, Hawai'i held "unique importance" due to "a position powerfully influencing the commercial and military control of the Pacific."[80] The 1898 US war with Spain in the Philippines sealed Hawai'i's fate. Over strenuous Native Hawaiian opposition, Congress annexed the former kingdom in support of the war and US interests in Chinese markets. Its military significance "was brought home to all of us once and for all on December 7, 1941," but the post–World War II debates over statehood extended its identity as a "bridge to the Asian world." Proponents argued that "Hawaii will be our center" for future "business and trade and social and political relations" with Asia, whose resources and markets represented "half of the people of the world."[81]

Hawai'i's mid-Pacific location was critical to establishing its identity as a bridge to Asia, but so were the territory's demographics. The "incredibly polyglot and racially integrated" population consisted of people of Asian, European, and Polynesian descent, with an Asian majority.[82] Proponents saw this as an advantage. A Washington senator argued that "Hawai'i in the Pacific" would be "not just . . . the 50th state, but . . . our diplomatic State": "a living example of the real fruits of freedom [to Asian observers]. Here the Occident and the Orient have met in a climate of mutual trust, understanding, and respect."[83] Observers marveled at the diverse population's ability to "live and work together amicably, democratically, and harmoniously."[84] In contrast to the contemporary mainland struggle for civil rights, Hawai'i appeared notable for its "racial aloha."[85] Certainly, daily life complicated this rosy view, stratified by ethnic divisions. More importantly, this view depended on the suppression of Native Hawaiians. According to the census of 1950, Native Hawaiians were 17 percent of the territory's population, while people of Asian descent were 55 percent; haoles, or white people, were numerically small at 22 percent but still held much of the political and economic power in the territory.[86] The celebration of interracialism and Asian assimilation therefore stigmatized Native Hawaiians, a demographic and socioeconomic "minority" in their own land, while hiding colonial conquest under a "melting pot . . . of interracial tolerance and affection."[87] As a result, the new state could be portrayed as a model of decolonization and friendly international relations between Americans and Asians, evidence "to all the peoples of the Pacific and of Asia that the US can still be the tolerant, hospitable melting pot of old."[88] This view was embraced in San Francisco, too, where anti-Asian attitudes had been "dominant . . . before the war": statehood was a "symbol to Asia of the desire of a predominantly white society to give equality to a predominantly non-white one."[89] Because of the subjugation of Native Hawaiians, Asians in Hawaiʻi could be used as a symbol of and support for good relations with Asia.[90]

People of Asian descent were also framed as a direct link with Asia. John A. Burns, the delegate from Hawaiʻi, argued for the significance of "Hawaii's citizens of Japanese, Chinese, Korean, and other oriental ancestries." The Cold War in Asia, he maintained, made them important: "the peoples of Asia . . . are a great question mark in our present battle with communism. They are numerous peoples with numerous resources and all signs indicate that they are going to be an even greater power in the future than they are now." Asian Americans in Hawaiʻi could help the United States win them over because of their "understanding of and sympathy for the culture and traditions of the Orient." Therefore, the "citizen of Hawaii, that new man of the Pacific, will be our most effective bridge to the Asian world."[91] Clearly, these sentiments evoke Hawaiʻi's Asian Americans as "alien," as *Newsweek* framed the islands.[92] Perceptions of their nonwhite, foreign status had in fact contributed to southern Democrats' consistent rejection of statehood, as part of their passionate upholding of white supremacy in Congress and the nation.[93] But statehood's ultimately successful proponents framed residents' Asian descent favorably, underscoring in positive ways their long-standing representation as foreign and "not-quite" American. The population's claim to full US inclusion lay in their "roots not in Europe but in Asia" and the transpacific connection they thus provided.[94]

When Ching and Tokioka trod the well-worn path of investment and commerce between Hawaiʻi and San Francisco, they were embedded in a broader postwar reconsideration of the islands. This included an economic reevaluation. As proponents had made abundantly clear during the congressional debates, the territory was no economic slouch. Its per capita income was higher than that of twenty-six states, its per capita tax burden was higher than thirty-three, and its gross territorial product twice that of any former territory to achieve statehood.[95] The secretary of the Interior further predicted a "considerable acceleration" in the new state's economy, based on what happened for other recently admitted states due to population growth, the migration of businesses, and outside investment.[96]

San Francisco observers, too, predicted heady, prosperous times for the new state. Increased land prices, heightened tourism, a sizable permanent migration from the mainland, more-frequent air travel, and frantic bids for new transpacific airline routes all attested to San Franciscans' confidence. The Chamber of Commerce was especially keen. The organization had consistently supported Hawai'i's statehood since 1946, when it claimed to be the first chamber in the country to do so. Now the chamber looked with optimism to "a great acceleration in the business tempo of the Islands" and a "strengthening of our economic ties with Hawaii."[97] Its members established the Hawaiian Affairs Section to research and promote its economic potential, the only one besides the Japanese Affairs Subcommittee that it created that decade. They also created a "Trade Development Tour of Hawaii" similar to the chamber's tours of Asia, conferences on business opportunities, and hospitality for visiting executives and officials from the new state.[98]

Tokioka and Ching represented a new state laden with economic and political meaning for all Americans, but especially for San Franciscans. Tokioka, the successful candidate, came to the project situated in an appealing set of ideas about his state as a "Pacific Crossroads" and especially about its Asian-descended residents, viewed as mediators with Asia.

THE DEVELOPER FROM THE CROSSROADS

Tokioka, a Honolulu-based Japanese American developer, exemplified the generalizations about Hawai'i in ways that resonated with the Redevelopment Agency officials' needs. Fluent in Japanese and English, well educated, experienced in business and urban development, he had built a career at the intersections of Japanese business, the Japanese American community in Hawai'i, and the white elite of Honolulu. However, despite US observers' tendency to abstract all those of Asian descent in Hawai'i as go-betweens with Asia, Tokioka was exceptional. His skills as a cultural and commercial mediator came from a background possible only in Hawai'i, and unusual

even there. Asian Americans and especially Japanese Americans in the islands had a very different history from those on the mainland. This historical, social, and geographic context shaped the skills that Tokioka brought to
San Francisco's Japanese Center.

Tokioka's background set him apart from most people of Japanese descent in Hawai'i, but his past made him familiar navigating between them
and the powerful white minority in the islands. Unlike the vast majority
of Japanese migrants who came as laborers for the sugar plantations, Tokioka's father had been a business manager in Japan and established a nurs-
ery in Hawai'i in 1895. Furthermore, Tokioka himself was born in Japan
and was thus a Japanese citizen: his mother, concerned about local hos-
pitals, returned to her parents' home for his birth. By the time he and his
mother returned to Hawai'i in 1909, his father had grown the business into
a thriving concern serving a wealthy white elite and, according to his bio-
grapher, deposed Hawaiian royalty. This relative prosperity allowed Tokioka to graduate from the University of Hawai'i at a time when less than
half of all Japanese Americans completed high school.[99] Tokioka completed
his education with a degree in international commerce from Harvard Business School, claiming to be the first Japanese to do so.[100] His class, citizenship, and education were unusual and gave him experience in worlds outside the migrant community in Hawai'i.

Tokioka's religion and specific congregation also gave him experience
moving between communities. Unlike most Japanese in Hawai'i and the
mainland, he was Christian.[101] Furthermore, his family attended the Makiki
Christian Church led by the Reverend Takie Okumura, an influential but
controversial congregation. Vocal critics in the Japanese community censured Okumura for his alliance with the Hawaiian Sugar Planters' Association, whom many workers saw as opponents. His church was built with donations from one of Hawai'i's white, elite sugar families, and Okumura had
created the Sugar Planters' Association–supported Americanizing campaign
that, critics derided, was mostly concerned with creating pliant plantation
workers. As a result, Okumura recalled, "the majority of the Japanese mis-
understood me" as "a traitor" and "a spy."[102] Throughout these controver-
sies, Tokioka remained a faithful and active congregant and a close friend
of Okumura.[103]

Tokioka brought his experiences moving between the Japanese American community and the local white elite to his lifelong career with the
International Savings and Loan family of companies. There he rose from
bank clerk to president because of his education and sharp professional

acumen, but also because of his ability to navigate the many eddies of society in Hawai'i.[104] The original company was founded in 1925 to serve the local Japanese American community by an unlikely pairing of prominent white and Japanese businessmen. Its board of directors included Clarence H. Cooke and John Waterhouse, men from the exclusive, missionary-descended plantation families that dominated Hawai'i politics and society. Wade Warren Thayer was the founding president, a former territorial attorney general (1912–1914) and secretary of Hawai'i (1914–1917). Thayer was also attorney for the Honolulu Japanese consulate, a position that likely introduced him to the executives of Sumitomo, one of the largest Japanese banks in Hawai'i, who contributed the foreign expertise and capital to International. Nakayama Goro, a former manager of Sumitomo Bank, served as International's vice president, and Harada Seiichi, a current manager, was its director; both were Japanese nationals. The collaboration was unusual—the boards and directorships of Hawai'i's Big Five were entirely white, although there was room for a man of mixed, aristocratic Hawaiian lineage—yet was understandable because of a series of developments in the early twentieth century.[105]

International Savings and Loan, and Tokioka himself, arose from the conspicuous position that Japanese held in Hawai'i's history since at least the turn of the century, due to their large numbers and their migration from a rising empire that competed with the United States. These conditions brought forth local responses to Japanese that toggled between virulent, racist xenophobia and laudatory acceptance. Japanese Americans in Hawai'i therefore had a contentious past, but it was one that made them, by the 1950s, prominent in Hawai'i politics and economy.

The white planter elite initially sought Japanese migrants as industrious and manageable labor for their sugar plantations after Native Hawaiian and then Chinese labor had proven undesirable in their view. Over 150,000 men and women arrived from Japan between 1885 and 1908, mostly to work on the sugar plantations, and quickly became the largest demographic group on the island and half of all sugar-plantation workers. But their large numbers spurred an anti-Japanese backlash. The same Hawaiian Sugar Planters' Association that had organized the tides of migrants began to fear their growing numerical leverage.[106]

A series of strikes in the early twentieth century gave plantation owners and other observers evidence of what they saw as a "Japanese conspiracy" and ignited full-blown anti-Japanese animus.[107] Workers of all origins had long struggled collectively and as individuals against violence from overseers, low pay, punitive fees, long workdays, and other poor working conditions; these conditions made up a paternalistic system that provided worker housing, basic supplies, and some medical care for workers as a means for labor control as much as welfare capitalism. In 1909, Japanese workers organized in the largest and longest strike to date, shaking plantations across Oʻahu for three months. This paved the way for a six-month successor that swept the territory in 1920. This strike was notable for its size and its deaths—over 150 people died when evicted and hungry families fell prey to the global influenza epidemic—but also for its departure from the "blood unionism" that had characterized previous organizing. Over eight thousand Japanese and Filipino laborers walked off the job. Despite the strike's interracial leadership and rank and file, however, employers and local media saw it as an attempted Japanese takeover. Filipino strikers, who had in fact initiated the action, were contrasted as naïvely deceived.[108]

Such inaccurate understandings of the strike were shaped by local conditions and Pacific geopolitics. Plantation owners had employed a long-standing strategy of ethnic stratification in everything from housing to strikebreaking that arose from and solidified stereotypes about plantation workers. The elite's ominous image of Japanese was also shaped by Japan's rising imperialism. Japanese colonial acquisitions in Korea, China, and the Pacific threatened US power in the region and stoked fears of Hawaiʻi's vulnerability. Japan augmented such fears by annually parading its naval fleet and by deploying its warships to the islands in ostensible protection of its citizens and their property, first in response to Japanese migrant exclusion and then to the US-led overthrow of the monarchy.[109] This context led mainstream dailies to compare the 1920 strike to Japanese colonization in China and to proclaim the "real issue": "Is Hawaiʻi to remain American or become Japanese?"[110] According to the territorial governor, locals held the "suspicion that the Japanese in Hawaiʻi, even though they be citizens of the United States, are acting under instructions from their own Government."[111] This view was deeply ironic since the consul general, Japanese banks, and migrant leaders such as the Reverend Okumura opposed not only the strikes but also most challenges to peaceable relations with white leadership.[112] Nonetheless, Japanese in Hawaiʻi were demonized as an imperial threat.

The 1920 strike birthed a robust anti-Japanese movement in Hawaiʻi. As on the mainland, the movement gained momentum from the xenophobia of World War I as well as concerns about foreign radicalism. Anti-Japanese activists in Hawaiʻi feared that a Japanese majority would cause the territory to "fall into the clutches of the Japanese in Hawaii and ultimately into the possession of the Japanese Government."[113] They therefore agitated for the territorial regulation of Japanese-language education, sought to repeal the citizenship of the US born, and energized Americanizing campaigns.[114]

The anti-Japanese movement struggled to stem the unmistakably rising influence of Japanese migrants and their children, who made up 40 percent of Hawai'i's population by 1920. They were also economically weighty, especially as they left the plantations in droves after the strike. Per capita real-estate investment among Japanese Americans rose from $2.61 in 1911 to $9.17 in 1920, and local savings increased about sixfold. Remittances, on the other hand, dropped dramatically. In the first decades of the twentieth century, Japanese migrants sent about $1 million annually to Japan. In 1922, this amount fell by almost half, an overall pattern of decline that continued through the decade. Additionally, there were approximately eighteen hundred retailers of Japanese ancestry in Hawai'i by 1930, almost 50 percent of the retail stores in the territory. Finally, the first wave of Nisei, US-born citizens, came of age in the 1920s. By the next decade, they would outnumber their parents as an even larger market.[115]

What seemed to some like evidence of a Japanese takeover struck the backers of International Savings and Loan as a goldmine. The company sought "to retain in Hawai'i a larger share of the savings of Japanese immigrants" and "make available a new source of longer-term loan funds" for those of Japanese descent.[116] They had little competition. The territory's major banks primarily served the white elite and did not loan to those of Asian descent, while Japanese banks such as Sumitomo or Yokohama Specie were primarily vehicles for remittances and trade financing. Nor could informal practices such as rotating credit associations or family borrowing match the rising demand. Two short-lived, pre–World War I examples had previously tried to tap into the Japanese American market, but International took advantage of the shifts in the 1920s to connect the resources of Hawai'i's white elite and Japanese banks to a burgeoning, underserved community.[117]

Tokioka's Japanese fluency, his education, and his relative comfort with the Japanese American community in Hawai'i fit perfectly with International's market and profile. Its original white and Japanese-national board members were joined by stockholders and customers of Japanese descent. Of the initial forty-four shareholders of the company, all but twelve had Japanese surnames, a pattern that continued for decades. All but four of the twenty-two mortgages held in September 1927 were under Japanese surnames. The Japanese American market, then, was crucial to the company, and Tokioka's ability to attract these customers impressed International's board and helped him rise through the company's ranks. When National Mortgage and Finance Company joined International in 1929, Tokioka was a vice president; ten years later, he was a manager and board member. By 1960, the family of companies included six financial institutions, and To-kioka was president of them all. This included the most recent addition, National Securities and Investment, added in 1959 for real-estate develop-ment and whose first project was the Japanese Center. Its nineteen direc-tors were all prominent men of Japanese descent. These included Robert Y. Sato, owner of a large local department store and a director of Dole, Inc.; Albert Teruya, owner of a chain of supermarkets in Hawai'i and director of another investment company; and Dr. Masaru Uyeda, a dentist involved with other local investment and real-estate companies. Their prominence extended beyond the ethnic community; their reliability as a group was vouched for by a vice president of the Bank of Hawai'i, the largest bank in the islands, and Castle & Cooke, one of the Big Five sugar companies. Like Tokioka, International and its officers had foundations in the Japanese com-munity that helped them grow to mainstream prominence.[118]

These Asian Americans in Hawai'i clearly occupied a radically different racial terrain than that of the mainland. Of course, actual conditions under-mined James Michener's claim that the islands had "practically no race dis-crimination."[119] Indeed, the plantation economy was built on a system of racial division that valued groups differently, hinging pay scales and ad-vancement on race and ethnicity. For most of the territorial period, a white oligarchy centered in the Big Five sugar factors' dynasties and boardrooms kept a tight hand on the island's politics and economy. However, the war, the declining fortunes of Hawaiian sugar, the new economic powerhouses of tourism and the military, a changing citizenry, and an active labor move-ment, among other factors, helped to alter, if not end, these conditions by the 1950s.[120]

Residents of Asian descent, and especially Japanese descent, could access far greater opportunities in postwar Hawai'i than on the mainland. This was in part due to the islands' lack of wartime mass internment. Although there had been some of the same anti-Japanese fears as on the mainland af-ter Pearl Harbor, local conditions demanded different steps, while logistics made mass internment nearly impossible: the economy would have shut down from such a rapid, huge loss of population. But Japanese Americans also engaged in overt displays of loyalty and patriotism, and local leaders, invested in maintaining peace with the Asian-descent majority, responded with interracial organizing. Eventually, territorial civilian and military officials overcame both local unease and federal pressures in order to prevent mass incarceration. Certainly, those of Japanese ancestry faced wartime hardships, but they did not have the psychological or economic setbacks of internment; some entrepreneurs even profited from the war. This was an enormous contrast with those on the mainland, where families had to unload businesses, homes, and property quickly and under duress. There were many stories of farmers selling equipment at a fraction of the purchase price, crops abandoned after planting, businesses sold for insultingly low offers, valuable merchandise given away, practices and careers suspended or abandoned. While such economic losses are impossible to calculate — and represent only a portion of the much larger loss — estimates range from $148 million to $400 million in contemporary figures. Internment, among other things, devastated "the accumulated substance of a lifetime" of those interned.[121] The vast majority of Japanese Americans in Hawai'i did not face such setbacks.

Changes in the postwar period amplified differences between Hawai'i and the mainland. People of Asian ancestry were the majority of the territory's population and electorate by the early 1950s. Japanese Americans alone accounted for about 40 percent of voters, significant given that the Issei were denied naturalization until 1952. These numbers helped them carve inroads in territorial politics, assisted by an insurgent Democratic Party seeking to unseat the islands' long-standing Republican monopoly and willing to reach out to Asian American voters to do so. Those of Asian descent were the majority of the territorial legislature by 1955 and ascended to the territorial supreme court in 1956. The new state's first congressional delegation included Daniel Inouye and Hiram Fong, both of whom had previously occupied leadership positions in the territorial legislature. In the decades following World War II, a select cohort of Japanese and Chinese Americans penetrated Hawai'i's once-exclusive spheres of power to destabilize more than a century of white-minority domination.[122]

These political elites joined networks of entrepreneurial, even wealthy, Asian Americans to pry open Hawai'i's once-segregated economy. Some, like Robert Sato, became important business owners in their own right, while joining the boards of the Big Five or other elite institutions. They also drove the islands' postwar land-development boom. Their wealth had been built in but exceeded their ethnic communities, and their alliances with other Asian American businessmen and politicians accrued them land and wealth once the "Democratic Revolution" overturned the long-standing, white, political oligarchy. Landownership had been highly concentrated; in 1959, twelve private landholders owned 30 percent of the new state's land while small property owners combined held only 27 percent.[123] Yet Asian Americans helped lead local development because the Big Five and their cohort maintained their historic land usage. Tradition and tax code favored large swaths of agricultural land, so even owners with the most exhausted or picturesque of properties had little incentive to develop. Small holders, on the other hand, had more. As a result, a well-capitalized class of entrepreneurial Asian Americans such as Hung Wo Ching, supplemented by mainland developers like Henry Kaiser (a Bay Area industrial mogul), drove real-estate development in postwar Hawai'i.[124]

Land was scarce and expensive in Hawai'i, however, so Ching, Tokioka, and many of their associates added mainland investments to their portfolios. The Japanese Cultural and Trade Center, indeed, was National Securities and Investment's first major investment. How the San Francisco project came to Tokioka's attention is unclear. Perhaps Franklin, Tokioka's son attending business school at nearby Stanford University, brought up the idea.[125] Tokioka himself narrated his involvement as a requested intervention: "three influential members of San Francisco's Japanese community came to his office 'out of the clear blue sky' and dropped the challenge in his lap." These unnamed men approached him because "Japanese groups in San Francisco wanted to undertake the project" but "couldn't quite swing it financially."[126]

Tokioka's telling accurately compared his resources with those of Victor Abe and the merchant-planners. Tokioka had accrued capital and experience unthinkable for the mainland Japanese Americans, necessary resources given the newly expensive and complicated parameters of the Japanese center project. By the time the call for developers went out in February 1960, the project delineated at least two, possibly three, blocks with a base price of $374,000 per block, almost double the price of the adjacent residential lots; at three blocks, the land alone would be more than the merchant-planners' initial capitalization. Projected construction costs, too, had ballooned with the required underground parking lot. These new expenses shocked Abe and his merchant-planners, who demanded to know "how any builder could comply with the requirements."[127] The group had pulled together a proposal that reflected a substantial investment of time and money, with architectural drawings, an incorporated investment company, and years of planning. Ultimately, however, they "decided that costs will stop them."[128] Abe and his group dropped out of negotiations without even entering a bid, despite all the resources and energy they had contributed to the project.

Tokioka, in contrast, saw no problem with the expense or the complexities that had daunted the other investors originally interested in the Japanese proposal. Joining with a Los Angeles developer, Paul Broman, who was relatively near San Francisco and was experienced in California redevelopment, Tokioka incorporated as National-Braemar to develop the Japanese Center. The company accepted the $1 million land price for three blocks and the terms for a $1.5 million garage without hesitation.[129] Tokioka's ability to do so reflected the prosperous career he had built in Hawai'i. Furthermore, his formidable professional skills and his experience navigating the interstices of American and Japanese, as well as white and Japanese American, interests made him comfortable with the intermediary tasks necessary for the project.

BUILDING CONNECTIONS WITH JAPAN

Tokioka's intermediary experiences in Hawai'i gave him "many contacts in Japan" that no other potential developer had.[130] These contacts were almost as valuable to the Redevelopment Agency as his financial resources and were essential to the kind of Japanese development M. Justin Herman desired. Although the statehood debates suggested that such networks were common to all those of Japanese descent in Hawai'i, Tokioka's arose from idiosyncratic circumstances. They were contingencies of his unique background, his stature in his local community, and the geography of transpacific travel.

By the 1950s, Tokioka had assumed a leading role in Hawai'i's Japanese American community that also introduced him to prominent Japanese nationals. He was an officer in several Honolulu-area Japanese American organizations, including the Honolulu Japanese Chamber of Commerce and the Economic Study Club, an ethnic civic organization with a weekly radio broadcast and monthly forums on territorial economic issues. When visiting Japanese businessmen, bureaucrats, and other dignitaries came through, he would, as an officer, help host them. Due to the contingencies of midcentury aeronautic transportation, this introduced him to quite a lot of people. Commercial air travel between Asia and the continental United States required stops, made most frequently in Honolulu. Japan Air Lines' first transpacific flight in 1953 from Tokyo to San Francisco went via Honolulu, for example. From 1949 through 1959, such stops added about

50 percent more passengers to those who made Hawai'i their primary destination, itself a growing number with the expansion of the islands' tourism. In 1953, the year that Tokioka was Japanese Chamber of Commerce president, through passengers accounted for an extra 60,455 visitors.[131]

Tokioka's leadership responsibilities therefore introduced him to many Japanese travelers going to the US mainland. When Tokioka was chamber president, for example, visitors included the twenty-nine–member Japanese contingent to the Japan-American Conference of Mayors and Chamber of Commerce Presidents in Seattle. The Japanese chamber had a presence in many of the official hosting activities, while Tokioka himself was a member of the "Mayor's special committee" that met the arriving delegation.[132] This visit, which lasted for three days, provided many opportunities for meetings, but even brief stopovers brought connections. When a former Japanese finance minister stopped in Honolulu for a few hours on his way to talks in Washington, DC, Tokioka greeted the envoy at the airport along with the Japanese consul general. Such contacts built upon the ones Tokioka made through his private activities, such as a recently engineered partnership with Daiwa Securities Company, of Japan, a business relationship that also bled into socializing. His business and civic activities situated him in numerous Japanese business and political networks.[133]

These connections proved critical to the Japanese Center. Tokioka's ability to call on them, and work knowledgeably with them, was crucial to his role as a developer and distinguished him from the others who had been interested in the development. It was clear that he and Broman understood this. National-Braemar's first meeting with the Redevelopment Agency Commission in August 1960, shortly before they made their formal bid, was a presentation of National-Braemar's architectural concepts, building assurances, and commercial goals as well as Tokioka's Japanese connections. Japanese speakers, appearing in Tokioka's absence, highlighted the transpacific trade and cooperation with Japan that would build the center. These included Japanese architects and executives from Bank of Tokyo and Japan Air Lines, two companies already secured as tenants. These speakers added little substance to the discussion with their brief, general statements of intent or support. Instead, they highlighted Tokioka's ability to summon their attendance. The Bank of Tokyo branch president hailed the "wonderful plan," explaining he was there because "Mr. Tokiyoka [sic] called me on the telephone and asked me to join this meeting."[134] The meeting suggested that Tokioka, who planned to "go to Japan immediately and obtain further commitments" from Japanese businesses, was capable of succeeding.[135]

International events amplified the importance of these abilities. In June 1960, shortly before National-Braemar announced its intention to bid, mass protests in Tokyo demonstrated the tenuousness of the United States–Japan alliance. The United States had pushed a revised security treaty through the Japanese Diet earlier in the spring, over much opposition. While the treaty did, in fact, include some concessions to a more balanced relationship, the formidable Socialist Party publicly opposed the treaty, and hundreds of thousands of Japanese took to the streets to resist what they saw as imperious terms. This furious opposition surprised US state officials who had organized a presidential visit to Tokyo to celebrate the presumed ratification. But when an administration representative arrived in advance of the president, he was met with an enormous, near-murderous demonstration. Eisenhower's trip was hurriedly canceled, while the treaty's ratification sparked the largest mass protest ever seen in Japan.[136] In San Francisco, the rumblings leading up to the massive protests were at least partly subsumed by the coverage of the *Kanrin Maru* centennial the previous month, but in other parts of the United States, it brought a revival of "Jap" to local headlines.[137] Japan's protests suddenly brought to the fore for Americans the obvious fact that Japan—even the generally sympathetic industrial sector—had aims beyond those of the United States. Indeed, the protests reflected domestic discontent with the current premier, who resigned shortly after, as much as chafing US-Japanese relations.[138] This "challenge to the US" underscored the need for continuing Japanese goodwill, even as it provoked anxiety about its stability.[139]

What journalists called a "crisis" added urgency to Tokioka's stated goal of "contributing to understanding between East and West."[140] He responded with proposals for a five-tiered "Peace Pagoda" for the Japanese Center's courtyard, conceived as a monument to the two nation's "firm friendship and good understanding."[141] The pagoda helped redefine his role from a commercial developer to a goodwill ambassador. His lobbying for and his emphasis on the pagoda's cultural nature attracted signatory support from some of the most prominent political and economic figures in Japan, including Prime Minister Ikeda Hayato, the powerful foreign minister Kosaka Zentaro, and Ishizaki Taizo, head of the Japanese Federation of Economic Organizations. Others included those with economic stakes in good US-Japanese relations, including the president of the Japan Export Trade Promotion Agency and the Tokyo Chamber of Commerce, and companies such as Ajinomoto Company, who were beginning to market directly to US consumers. The pagoda was explicitly cultural, rather than commercial, in nature and entailed relatively low-stakes backing. This smoothed the way for the participation and donations from a much broader swath of Japanese political and economic institutions than would be directly invested in or occupying the center itself. This balance in cultural and commercial representation contextualized the city's economic aims with the Japanese Cultural and Trade Center, and showcased Tokioka's ability to facilitate and cultivate Japanese networks.

CONCLUSION

By the time the Redevelopment Agency had embraced the idea of a Japanese center, incorporated it into the Western Addition plans, and secured a developer, the Japanesetown project had come a long way from Victor Abe and the merchant-planners' proposal for a "Japanese Village." The merchant-planners had envisioned an expansion from an enclave into a touristic and ethnic marketplace, a small shopping center with Japanese goods, foods, and cultural elements run by local entrepreneurs. But city officials reconceived their project into one better aligned with citywide goals and programs of Japanese relations. The Japanese Center had become a transnational production seeking Japanese businesses and capital in order "to revitalize San Francisco's international character."[142]

The trajectory from "Japanese Village" to Japanese Cultural and Trade Center depended on a number of contingencies, including Japanese domestic legislation, explosive international relations, local ethnic change, Pacific decolonization, and San Franciscan leaders' never-ending search for urban revitalization. The merchant-planners forwarded their proposal while San Francisco officials sought new ways to embed their transpacific ambitions into the built environment. Officials were therefore far more interested in the project by the late 1950s than they had been just a handful of years earlier, and Japanese retailers were primed for participation. But the Redevelopment Agency's new ideas for a Japanese center were far more expensive and complicated than those of the merchant-planners, and made far more demands of the potential developer. These new parameters winnowed down what had been a robust cohort of interested parties to Tokioka, the last man standing. By coincidence, the United States accepted Hawaiʻi, a former territory, into the union with a discourse that ran parallel to San Francisco's goals. This discourse highlighted the abilities of Tokioka, who brought the full weight of his unique, Hawaiʻi-cultivated professional and personal experiences to the project.

Tokioka emerged as a prototypical intermediary with Asia, shining with all of the promise of his home state. This was possible only because of an exceptional background that helped position him within a host of networks linking Japan and the United States. Despite its singularity, this same intermediary status would come to color all those of Japanese ancestry working toward San Francisco's Gateway to the Pacific identity.

Intermediaries with Japan: The Work of Professional Japanese Americans in the Gateway

Masayuki Tokioka, the president of the National-Braemar corporation and developer of the Japanese Cultural and Trade Center, epitomized an intermediary role between Japanese and US interests. His job as developer was predicated as much on his connections with Japan as it was on his company's experience with urban development and financial backing. This was in part economic, to attract the Japanese business and investment that San Francisco officials could not. But there was also a cultural function in his position. Tokioka was fluent in Japanese, familiar with Japanese business networks, and able to work between the city's and the foreign companies' desires in mutually acceptable ways. These intermediary skills were unique and forged in his relatively privileged position in Hawai'i's idiosyncratic social terrain.

As the Japanese Cultural and Trade Center went from an accepted bid in 1960 to a realized commercial development eight years later, other Japanese Americans joined Tokioka in his intermediary role. These were the architects, designers, cultural practitioners, and negotiators who mediated Japanese and US interests to construct the center. In their work of finding occupants, working with the city, designing the building, and promoting it to the public, these men and women interpreted Japanese traditions and practices for the Redevelopment Agency and other city institutions. Their work applied not only their professional skills and training but also domestic racial thinking and US postwar expansionism in Asia. Scholarly and popular thinking, animated by hostilities and intervention abroad, understood Japanese Americans as uniquely suited to interpreting the alien but critical nation of Japan. While such views did not automatically lead to employment, it gave Japanese Americans a basis to claim cultural knowledge that augmented their professional skills. Perceptions of cultural fluency, and their actual hard-won fluency, allowed them to inscribe US-Japanese exchanges into the city's landscape and to use those exchanges to pry open once-segregated spheres of employment and civic life. Ironically, in doing this they embraced popular understandings of their foreignness that had supported prewar Japanese exclusion and their internment during the Pacific War. The networks and pressures of the Cold War, however, gave Japanese Americans the opportunity to use that perceived foreignness to promote their inclusion.

Japanese American work supported San Francisco's Gateway to the Pacific identity in crucial ways outside the Japanese Center as well. The Japanese interests of local employers and cultural institutions opened a number of doors in San Francisco and the Bay Area. Japanese American professionals, cultural workers, and entrepreneurs helped create an entire geography of Japanese connections in the metropolitan area. Their work supported Japanese connections in the city's built environment, homes, and offices, thereby expanding and strengthening the city's place in Pacific networks. These activities, too, relied on popular preconceived notions of Japanese American cultural knowledge, while perpetuating and strengthening them.

These transformations were based in racial thinking about Japanese Americans that positioned them as figures detached from, yet inherently connected to, their ancestral Japanese culture. The Japanese Center was just one example of the burgeoning commercial, cultural, and personal networks increasingly connecting the United States and Japan. As Japanese companies expanded and invested in the states, US businesses and boosters sought ways to engage and encourage them. Japanese Americans found ways to work between those interests, augmenting professional skills with the racialized perceptions of them. Increased friendly US-Japanese contact created new opportunities for Japanese Americans. Yet while these ideological and economic conditions offered opportunities, they also held limits, contradictions, and risks.

JAPANESE AMERICANS AS TRANSPACIFIC BRIDGE

The popular and scholarly imagination established a context for Japanese Americans to act as intermediaries. Interwar social science established a framework for understanding migration as productive of interpreters and intermediaries. Robert Park, father of the Chicago school of sociology, theorized this process in his concept of the "marginal man." According to Park, mobility and migration made this idealized type a product of "two worlds, in both of which he is more or less of a stranger." This detached connection made him a keen observer with "more general, more objective standards."[1] This was the ideal of a professional sociologist, and therefore carved out a valued place for Asian Americans in the social sciences when such professional opportunities were rare. These men and women — despite Park's gendered language, women also employed this dynamic — made careers out of studying and interpreting Asian Americans for their white colleagues.[2] The Nisei as interpreter had a place in Japanese and Japanese-migrant thought as well. Shaped by rising social and legal exclusions, prewar Japanese migrants also viewed their US-born children as a product of two cultures. This, they believed, naturally situated the Nisei to promote a peaceful, productive US-Japanese relationship and to smooth relations for Japanese in the United States. The idea of Nisei as a transpacific "bridge" was even encouraged by Japanese imperial officials, who saw them as ideal propagandists for the expanding militarist state. Most Nisei ignored this idea, but it nonetheless demonstrated a consistent yet flexible view of the Nisei as interpreters between two nations.[3]

Japanese and white American observers both understood Nisei as interpreters, but there was a crucial difference. Japanese and Japanese Americans believed Nisei required schooling in Japanese mores, language, and culture to properly serve as intermediaries. Theirs was not an inherent knowledge, and any understanding that Japanese Americans might have had would likely have been based in a Japan of decades prior, when their parents emigrated. In fact, an entire educational apparatus aimed to rectify this presumed deficiency: tours of Japan, periods of education abroad, language schools, and even festivals in the United States. White American scholars and writers, in contrast, assumed Nisei's Japanese knowledge. Japanese culture, they suggested, was transmitted untouched to the US-born. Concordant with contemporary racial thinking, this was a question of culture, not blood or "race," but was almost as unchanging. Law and practice compounded a cultural understanding of Japanese Americans, along with other Asian Americans, as "aliens ineligible for citizenship." This thinking culminated in their wartime incarceration. General John DeWitt expressed this stamp of foreignness while arguing for their detention: "The Japanese race is an enemy race and while many second and third generation Japanese born on United States soil, possessed of United States citizenship, have become 'Americanized,' the racial strains are undiluted."[4] Japanese loyalties and identities, in his formulation, remained across the generations.[5]

The prewar understanding of Japanese American foreignness contin-ued into the postwar period, as scholars described Nisei's persistent Japa-nese culture. Postwar social scientists identified and sought to explain the intriguing puzzle of Asian American integration into the socioeconomic mainstream. Chicago sociologists provided the foundation for this research with their exhaustive examination of Japanese American resettlement af-ter incarceration. As sociologists Alan Jacobson and Lee Rainwater stated the problem, Japanese Americans "have, in less than ten years, achieved a socio-economic position more favorable than many European groups which have been in [Chicago] much longer. What accounts for the appar-ent rapid success of Japanese Americans, a highly visible ethnic minority?"[6] Their answer and those of other contemporary scholars largely disregarded the economic conditions of the military-industrial city and the particular history of race in the Midwest. Instead, they identified a convergence in the "value systems found in Japanese culture and value systems found in American middle class culture."[7] As Chicago-trained sociologist Setsuko Matsunaga Nishi maintained, "Japanese cultural values functioned to de-fine, support, and maintain for its members certain behaviors essential to the emergence of upward mobility in American society."[8] The answer, in other words, was the presence of Japanese culture in American-born Japa-nese Americans, which endured despite American birth, education, mores, and language, or what sociologist William Caudill called their "superficial external characteristics."[9]

This idea filtered out to the popular media in the 1950s and 1960s and was applied broadly to explain what was described as the entire popula-tion's assimilation. Journalists and commentators outlined a trajectory that took Chinese and Japanese Americans from a segregated minority to assimi-lating, middle-class "success stories." A 1966 article for the *New York Times Magazine*, later expanded into a book, famously exemplified this pattern. William Petersen stated that Japanese Americans might well have "been subjected to the most discrimination and worst injustices" of any minority in the United States, yet their educational attainment was higher and "social pathology" was "lower . . . than for any other ethnic group in the Ameri-can population." Petersen, himself a Berkeley sociologist, argued that this puzzle could not be explained by protest, lawsuits, or aid. Instead, he cited sociologist George DeVos to assert the causal role of culture. Meiji Japan had impressed a "diligence in work, combined with simple frugality" on migrants that "had an almost religious imperative, similar to . . . 'the Protestant ethic,'" qualities they passed on untouched to their US-born children. This was phrased in more explicitly laudatory terms than in the scholarly literature, but the conclusion was the same: a retained ancestral culture explained Japanese Americans' successful trajectory (and neither scholars nor journalists questioned its success).[10]

It is important to note that culture was not the only explanation for Japanese American success, although it emerged as the most salient in the national imagination. Observers also briefly identified the internment as cause. This first emerged from the assimilationist strictures of the War Relocation Authority officials, many of them New Deal liberals. As historian Scott Kurashige notes, officials "argued that the [postwar] status of Japanese Americans improved not *in spite of* their having been interned but *because of* their internment."[11] Popular media from the *Nation* to the *Saturday Evening Post* continued the WRA's story. Journalists reporting on the internees' return to the West Coast uncovered isolated instances of racism but argued that a racist discriminating against resettlers "found himself fighting not a helpless, discouraged Japanese family but the WRA plus the state and federal governments."[12] Furthermore, their incarceration had proved Japanese Americans to be "useful, conscientious citizens" to their fellow Californians, who fully appreciated their contributions to the state's economy in their absence.[13] Within a decade, this coalesced into a triumphant narrative of government-assisted assimilation. *Newsweek* was among a number of outlets that framed their incarceration as a "disguised blessing": "pushed into the mainstream of American life, Japanese-Americans entered new occupations, improved their economic status, and helped pull down the racial barriers against them."[14] Their incarceration was interpreted as a salient explanation for the community's monolithic socioeconomic achievement.

Despite this compelling story and its telos, Japanese culture overshadowed the internment as the explanation. The latter story necessarily spotlighted a deep-seated US intolerance. At best, the government was reluctantly unjust, while the American populace had progressed from prejudiced and undemocratic impulses. Japanese culture, in contrast, explained achievement by highlighting complementary and positive US ideals. The internment did not disappear in this story. Instead, it emerged as another hardship successfully defeated. Furthermore, it dovetailed with the similar story told about Chinese Americans, who were similarly framed as "model minorities" who succeeded because of cultural attributes, not legal or social protective measures.[15]

Additionally, as scholars have noted, the explanation of retained ancestral culture for Asian American "success" served another important cultural function. Because this narrative was completely detached from socioeconomic context or historical inequality, it disciplined and cautioned other minorities. Hence, if Japanese Americans could successfully rally from America's "worst injustice," then by implication African Americans could as well and without their grating protest movements. The story of Asian American achievement resonated with criticisms of contemporary civil rights movements and racial nationalisms.[16]

By the 1960s, Japanese culture became the prevalent explanation for Japanese American socioeconomic success. This flexible and broadly useful postwar story reinforced the notion of an inherent "Japaneseness" to American-born Japanese Americans. Unlike Japanese understandings of the Nisei as intermediaries, American observers presumed an inherent fluency with Japanese culture. This cultural knowledge positioned Japanese Americans as ideal cultural intermediaries with the growing economic power of Japan.

ERASING DIFFERENCES AMONG
JAPANESE AMERICANS

The shift from internment to Japanese culture as an explanation for Japanese American success was not preordained, however culturally useful. In San Francisco, this transition was aided by Japanese Americans on the ground who flourished their Japanese associations, such as Masayuki Tokioka. He underscored his Japanese knowledge, particularly when progress lagged or the project needed extra support. But while his Japanese fluency was hard earned and contingent, he developed partnerships with local Japanese American merchants that framed his fluency as an ethnic trait. Tokioka and the neighborhood merchants used both foreign and local knowledge in ways that homogenized their gaping disparities, and added to their ability to contribute to San Francisco's Gateway to the Pacific identity. In San Francisco, this helped to move ideas about Japanese American innate ties to Japanese culture into local politics and institutions in concrete ways and to emphasize their utility for the city's Gateway to the Pacific ambitions.

Tokioka's uniqueness endeared him to San Francisco officials because he connected the Japanese Center to what appeared to contemporary Americans to be an especially opaque country. The decades from World War II through the 1960s saw an upsurge in scholarly and popular writing about

Japan, and much of it defined Japan and the United States in oppositional terms. Ruth Benedict's *The Chrysanthemum and the Sword* is perhaps the most enduring example.[17] Benedict wrote the 1946 book at the behest of the Office of War Information, and it had shaped military action, occupation policy, and popular thought. The book began straightforwardly: "The Japanese were the most alien enemy the United States had ever fought."[18] What followed was an interpretation of a society fundamentally organized by alien concepts such as shame, saving face, unstated intentions in social interactions, and hierarchical social structures. As anthropologist Clifford Geertz has noted, Americans long viewed Japan as the most "other" in a long history of intercultural interactions with other nations. Benedict, he argues, took that tradition to an extreme by "accentuating" Japan's difference through continuous contrast with the United States.[19] Benedict's ideas thus resonated with and helped frame US understandings of "the Japanese" as a quite uniform, static national group very different from—and often unintelligible to—Americans. Some contemporary scholars noted that the book's flat portrayal of Japan "assume[d] that all Japanese conform to [Benedict's] pattern" and avoided the historicity of current conditions, yet most agreed that the book was "the most important contemporary book yet written about Japan."[20]

The book's status was apparent in its influence on other scholars and writers, including some of the postwar social scientists studying Japanese Americans. Caudill, for example, used Benedict's work extensively to support his understanding of Japanese culture; although he claimed to have considered scores of titles for his study, *The Chrysanthemum and the Sword* was almost his sole citation for the subject.[21] Such social science made for an ironic full circle: Benedict had used Issei as informants because US ethnography in wartime Japan was impossible. The book contained incongruity, but it was an important part of a larger contemporary conception of Japan as a particularly alien and illegible culture.

Tokioka's role was therefore doubly important, since he had to act as not only an economic intermediary but also a cultural one. His economic responsibilities were the most straightforward, as demonstrated in his relations with tenants. His friend at the Bank of Tokyo arranged for a series of introductions with "many of the principle manufacturing and service organizations in Japan" immediately after Tokioka was awarded the project.[22] This trip resulted in two tenants, the Sentochi Theatrical Company, run by Matsuo Kunizo, a "relative newcomer" to the business but, according to Bank of Tokyo, "Japan's top theatrical entrepreneur."[23] Meitetsu Department Store, a Nagoya-based company that was one of the largest department stores in Japan, also signed a preliminary contract to "develop and coordinate the rentals in [one portion of] the Japanese Cultural and Trade Center."[24] Tokioka filled the remaining commercial and exhibit space with Kintetsu, another large department store. Later, he retained Tokyo Kaikan, a large and popular Tokyo restaurant, as the center's sole Japanese-food purveyor.[25]

Tokioka's central task appeared accomplished. In August 1961, National-Braemar officially purchased the land from the city for the Japanese Center.[26] By early the next year, Meitetsu had secured subleases with companies such as "Soni," "Nicomoto," "Hoya glass," and others.[27] Sony transistor radios, Mikimoto pearl jewelry, and Hoya crystal were becoming increasingly familiar brands in the United States, but the flagrant misspellings of agency staff suggested they had not yet gained complete US consumer recognition. The Japanese Center appeared set to introduce new Japanese products and showcase the city's transpacific commerce.

The rapid pace of the project up to that point, however, quickly tapered off, and Tokioka and Broman frequently referred to cultural differences to explain or elide weaknesses in their project. Right after the signing ceremony, for example, the Japanese tenants became mired in the red tape of Japan's strict financial oversight. Japan's postwar government tightly regulated capital as a way to shape industrial development and speed economic growth. All overseas investment had to meet Ministry of Finance approval before it could be released. Tokioka imparted the seriousness of the situation to the Redevelopment Agency, warning that "it easily takes three to six months to receive proper action." However, he assured them that he had the situation in hand: it would be a "great help . . . if a letter were written by proper officials, both of the municipal and state governments of California, to the proper officials of the Japanese Government." He closed with the convoluted request that these letters "be sent to me so that I can have my representative in Japan deliver them in person to the proper officials who will see to it that they are delivered to the individuals addressed."[28] As the project's pace slowed, Tokioka painstakingly explained the situation to agency officials in terms of cultural divides and highlighted his own significance as an intermediary.

National-Braemar used this dynamic again when Tokyo Kaikan dropped out of the project. The enormous costs for a Japanese investor to export capital, sink money into a risky investment, send over Japanese managers and workers, and sign a multiyear lease on a building they did not own and therefore could not fully control, required concessions that weakened the overall project. The company had insisted on being the only Japanese-food concessionaire in the three-block center. This meant the hotel could sell only "American" food, and the theater would peddle Chinese food.[29] Conflicts with Kintetsu management further discouraged Tokyo executives, who eventually dropped out. After their departure, Broman assured Herman, always concerned about delays, that there were "numerous other tenants prepared to take the restaurant." This may or may not have been true; his failure to name prospective companies suggested that it was, at best, a stretched truth. Yet Broman explained the lapse with culture: Tokioka understood the "delicate situation regarding saving face in Japan" and therefore felt it better to delay any new agreement.[30] Eventually National-Braemar did secure another restaurant, but Broman notably chose a widely accepted, stereotypical reason to smooth over the problem.

Transpacific hurdles of tenancy, capital export, and foreign regulations provided just one set of obstacles to the development, which ultimately took eight years to complete; domestic complications proved almost as difficult. Executives of Meitetsu, the project's earliest Japanese tenant, grew exasperated with the arduous process of domestic redevelopment financing in the United States. Herman's scheme for an underground parking lot depended on a novel financing mechanism, one involving almost a year of tangled negotiations between agency officials, the Board of Supervisors, and the parking authority. These "many delays" threw the project into disarray. The Japanese companies were forced to extend their preliminary contracts three times. Drawing the line at the fourth, Meitetsu withdrew. Not only did the project lose an experienced and respected retailer, the company took with it all their sub-lessees. This left Matsuo's company with the theater and Kintetsu as the default central commercial presence by 1964. National-Braemar had to retain the center's final wing.[31]

With fewer assured tenants, local Japanese Americans were crucial lessees for National-Braemar's wing. They also provided a local foundation for the Japanese Center. Tokioka's claims to Japanese fluency had come from all the things that made him unique and distinct from local Japanese Americans: his Issei status, class background, Christianity, Hawai'i location, and the new state's idiosyncratic opportunity for Asian Americans. But the development's progress helped elide his difference. National-Braemar's wing became the "San Francisco tenants area."[32] This relieved Tokioka of finding the scores of businesses he would have needed to secure from Japan, attesting to the difficulties even such an experienced intermediary had with the tedious process of securing overseas leases. But it also helped ground the project in the surrounding neighborhood, or, as Tokioka argued, "secure the goodwill of the people in the community." This drew on their original investment in the center: "this project was originally intended to be developed by the San Francisco Japanese community," but they abandoned it "due to various reasons."[33] Left unstated, of course, was Tokioka's part in precipitating their departure. Local Japanese American merchants' participation helped paper over what could have been an open conflict between them and Tokioka, one that would have undermined both the logic of the project's location and the utility of the city's migratory transpacific connections. Furthermore, these small local businesses shared space with the Bank of Tokyo of California, a US-based subsidiary headquartered in San Francisco that had supported the project from the beginning and signed on as one of the earliest tenants. Initially, its president had served to witness Tokioka's transpacific networks at the Redevelopment Agency hearing. But the branch moved to what became the San Francisco Building, where it would nest alongside local Japanese American businesses such as a drugstore, a shop selling carp, and a bonsai store.[34] These diverse but proximate shops helped elide the differences between Japan, elite and outsider Japanese Americans such as Tokioka, and local Japanese Americans.

The San Francisco Building's tenancy made manifest the cooperation that had begun since Victor Abe had first approached the Redevelopment Agency in 1953, and continued even after the merchant-planners abandoned the project. The Bank of Tokyo of California president was not the only witness on Tokioka's behalf at the 1960 Redevelopment Agency hearing. Abe also spoke. Describing himself as "formerly head[ing] the group which you might say in a way initiated this project," he testified that local merchants were "definitely interested in coming in" as renters in Tokioka's center. He painstakingly denied any hard feelings: "Our group is wholeheartedly behind Mr. Broman and Mr. Tokioyoka [sic]." In fact, he was "sure that all of the members of the Japanese community are wholeheartedly behind this project," viewing it as "an asset to San Francisco and to their own community."[35]

Local Japanese American merchants indeed consistently supported the project. Abe, the Japanese American Citizens League (JACL), and the Japanese Chamber of Commerce had representatives at all of the Japanese Center's milestones, including the contract signing in 1961, the groundbreaking in 1964, and, eventually, the grand opening in 1968.[36] When ground was finally broken for the center in 1964, the daylong ceremonies expressed relief as much as celebration. Both the Redevelopment Agency and National-Braemar had been heavily criticized by the press and residents for the protracted progress, and this was the first visible step on a site that had been an empty, fenced-in eyesore for four long years. Japanese Americans supported the embattled project. Nearby shopkeepers hung agency-provided celebratory signs in their windows, while Masao Ashizawa and other merchant-planners hung paper carp around the neighborhood to add "color and a festive air for the occasion."[37] Community priests sanctified the site with a Shinto groundbreaking ceremony, "a tradition in Japan for more than 1200 years" that added cultural portent and color to the common ritual.[38] Local Japanese Americans had a symbolic role, too: at the ceremonies, the Redevelopment Agency Commission chair opened his speech praising the "historic relationship to Japan of San Francisco's old Japanese Town" as "prompt[ing] the Agency to encourage the establishment of a Japanese Cultural and Trade Center."[39]

The cooperation between Tokioka, local Japanese American merchants, and the Redevelopment Agency homogenized Japanese Americans and helped cement their postwar associations with Japan. This was useful for all participants. It allowed Tokioka and the Redevelopment Agency to demonstrate the Japanese Center's positive community relationships at a time of growing anti-redevelopment resistance nationwide; there would be no boycotts or resentments to spoil its commerce. Abe and his group also gained through continuing good relations with the city, relations they would eventually leverage into their own long-desired commercial project. Together, the different actors gave institutional form to San Francisco's transpacific ties as well as a portrayal of Japanese American cultural mediation.

INTERPRETING JAPANESE DESIGN

As a "symbol of the cultural and commercial ties that unite us to Japan," the Japanese Cultural and Trade Center's form as much as its content was a crucial expression of the city's Gateway to the Pacific identity.[40] Its appearance was so critical that Redevelopment Agency officials had specified the development's aesthetics in the district's plans, well before Tokioka entered the project. From the start, M. Justin Herman and others rejected the idea of a "crudely imitative" or pleasing image of Japan similar to the Japanese Tea Garden in Golden Gate Park. Instead, the center was meant to evoke the "contemporary and traditional architecture" of Japan.[41] Officials demanded a "contemporary" building "sympathetic to the principles of traditional Japanese architecture."[42] Contemporary design reflected the modernist ideals of urban redevelopment: professionally and scientifically engineered buildings and cities could improve the life of their inhabitants and society more broadly; all of Herman's redeveloped landmarks were pointedly modern. Moreover, the center's modern design reflected the specific goals of the project. It articulated the city's ties to the "true, modern Japan," the industrialized partner that boosters sought.[43]

Japanese American architects merged the modern and Japanese aesthetics so crucial to the Japanese Center's function. The lead architect was Minoru Yamasaki, a well-regarded young architect based in Troy, Michigan, who would go on to design the World Trade Center in New York City.[44] He and the architect of record, the Oakland-based firm Van Bourg/Nakamura where architect Noboru Nakamura was partner, acted not only as architects but also as interpreters of Japanese design. Yamasaki and Nakamura both brought professional stature and skill to their task, supplemented by hard-earned Japanese knowledge and demonstrating another form of Japanese American mediation with Japan.

Agency officials presumed a distinction between modern and Japanese aesthetics that shaped their need for an architect who could bridge them. Norman Murdoch, the head of the agency's planning division, made this clear when he devised a metric for evaluating the designs of the initial lead architects, the Southern California design team of Millard Sheets and S. David Underwood that was National-Braemar's first choice. Murdoch produced a quite scholarly aesthetic framework that went well beyond the usual scrutiny of form and function. Drawing from diverse US publications on Japanese design, he isolated key "principles of traditional Japanese architecture," such as an emphasis or exposure of functional structures, flexible and modular space, and, rather vaguely, "respect for nature."[45] Adherence to these basic principles, he argued, could unite modern materials and dimensions with Japanese design without slavish imitation or parody. This framework filtered out much of the heterogeneous Japanese built environment in ways that, while neatly aligned with the redevelopment program's modernist sensibilities, defined Japanese design in opposition to modern styles. His principles also provided a seemingly objective measurement for agency officials to assess the Japanese content in the architects' designs.

Murdoch and Herman quickly found Sheets's designs lacking. Sheets had produced a two-story, horizontally oriented building with a large decorative entrance reminiscent of a torii, a gate-like entry to Japanese shrines. Murdoch's critique noted the awkward traffic into and within the buildings and some instances of poorly integrated functions. He based his most pointed critique, however, on his aesthetic framework. He found a systemic failure to engage Japanese design beyond surface and artificial elements. Among other points, he noted, the plans lacked natural materials, a prominent roof, or a visible structural framework. Based on Murdoch's analysis and undoubtedly his own dislike of the drawings, Herman refused to approve the proposal and demanded "fresh design lines."[46]

In contrast, the portions of the Japanese Center assigned to Nakamura's firm were accepted as a more "successful adherence" to Japanese aesthetics. Murdoch generally praised the visible structural system, the modular units, and transitions between the buildings with little attendant critique. However, the aesthetic analysis was not evenly applied to the Van Bourg/ Nakamura designs. For instance, the agency's evaluation observed that Nakamura's high-rise hotel was a form "foreign to traditional Japanese architecture," and so Murdoch's aesthetic analysis was "no longer applicable."[47] This did not apply to Sheets's mall, a building similarly foreign to "traditional" Japanese architecture but nonetheless scrupulously subjected to and rejected on the basis of cultural scrutiny. This evaluation gave the agency a way to reject Sheets and pressure Tokioka and Broman to find a new architect. Once Yamasaki agreed to the project, Murdoch and Herman appear to have dropped the aesthetic framework completely.

Herman and Murdoch may have felt little need to assess the Japanese principles in Yamasaki's work because by this point the architect's aesthetics so well paralleled that of the desired Japanese Center. He had embraced Japanese design, assiduously studying its history and forms in order to incorporate it, as well as other non-Western forms, into his work. Indeed, it had become a passion for him, one lauded by contemporary critics and popular audiences enamored with Japanese design. These ideas supplemented his original, quite conventional education. Yamasaki had been born in Seattle and trained in the European-American tradition at the University of Washington and New York University. The buildings that had garnered him his earliest recognition were two examples of massive, modern form: the notorious Pruitt-Igoe housing project (1954) and the Lambert–St. Louis Municipal Airport (1956) in Missouri.[48] From the buildings' scale to his use of materials such as steel, glass, and concrete, he was in many ways well inside the mainstream of modern, midcentury US design.[49]

This is not to say that Japanese-design ideas were absent from the modernist lexicon of Yamasaki's training and influences. Modern architects had expanded the European and US vocabularies decades before with Japanese influences. Frank Lloyd Wright was a well-documented example: the architect collected Japanese art and used roofs, natural context, and interior spatial relationships in ways influenced by Japanese models. Midcentury architecture and art referenced Japanese ideas overtly; Murdoch could freely cite from literature on Japanese architecture because it was readily available. The two decades following the Pacific War saw an explosion in English-language reprints and new studies on Japanese architecture.[50] For example, Arthur Drexler, one of Murdoch's citations and the influential curator of architecture and design at the New York Museum of Modern Art, supplemented his published work on Wright, Ludwig Mies van der Rohe, and postwar US architecture with a study on Japanese buildings. In 1954, he even displayed a complete Japanese home at the museum, a curatorial choice that reflected "the relevance of Japan's architectural traditions to contemporary Western building."[51] Japanese design was an accepted and valued influence on 1950s US architecture and interior design.

Very much in line with these developments, Yamasaki had also begun to consciously integrate Japanese influences into his work about five years before his involvement in the Japanese Center. As he noted in 1957, contemporary architectural thinking had begun to "encompass both Saracenic and Asiatic historical architectures to round out the European interests of our architectural fathers."[52] Yamasaki himself began to abandon stark modernism in favor of "human scale" and "delight, serenity and surprise."[53] This emerging sensibility was apparent in his McGregor Memorial Conference Center at Wayne State University (1958). He softened the modernist form to build a "dream palace" with "visual delight" evoked by, among other influences, "Japanese temples."[54] His new "approach to architecture" was awarded an American Institute of Architects' First Honor Award and was warmly received by critics, who saw this as part of an "emerging assurance of a mature artist."[55] Yamasaki attributed this aesthetic departure to his formative experience designing the US consulate in Kobe, a project he received in 1954. He did not seem to consider the building a major legacy; he left it out of his 1979 autobiography, *A Life in Architecture*. Yet the process of designing the consulate introduced him to Japanese design traditions and "crystallize[d] his philosophy of architecture."[56]

The Kobe consulate was an education in Japanese architecture for Yamasaki. His commission was part of the State Department's massive program of postwar embassy building. The new program's rigorous design standards specified modern architecture in order to project the image of the United States as modern, open, dynamic, and democratic; the program also demanded attention to local conditions and culture in order to integrate buildings into their context and promote indigenous goodwill. With no previous background or experience with Japanese buildings, Yamasaki had to familiarize himself with the local architecture through what *Architectural Forum* described as the State Department's "enlightened policy" of encouraging the "serious study of climate, culture, and local materials" to produce its consular buildings. His consulate was described as a "reward" of that policy.[57] His plans incorporated a Japanese garden and adopted local building techniques to protect the structure from Kobe's recurrent flooding. The buildings borrowed a "Japanese look," as Yamasaki described it, from such forms as shoji screens but rendered it in materials such as concrete, fiberglass, and marble.[58] Such components offered what a critic called a "graceful acknowledgement of U.S. appreciation for Japanese culture."[59] Designing the Kobe building gave Yamasaki direct experience with bringing together the materials and structures of modernist architecture with Japanese forms and ideas.[60]

Yamasaki was as much goodwill ambassador as project designer. As the *Architectural Record* noted, "The Japanese have been tendered a gracious double compliment" by the State Department's selection of Yamasaki, "a native-born American of direct Japanese descent."[61] As the consulate's architect, he was a US representative, demonstrating US racial liberalism at home and in relations with a former enemy. Yamasaki's role thus paralleled that of the State Department's official goodwill ambassadors, prominent artists or musicians who toured racially similar countries to highlight US cultural innovations and racial democracy.[62] Yamasaki was not the only architect selected for such a role; Finnish American Eero Saarinen designed the Helsinki consulate and Argentinian American Eduardo Catalano the Buenos Aires one. There was also another Japanese American selected for a second Japanese consulate, in a country with an enormous postwar US military, political, and business presence: George T. Rockrise, whom the committee in charge of choosing architects baldly identified as "1/2 Jap."[63] Yamasaki's rising prominence explains his selection by the committee, which favored well-known architects, but the strategic significance of his goodwill role was possible only because of his Japanese ancestry.

Well before the Japanese Center commission, then, Yamasaki had learned that Japanese traditions and his Japanese ancestry had value for him as an architect. Art historian Bert Winther-Tamaki argues that the Kobe project provided Yamasaki a positive view of his Japanese identity after a lifetime of discrimination.[64] Yamasaki himself attributed a "nearly fatal ulcer attack" in 1954 to his "complex" about being Japanese. However, "the two visits to Japan [for the Kobe project] and the return trip around the world changed my life—maybe saved it." His plunge into Japanese architecture left him "stunned" and inspired.[65] He came away with not only a new appreciation for Japanese design, but also a new professional significance. Japanese design itself, he argued, had social import. "Qualities in Japanese architecture" such as "serenity" or "scale" could "help U.S. architects shape the kind of environment necessary to a better life." Furthermore, their incorporation could strengthen US ideals of freedom and human dignity crucial for global leadership. He himself, he argued, was uniquely positioned to introduce this tradition to US audiences: "[B]eing a Nisei probably makes me a logical candidate for this kind of discussion."[66] His embrace of Japanese architecture gave him a new aesthetic sensibility and claim for expertise aligned with his racialized identity.[67]

By the time that Yamasaki was appointed as lead designer in 1960 to replace Sheets, he had developed an aesthetic incorporating "Japanese qualities" into "American" architecture. His experiences suited him to the cultural mediation considered necessary for the Japanese Center. The complex he produced reflected a "framework of contemporary architecture" while expressing "the serenity and dignity of characteristic Japanese architecture" (fig. 5.1).[68] And while he referenced buildings such as the "Imperial Palace of Kyoto built during the 11th Century Heian Dynasty," he did not want "purely Japanese-style buildings": "formal Japanese temple-style structures for a building complex to be constructed in an American city would be unreal." Such structures would also be out of step with the "true modern Japan" agency officials wanted him to express. Instead, he used materials and technology associated with modern architecture, such as glass and concrete, to evoke Japanese "characteristics."[69]

Despite the expectations placed on Yamasaki and his own attempts to fulfill them, there were limits to his self-taught Japanese knowledge. At one point, Herman asked Yamasaki to make the hotel rooms "as pleasant and as 'Japanese' as possible." Yamasaki replied, somewhat testily, "As I have told you before, I really do not know enough about Japanese detailing to feel adequate in this respect." And to prevent Herman going down another fruitless path, he added, "Neither we nor Van Bourg-Nakamura are suited to this kind of Japanese detailing."[70] Yamasaki had expressed similar sentiments before, telling a reporter, "I couldn't build a Japanese house if I tried. . . . I haven't the training or the background." Yet, he grumbled, "some people forget that I am American, not Japanese. . . . They can't understand why I

FIGURE 5.1 Minoru Yamasaki's model of the Japanese Cultural and Trade Center, 1965. His project evoked shoji screens in its black-and-white exteriors, while the repetitive rectangularity nodded to a Japanese modularity celebrated by contemporary critics. Other flourishes included the buildings' jutting eaves, most evident on the high-rise hotel but also present on the low buildings, which referenced a Japanese tradition of prominent roofs. Photo ID# AAZ-0966, San Francisco Historical Photograph Collection, San Francisco History Center, San Francisco Public Library.

can't."[71] Yamasaki had self-consciously and determinedly assumed a knowledge of Japanese design, but he was unwilling to accept any indication of innate, easy, or boundless cultural knowledge.

The Japanese companies had no such expectations, as Noboru Nakamura's experiences demonstrated. Part of the responsibilities of Nakamura's firm was the center's dinner theater, planned to house kabuki-influenced shows. This building required acquaintance with the kabuki form and its elaborate costuming and staging, as well as the needs of the extravagantly commercial character of Matsuo's Sentochi Theatrical Company (incorporated as Dream Entertainments in the United States), which controlled the building. Mirroring prewar Japanese intellectuals' assumptions about Japanese Americans' cultural understanding, Dream executives presumed Naka-mura's need for education. They brought him to Japan multiple times to study their theaters, layouts, and facilities. Ironically, Nakamura had early exposure to Japanese architecture, far more so than Yamasaki. His grand-father had been a master builder in Japan, and his mother, who encouraged Nakamura's career, had done drafting work for her father. Nonetheless, Matsuo recognized the need for training in the specialized building they required.[72]

Nakamura's work on the center was the result of conscientious study of more than just theater design. Traveling to Japan, learning from his Japa-nese counterparts, and engaging the executives, his work also benefited from his fluency in the Japanese language. His linguistic skills might not appear particularly surprising in a child of Japanese migrants who had him-self been born, although not raised, in Japan. His father had emphasized the language for his children, and Nakamura indeed grew up speaking Japa-nese with his family. Yet his abilities had been supplemented by the tute-lage of the US government through the World War II Military Intelligence Service.[73]

Nakamura and about five thousand other Japanese Americans became fluent in Japanese not through their families or upbringing, but through intensive wartime language training. Like Yamasaki's experience with Jap-anese design, Nakamura's Japanese fluency was the product of assumed Japanese American knowledge of Japanese culture. Prior to World War II, few Americans spoke Japanese: in 1934, only eight colleges and universi-ties offered the language. Facing a need for interpreters as war with Japan loomed, the Military Intelligence Service searched in vain for soldiers who spoke Japanese, a language the army supposed was particularly difficult for native-English speakers. According to *Life* magazine, this was "one of the most troubling war shortages faced by the United States since Pearl Harbor."[74] Disregarding the suspicions in other realms of government, in-telligence officers turned to the population they assumed would be fluent. But even Nakamura's abilities, which still needed the assistance of the full six-month intensive course, were rare.[75] As it turned out, of all the Japanese Americans who had attended Japanese-language schools as children or who grew up with Japanese-speaking parents, a 1941 survey of enlisted Japa-nese Americans found that only 3 percent were fluent.[76] The military there-fore established a language school to teach them their ancestral tongue. This select group went on to translate intelligence, interrogate prisoners in the Pacific theater, and serve in the US occupation, where they operated as interpreters and, as historian Eiichiro Azuma argues, cultural brokers. Nakamura gained his cultural knowledge at the hands of both Dream executives and the US Army.[77]

The Japanese American architects brought a unique facility with Japanese culture to their jobs with the Japanese Center. A press release noted, "Naturally, the leading Japanese architect of the day, Minoru Yamasaki, did the planning along with the firm of Van Bourg and Nakamura."[78] But cultural fluency was "natural" for neither Yamasaki nor Nakamura. Both architects had gained familiarity with Japanese design or language through purposeful education as part of wartime or postwar US expansionism in Japan. US assumptions of their connections with Japan became a self-fulfilling prophecy of cultural knowledge.

DEFINING JAPANESE

The Japanese American architects and Tokioka adopted interpreter roles, distanced yet privileged translators of Japanese culture for US audiences. But there were others working in the Japanese Cultural and Trade Center who represented Japan directly and were understood to bring aspects of unmediated Japanese culture to San Francisco. The cultural representation of these men and women was reserved for interior, private, or decorative spaces and roles. Their work boosted the authenticity of the center, supporting its pretensions to Japanese experience and its claims to natural, San Francisco–Japan connections.

The Japanese Center, where "Japan awaits you," was built to give visitors an experience of Japan.[79] This made the Japanese character of the interiors critical. Although the complex was composed of three separate blocks of buildings, the visitor was almost completely enfolded. A covered outside path connected two of the buildings, while the third was attached by an "enclosed bridge" over a dividing street that allowed for "a continuity of experience." The sightseer entered from the street, or perhaps directly from the underground parking lot, and wandered from store to store, hotel to restaurant, without any outside exposure except for glimpses of the sky through high windows and skylights. The visitor meandered along an interior "labyrinth of streets" lined with varied storefronts designed for a "festive and delightful" experience (fig. 5.2).[80] Some shops had the modern glass fronts found in any mall, but their goods and shop staff helped cast a Japanese effect. Other interiors enveloped the guest in Japanese atmosphere. The Miyako Hotel, for instance, had a floor of "Japanese-style" rooms with

FIGURE 5.2 The interiors of the Japanese Cultural and Trade Center were designed to enfold the visitor and focus their attention on the individualized shops, inside greenery, and decorations. The varied storefronts and enclosed walkways aimed to create a "festive and delightful" mood, thematically connected by Japanese foods, design, and people. Photo ID# AAZ-0967, San Francisco Historical Photograph Collection, San Francisco History Center, San Francisco Public Library.

sunken baths, tatami mats, and futon beds.[81] Economizing shaved off some embellishments such as tokonoma, or display alcoves, and sliding shoji panels intended to provide "spatial flexibility." Norman Murdoch in the agency's architectural division acceded to these changes because "how [the rooms are] detailed and furnished" would determine "the 'Japanese' character" more than the elements such as modularity he had once deemed so critical.[82] These decorative touches were, in fact, judged more essential than the cultural exhibits that lent the Japanese Cultural and Trade Center part of its name; gallery space was converted to retail in the interest of more income. Murdoch deemed the mall's enlarged paths and public areas "adequate to serve as exhibition spaces," despite the challenges for displaying valuable or rare materials.[83]

Such evocative interiors required the "proper presentation of Japan." The décor was therefore "designed and manufactured in Japan" and shipped to San Francisco.[84] Similarly, Tokioka and the Redevelopment Agency secured a succession of Japanese firms for these spaces. These included the Japanese firm of Takenaka and Associates, a Japanese company "whose history dates back prior to 1600." The firm, described as a "specialist in traditional Japanese architecture," employed Japanese nationals at a branch in San Francisco.[85] After the first years of delays, Takenaka was replaced by the Tokyo-based Shimizu Construction Company, "one of the largest Japanese concern[s] in its field."[86] Authenticity was so key that Herman pursued a Japanese consultant before Yamasaki signed on. Herman contacted the well-regarded Japanese urbanist and architect Tange Kenzō, a professor at the University of Tokyo, about the possibility of providing "architectural advice to the Agency" in order to "assure . . . an aesthetic quality and integrity which I consider so important."[87] However, Tange and other Japanese architects worked in the same international circulation of ideas as any contemporary architect, and they were by no means invested in any kind of pure, unmediated Japanese sensibility. Tange was in fact known for merging Japanese design with modernism, as evidenced by his famous Hiroshima Peace Memorial Museum. Echoing Yamasaki, an architect with Takenaka further argued that the "qualities" of Japanese traditional architecture could "be produced in reinforced concrete just as in the traditional materials of wood."[88] Shimizu had already translated Japanese design for enthusiastic San Franciscan audiences; the firm had remodeled the popular Yamato Sukiyaki House.[89] Although they were hired for their Japanese expertise, the Japanese architects were no more invested in traditional purity than their Japanese American peers.

Even if they were loyal to the idea of cultural purity, the Japanese architects did not have complete authority to decide what constituted "Japanese." Herman in particular had his own ideas on the subject. He urged the developers, for example, to include "stone water basins and stone lanterns, used in both traditional and contemporary Japanese gardens" in the inside landscaping; this, he argued, would "strongly identify them as Japanese gardens" and "do much to establish the character of the Center."[90] The entire mall form was his insistent concept. The first tenants, Meitetsu, ran a large and successful department store and initially envisioned a similar concern for the center. Herman, however, argued that a department store "has very limited cultural or tourist interest."[91] Instead, he insisted on "typical Japanese Ginza," referring to the bustling Tokyo shopping area: a "series of shops each one devoted to a special product such as pearls, cameras, silks, etc. from Japan."[92] There was some irony to his insistence. Department stores were a tradition in Japan about old as in the United States. Born from Meiji Japan's westernizing efforts, they were crucial for postwar Japan's consumer economy and industrial production.[93] Japanese department stores were therefore evidence of the modern, burgeoning Japan that San Francisco sought connections with. Yet they did not fit Herman's thinking. Clearly, what was defined as "Japan" was the product of redefinition and some conflict.

This was true among US critics and designers more broadly. The understanding of Japanese "traditions" in architecture was defined and codified in part by American architects and critics seeking usable ideas for contemporary structures. These observers distilled centuries of a highly heterogeneous built environment into a set of principles that complemented modernist ideas. This left out a host of castles, temples, and homes with ornate design, mammoth size, or other "idiosyncratic" elements that did not fit into minimalist, modular, or horizontally oriented frameworks.[94] This reflected almost a century of US notions of Japanese simplicity—what some observers called a "civilized emptiness"—that emerged through a contrast with the contemporary US home, cluttered and filled with ornaments and curios.[95] US observers originally defined Japanese aesthetics in opposition to familiar sensibilities, and this was refined further by midcentury designers seeking useful models.

Japanese tenant companies, like the architects, were less interested in cultural authenticity. Matsuo Kunizo of Dream Entertainments, the company behind the center's theater, had made a career of flamboyantly commercial and adapted institutions. His "famous Kabukiza Theater" in Osaka was "the only multi-stage system in the world which provides for five separate and complete stages along with specially designed elevator equipment for rapid set changes." As this suggests, he had no attachment to traditional forms. His flashy and popular productions used elaborate lighting, tiered staging, and even fire effects that drastically departed from the kabuki tradition, which was performed by an all-male cast in elaborate costuming and a highly stylized manner. But a more traditional theatrical group would not have been interested. The Murayama family, for example, was an "old established family" in theatrical building and had refused the overseas project. Matsuo's "resourceful progressiveness" bent him toward such an undertaking; a few years earlier, he had taken a kabuki troupe on an acclaimed US tour and he built a Disneyland-type amusement park in Nara during the Japanese Center's development. Given these proclivities, he en-visioned a self-consciously reinterpreted theater. Concerned that kabuki, while very much attached to ideas of Japan for US audiences, would appear too esoteric and unpopular for the theater's 750 seats, Matsuo instead of-fered "Kabuki" as a familiar, dinner-revue show with a revolving stage, waterfalls, and pretty female dancers.[96]

The interiors of the center were defined as Japanese and so were given over to Japanese companies and designers, but there was another set of Japa-nese cultural practitioners involved in the center's development: Japanese American women. Unlike Japanese or Japanese American men, who occu-pied all the professional roles interpreting Japanese culture and interests, Japanese American women held a different set of roles and embodied a dif-ferent set of assumptions. The different types of female positions can be seen at the ceremony for signing the purchase agreement at City Hall in 1962. The male executives and city officers attended in their official capacities, but Hatsuko Kusano, a Japanese American woman, was pure decoration. She provided cultural color, dressed in a kimono and handing pens to the male signatories. The documents the men signed reflected another female role. To underscore the transpacific symbolism of the event, they signed an official, English-typescript document and then a Japanese translation in flowing cal-ligraphy using a "Japanese traditional brush, called a fude, dipped in special Japanese ink."[97] The latter document was rendered, for a small fee, by artist Masae Yamamoto, who did subsequent work for agency publicity and cer-emonial materials. Her work offered authenticity to the city's claims of Japa-nese connections, as the Japanese architects did. She employed skills honed over time—she was a graduate of the Women's College of Art, Tokyo—that she exercised in public forums in response to growing public interest. Ku-sano, on the other hand, was as decorative as the brush and calligraphy itself, simply there to provide color to the events. Other women took on similar positions at Japanese Center events, embellishing ceremonies or per-forming classical dance.[98] These positions, diverse in themselves, differed in substantial ways from those of Japanese American men. The women were seen not as intermediaries but as direct transmitters of Japanese culture, and were employed in occasional and less-remunerated ways.

The women's "traditional" practices and objects displayed in the Japa-nese Center conjured an essential, timeless, and homogeneous Japanese culture. The calligraphy, women in kimonos, flower arranging: these were decorative elements that harkened back to a timeless, traditional culture that was both avoided and evoked in the center's insistence on modern form

and function. This awkward balance of the modern and the traditional es-
sentialized both into recognizable markers in order to bring them together.

Japanese Americans found a place in that awkward balance, demonstrating through their hard-earned knowledge of Japanese culture both their fluency with it and distance from it. In fact, they themselves served to integrate
the modernity and tradition much more easily than the Japanese Center's
portrayal of Japan. They embodied a mixture of Japanese and American
cultures: Japanese bodies dressed in suits or dresses, speaking clear US English. This paralleled the mixture that the center represented and mani-
fested: the modern technologies of industrialized, mass production that
Americans were used to (cameras, jewelry, cars, cast concrete, malls) with
a distinct Japanese sensibility. Familiar yet unfamiliar: enough of the exotic
to spur interest, but enough of the ordinary to spur consumption.

This dynamic also suggested that, despite San Francisco boosters' calls
for partnership with Japan, there were lingering inequities that US actors
took for granted. The decorations, "color," and aesthetics were comfortably Japanese. But the structural, foundational, and external elements—the
reliable substance of the building—were "American." As Yamasaki emphasized, the Japanese Center was intended to "produce a Japanese feeling," but
with "American products, knowledge, and equipment."[99] This ultimately
provided the solid foundation that the developers and agency trusted. San
Francisco's increased ties with Japan rested on a belief in the nation's maturity and development, but in practice US representatives, Japanese American or white, were unwilling to grant Japan complete equity.[100]

INTEGRATING JAPAN THROUGHOUT SAN FRANCISCO AND THE BAY AREA

Japanese Americans added to San Francisco's transpacific urbanity by
building and centralizing municipal transpacific connections. But other
Japanese Americans in San Francisco, and elsewhere in the Bay Area, added
to those connections as well. They also either employed hard-earned familiarity with Japanese culture or trafficked in their associations with the
nation. Their work supported the Japanese Center by contributing other
forms of Japanese connections throughout the city. They showed the
breadth of the city's Japanese institutions and demonstrated their popular
support, while creating professional and civic opportunities for Japanese
American advancement.

One notable cohort of Japanese Americans who built on and expanded their city's growing interest in Japan was restaurateurs. As noted in chapter 2, Japanese restaurants, especially those featuring sukiyaki, enjoyed a booming business in postwar San Francisco. These restaurants propelled the owners of the most popular into thriving careers while embedding Japanese icons throughout the city. The most prominent were Yamato Sukiyaki House in Chinatown, Tokyo Sukiyaki in Fisherman's Wharf, and Nikko Sukiyaki near downtown. The first two were in well-known tourist sites, and the frequent advertising of all three indicated their owners' ambition to attract more than ethnic patrons. The restaurants became popular staples in newspaper columns and fashionable destinations for local and visiting luminaries. The owners themselves gained local fame and some wealth by cultivating San Franciscans' interest in Japanese cuisine and culture.[101]

The sukiyaki restaurants served not only food but, as the Japanese Center aimed to do, an entire "aura of Oriental serenity and beauty."[102] The dishes, which included "*sushi* (raw fish) and *sake* (rice wine)," were only one novelty. For example, all three had floor seating, a popular and unusual accommodation: one reviewer noted that guests forwent ordinary seating for cushions and low tables "nine times out of ten" despite "cramped legs."[103] Such touches proved so popular that the owners engaged in a kind of décor arms race with each remodel. Edward S. Ishizaki, the owner of Yamato, opened his first restaurant shortly after he returned from Topaz, but it "lacked the proper Oriental atmosphere" and did poorly. Forced to move in 1950, Ishizaki presciently hired a Nisei architect to design a restaurant with floor seating, tatami mats, and paintings by the famed artist and Berkeley professor Chiura Obata (Ishizaki's art instructor at Topaz).[104] Enabled by his booming business, five years later Ishizaki took "a trip to Japan to study current Japanese recipes as well as to procure objets d'art" for a remodel, claiming to have "the most elaborate, truly authentic Japanese type restaurant in America."[105] Shotaro Yasuda opened Tokyo Sukiyaki a year after Yamato, with "private Japanese dining rooms" and an "exotic atmosphere."[106] Within three years, the restaurant had done "such a fine business that the establishment [was] enlarged" with rooms for parties of over one hundred and "one of the best views of the Bay."[107] The atmosphere was also amplified: "every inch of material in this fascinating place came directly from Japan," one journalist enthused.[108] The "$250,000 Nikko Sukiyaki restaurant" adopted the model set by its successful predecessors when it opened in 1958. It had "nine Japanese tatami rooms" and interiors reminiscent of a "picturesque Japanese village, with smooth water-polished stones, imported from a river bed in Japan, and a miniature waterfall tumbling down a rock bank into a lily pond."[109] Like the others, Nikko had "attentive" and "pretty hostesses, garbed in traditional Japanese robes." There were some adjustments, like "holes cut in the floor for added comfort" in the "Japanese style" seating.[110] Yet even this was understood as authentic: "many private homes in Japan have them."[111] These restaurants served a Japanese environment as much as cuisine.

Patrons were drawn to the restaurants' authenticity and viewed the restaurateurs as authorities on Japanese cuisine. Frank Dobashi, general manager of Nikko Sukiyaki, gave public demonstrations on "authentic Japanese dishes."[112] The Japan Trade Center, the Japanese trade-promotion agency, used Nikko's and Yamato's chefs and owners to showcase Japanese cookery.[113] The *San Francisco Chronicle*'s "Gourmet Guide" praised the "ancient Japanese recipes."[114] But their authenticity was as modified as the floor seating. Despite the claim to tradition, the restaurants' signature dish was a relatively recent innovation. Sukiyaki, as a beef-based dish in a habitually pescatarian nation, was a product of Meiji-era reforms. Officials encouraged the consumption of a once-taboo ingredient as part of a self-conscious, selective adoption of Euro-American knowledge and behaviors. Sukiyaki, then, was a determined departure from "traditional" food customs and a millennium of practices and beliefs, shrewdly marketed by Japanese American entrepreneurs. Like Yamasaki or Tokioka, the sukiyaki restaurateurs benefited from popular and professional interest in Japan, and cultivated a crafted and responsive set of cultural skills in dialogue with surrounding tastes.[115]

Other Japanese Americans built careers not on popular taste but on the transpacific economic interests enlarged by San Francisco's Gateway to the Pacific ambitions. A number of Japanese American brokers built successful careers at the interstices of Japanese and US interests, furthering the city's Gateway identity and enhancing their own status. One example was Mas Yonemura, who rose to Japanese and US prominence but began as a struggling Bay Area lawyer. Fresh out of the University of California, Berkeley School of Law, Yonemura was denied job after job at prestigious white law firms. So his first position in 1947 was in an office of African American lawyers in West Oakland, where he primarily served black clients. This gave him the foundation to begin his own practice some years later.[116]

Yonemura subsequently transformed his small practice through the networks of US-Japanese commerce. This change began as a result of a grisly but straightforward case: a 1954 murder. A Japanese marine engineer on leave engaged in a drunken fight that resulted in his stabbing a local Japanese American to death. The Japanese consulate was responsible for his defense and sought a Japanese-speaking lawyer. Yonemura, trained by the Military Intelligence Service, was bilingual. Yonemura lost the case but began a career navigating Japanese interests in the United States. The murder happened to occur just as Japanese companies reestablished US business and cracked new US markets. He was consequently in on the ground floor. He quickly obtained more work with the consulate and then, based on his language skills and their recommendation, represented the still young Japan Air Lines in labor negotiations in Hawaiʻi and San Francisco. Later he took on more Japanese clients, including Kintetsu beginning in the 1970s. Thus, Yonemura's professional development mapped onto the postwar expansion of Japanese businesses in the United States and helped establish their presence in San Francisco and—as one would expect from the "Gateway to the Orient"—the United States more broadly.[117]

Yonemura worked not only with Japanese economic networks, but also with civic ones, bringing them to San Francisco's neighbor across the Bay. Oakland also had an international harbor connected to ports and industries in Japan, and was sought out as a sister-city by Sakai, Fukuoka, and Kamagaya. He and Frank Ogawa, who went on to serve on the city council and became a beloved Oakland civic leader, were founding members of the city's sister-city committee and were integral to their city's decision to affiliate with Fukuoka. As the representative for Japan Air Lines, Yonemura frequently flew to Japan and was therefore able to study the city, meet with Fukuoka officials, and report back to his mayor. He and Ogawa then served in the official delegation to Fukuoka, traveling with their mayor and other municipal dignitaries for the affiliation celebration. This led to more civic opportunities, including statewide positions with the Democratic Party. His earned prominence was demonstrated most vividly by a 1965 invitation to a White House dinner in honor of Premier Satō Eisaku; nine other prominent Nisei "and their wives" also attended, including Minoru Yamasaki and the Japanese American legislators from Hawaiʻi. Certainly, Yonemura's position did not come out of nowhere. He had been an officer with the Japanese Chamber of Commerce of Northern California, for instance, and had worked with Japanese American political groups in support of local political candidates.[118] However, his language skills, his Japanese links, and the sister-city tie were crucial in his rising career and civic life, providing him a far more prestigious livelihood than his early opportunities had indicated.

Other Japanese Americans facilitated additional forms of US-Japanese connections in San Francisco. Transportation was one particular field. San Francisco was the primary hub of transpacific traffic during this period, both air and surface. With growing competition from airlines, American President Lines aggressively pursued an enlarged Japanese market as transpacific tourism and business grew. The San Francisco shipping company had roots in nineteenth-century mail and passenger steamship firms with some of the earliest US routes between the United States, Japan, and China. Like the city of San Francisco itself, American President Lines engaged in cultural activities to supplement its business ambitions; its president, George Killion, was a leading member of the Japan Society and an officer in San Francisco's sister-city affiliation committee with Osaka. His business innovations included partnerships with Japan Air Lines to encourage travel to Asia, and new, luxurious passenger ships to tempt travelers. Another strategy was creating the new position of Japan traffic manager to drum up new business. Executives hired another former intelligence linguist, Marvin Uratsu, who had served in the US occupation of Japan. He and a second Japanese American traffic manager pursued Japanese and Japanese American contacts through the Japanese Chamber of Commerce, churches, and the vernacular press. Airlines took similar steps, and Northwest Orient Airlines and Pan Am hired their own Japanese American "special Japanese representatives."[119] Their job was not only to cultivate general traffic, but also to tap into the market of Japanese American travelers visiting their nation of origin. Japanese Americans' presumed links to Japan made them valued potential customers as well as intermediaries.[120]

Women also worked in transpacific transportation from San Francisco, but they serviced transpacific travel instead of facilitating or cultivating it. Passenger airlines such as Pan Am and Northwest Orient Airlines began hiring Japanese American women as attendants on the "Orient portion" of their flight systems in the mid-1950s, lines that connected San Francisco to various parts of Asia. The airlines actively recruited what they called "Nisei stewardesses" (although not all were technically second generation, and later, not all were of Japanese descent), advertising for them in vernacular newspapers and holding interviews at, for example, a Buddhist church. Like the military intelligence with their male counterparts, the airlines courted Japanese American women for their presumed bilingualism; because the company could not legally hire Japanese nationals, the airline turned instead to Japanese Americans. Yet, as with the military, Pan Am found that while "most of the Nisei speak some Japanese," it was limited and "ordinary family style Japanese" and not the "polite phrases" appropriate for the work.[121] Mannerly, Tokyo-style language lessons were therefore an impor-tant part of their training. The women required education in the presumed "natural" demeanor and language skills that made them desirable as flight attendants.[122]

Airlines made assumptions about Japanese American women's attitudes as much as their language abilities. Like the waitresses at sukiyaki restau-rants, Japanese American flight attendants' appeal came from their racial-ized sexuality; as one contemporary journalist wrote, "A Japanese woman knows how to serve and desires to serve." Anthropologist Christine Yano argues that their "'look' of exotic cosmopolitanism" was particularly ap-propriate to the flight industry because it also embodied the sophistication of international air travel.[123] Most of the passengers they served were in fact Americans, so their language ability, however developed, was far less use-ful than their exotic face. This reflected the rising feminization of in-flight work more broadly. As historians Kathleen Barry and Phil Tiemeyer have shown, airlines shifted from stewards to lower-paid stewardesses during this period and redefined the work as feminized and sexualized; require-ments for gender, age, weight, and marital status overshadowed the medi-cal or safety training favored in the past.[124] The Nisei stewardesses were so successful in this context that they were eventually employed on domestic flights as well. But "Nisei stewardesses" had begun on itineraries that physi-cally connected San Francisco—and through that city, the United States as a whole—to the Orient, adding the "color" that made the route as distinct and attractive as the destination.

Contributing to San Francisco's Gateway ambitions could advance ca-reers, but capitalizing on popular racial thinking also came with risks. The JACL, headquartered in Japanesetown, was well aware of both sides. Both the national organization and the local chapter had helped illuminate Ja-pan's connections with San Francisco in various ways, as we have seen. At the national level, too, policy makers and politicians had begun to look to the JACL for connections with Japan. As the JACL member most likely to engage white politicians and bureaucrats, Mike Masaoka, the organization's famed wartime spokesperson, had seen for himself that "Americans gener-ally expect us to speak out on things which relate to Japanese-American relations and that they are interested in what we have to say."[125] Indeed, when Secretary of the Cabinet Maxwell M. Rabb spoke at the National JACL convention in 1956, he advised members of their "unique position" to con-tribute to US-Japanese relations.[126] Their ancestry, he argued, gave them "more information and more concern for the continuance of good Japanese-

American relations than any other group."[127] Such views suggested to some officers that the organization could gain stature as an international intermediary: the group's "greatest contributions to present-day America can be in the field of Japanese–American relations."[128] This would expand the group's field of recognized expertise and win them "new friends."[129] It could also mark their territory. As more people and institutions entered the growing field of transpacific relations, such a role would avoid "some other group taking over by default."[130] As a result, as historian Ellen D. Wu has shown, the JACL began reconsidering its official policy toward Japan, which had been, officially at least, "strictly hands off" since 1940.[131]

Masaoka was supportive of this policy because he made his living by mediating between US and Japanese interests. He worked part-time as the JACL's Washington, DC, representative, but his most remunerative work was representing Japanese business interests in US politics. Beginning in the 1950s, he leveraged his political contacts and knowledge about the US political system, as well as his foreign associations, into a livelihood. According to his autobiography, he "did not speak the [Japanese] language, and knew almost nothing of its culture." Nonetheless, he found that US officials regularly assumed that he had special insights into "the Japanese mind." Japanese executives also regularly consulted him as they expanded their economic presence in the United States. He leveraged this mutual dearth of information into a career lobbying against tariffs or in support of trade agreements in Japanese interests. Masaoka had initially put off lobbying directly for Japanese businesses. So while Japanese textile manufacturers were among his first clients, he represented their US importers rather than Japanese companies themselves. To do otherwise would have required registering as a foreign agent, a title with an "unfortunately sinister ring" and uncomfortably reminiscent of wartime assumptions about Japanese American disloyalty.[132] But after many years, his clients included Toyota, Sumitomo Corporation of America, the Japan Trade Center, and the Japan Telescopes Manufacturers Association.[133]

Many JACL members had reservations about Masaoka's representation of both Japanese and Japanese American interests that reflected discomfort with any JACL relationship with Japan. As one Fresno chapter member angrily pointed out, Masaoka had been previously identified as a representative of the JACL and Japanese Americans when he was in fact arguing on behalf of Japanese clients over sensitive trade issues. There was "inherent danger" in such mistakes: "our present position in American society is not so secure that we can risk the censure of any considerable portion of the

American public."[134] Similarly, many members rebelled when the JACL reconsidered its policy toward Japan. As one leader noted, it became "one of the most controversial issues we have ever faced as an organization."[135] This was an "emotional" response based on "old fears": adult Japanese Americans had already experienced the dangers of too close an association with Japan.[136] Friendly US-Japanese relations could easily change and with it their fortunes at home. Many JACLers were therefore "adamantly opposed to laying [the JACL] open to any future charge that it acted as a spokesman for Japanese interests."[137] These concerns sobered even the supporters of a revised JACL policy. As the group's national president admitted, too close an association with Japan could result in "the same doubts as to our loyalty as occurred during the evacuation. It works both ways!"[138] A role in US-Japanese affairs could bring immediate benefits, but it also came with risks.

The JACL finally adopted a formal position on Japan in 1958: it hedged. Chapters voted overwhelmingly to create the internal National Committee on International Relations to consider the organization's response to all major questions of US-Japanese relations as they arose. As a committee within the JACL, it was charged with studying and making recommendations for the national board, who could then act if it so chose. At the same time, members created a separately incorporated group, the American Committee on Japan, to "prevent any group from taking over" the JACL's potential role in expertise and consultation on US-Japanese relations. With Masaoka as executive secretary and Saburo Kido, the JACL's former wartime president, as chair, it was clearly a JACL affiliate. This new partner association allowed the JACL to retain its role as the preeminent representative in all things related to Japanese Americans and also acknowledged that "our own status in American society is dependent to a very important extent upon the degree of good relations between" Japan and the United States.[139] The committee provided a voice on Japanese issues while neatly segregating them—related decisions or statements could be carried out in the name of the "special committee" without implicating the entire organization—thus providing critical distance when international events soured.[140] This proved to be important, for instance, when the 1960 antitreaty riots in Japan brought nasty letters and phone calls to the organization as well as worrying hostility in the national press toward Japan. Fearing a crisis, the JACL was able to publicly renew its "Americans first" stance.

The JACL's American Committee on Japan also carved a distinction between the institutional, official representation of Japanese Americans—"Americans first"—and discrete responses from those "with a special interest and competence in Japanese-America relations."[141] This allowed individuals to further their interests and careers through associations with Japan. Masaoka, for instance, could continue his lobbying business while keeping a coherent relationship with the JACL. Others like lawyer Mas Yonemura or the calligrapher Masae Yamamoto might or might not have been a member of the JACL or agreed with their politics. Still, the JACL's all-American identity gave cover for individual Japanese Americans to cultivate beneficial links with Japan.

This awkward reconciliation was increasingly unnecessary as individuals and organizations moved toward socioeconomic integration. By the 1960s, the JACL felt less compelled to maintain the official distinction with Japan. In 1966, members of the JACL were buoyed by President Johnson's declaration of a "Pacific Era" in which Asian concerns would receive equal consideration as those in Europe, a "historic crossroads in United States overseas policy" in which the "JACL inevitably will have a leading role in the coming era of United States–Asia relations."[142] The JACL then began integrating Japanese items into their regular activities. For example, the group partnered with Japan Air Lines to sponsor scholarships for Japanese Americans to attend Japanese universities.[143] The following year, the organization sponsored its first "JACL Goodwill Tour to Japan," while the Cultural Heritage Committee was organized to "encourage participation in programs toward learning and understanding the culture of Japan," including sister-city affiliations. These smaller initiatives signaled a larger institutional shift: the JACL Planning Committee argued that the organization should "function more directly on specific issues affecting United States–Japan relations" and be more "vocal and active to influence our government . . . for maintaining friendly relations between Japan and the United States."[144] At institutional and individual levels, Japanese Americans paradoxically increased their affiliation with Japan while increasing their socioeconomic integration in the United States.

Notable successes at the individual level helped pave the way for such institutional change. But changes in US culture may also have encouraged a shift. An exploration of "heritage" was itself a form of assimilation with mainstream patterns. Italian or Irish Americans also toured ancestral homelands and explored their roots. As historian Matthew Frye Jacobson has argued, the 1960s saw an ethnic revival as whites attempted to shed the burden of white privilege and prejudice through claims to "ethnic" identification.[145] Japanese Americans argued for their own place in the nation of immigrants. The assertion was not easy, as the hyphen fit most comfortably between a European ethnicity and American nationality. Yet it was another realm in which they could struggle for inclusion and one with particular resonance for San Francisco's Gateway identity.

Claims to inherent or learned Japanese knowledge provided an avenue into mainstream employment and even sociability for a select few. It is also possible that related ideas opened doors for others. While sexualized and racialized notions of Japanese American women opened doors into the enviably sophisticated and cosmopolitan work of transpacific in-flight service, for example, such ideas may also have informed more mundane employment. In 1960, for example, the JACL noted that non-Japanese businesses had a "great demand for Japanese girls as clerical workers. As a matter of fact, there is discrimination the other way."[146] Almost 40 percent of Japanese American female workers, in fact, were clerks, almost double the next occupation, domestic workers. This was a significant and rapid change from earlier years, when a 1941 study found that Japanese American women had little access to white-collar office employment.[147] Popular portrayals of Japanese American women, subservient and sexualized, nonthreatening and attractive, may have smoothed this particular avenue of integration into offices and businesses outside the ethnic economy.

These ideas about Japanese American foreignness or associations with Japan worked in conjunction with other trends and patterns identified by scholars in contemporary Japanese America and Asian America more broadly. Historians such as Charlotte Brooks, Ellen D. Wu, and Cindy I-Fen Cheng have demonstrated the ways in which Cold War international pressures encouraged whites to hire or reside near Asian Americans. Other factors pushed acceptance of Japanese Americans in particular. As T. Fujitani has shown, a flurry of magazine articles, government pamphlets, newsreels, and, by 1951, even a film (*Go for Broke!*) resolutely celebrated the exploits of the all–Japanese American 442nd Regimental Combat Team. The JACL also picked up the story of Nisei soldiering, as Wu has argued, and used it to promote the organization's standing among Japanese Americans and that of Japanese Americans generally. Additionally, while interwar Japanese Americans had high levels of education, Richard Alba and Victor Nee argued that such education aligned them especially well with postwar economic transformations. This was true at the national level and particularly so in San Francisco. In California, Japanese American men and women had a higher level of education by 1960 than for all other ethnoracial groups. Unlike in prewar years, they could translate this into employment: nearly 20 percent of Japanese American men in San Francisco were in technical and professional occupations, higher than for all other groups, while far fewer engaged in blue-collar work. This was a profound change from twenty years before. In 1940, despite comparable levels of education, less than 7 percent of Japanese American men were in professional fields while 36 percent of Nisei men were laborers or domestic workers. Such changes reflected the general expansion of government, services, and the finance, insurance, and real estate industries in San Francisco and California. Not only did employment barriers decline, but jobs requiring high levels of education also grew in number. Japanese Americans, suitably trained, were needed by employers as never before.[148]

These kinds of achievements were not simply the result of advanced education or other forms of striving. Lawsuits and negotiations in the Bay Area helped to expand postwar fair-employment and housing regulations, while civil rights activism all over the nation chipped away at racial covenants in real estate, bigoted employer practices, and other forms of private and public racial discrimination. In San Francisco, such activity resulted in a voluntary fair-employment provision enacted in 1950 and a fair-employment commission created in 1957. Japanese Americans, as with other Asian Americans, were able to benefit to a greater degree than black workers. A significant gap in education explained some of the disparity, yet educated and skilled African Americans continued to be turned away from jobs granted to Asian Americans. Japanese Americans, after all, had been supported by federal and private agencies since the Pacific War and into resettlement, efforts that had connected them to the mainstream job market and resources in ways unavailable to other minorities.[149]

There were therefore numerous reasons for Japanese American postwar advancements, but they remained more equivocal than popularly portrayed and did not denote full racial inclusion. In part, opportunities in the ethnic economy declined. Ethnic establishments had been a pillar of the prewar Japanesetown economy, second only to domestic service. Such small businesses steeply declined by 1960. Redevelopment in Japanesetown uprooted some, while many young workers undoubtedly happily left the long hours of proprietorship for more stable employment. Competition from big chains and rising integration also hindered their continuation. Additionally, advancements in the primary labor market were not without ambiguity. The largest number of Japanese American women were clerks, a respected white-collar occupation; the next most numerous trade was private domestic labor, at a percentage in 1960 comparable to 1940. Furthermore, while Japanese American women actually averaged higher incomes than women in other groups (although less than men of all races), Japanese American men had incomes 15 percent less than white men despite higher rates of professionalization. And less had to do more: incomes had to accommodate their larger families.[150]

Japanese Americans, then, found avenues in the ethnic market lessened, but access to the mainstream job and housing market on the rise during the 1960s. Not only were they afforded more opportunities; the structure of employment and industry had shifted in accommodating ways. Among the many reasons for this pattern, and not easily separated out, were the expanded opportunities offered through the nexus of US-Japanese relations. These international and domestic bases of Japanese American integration worked in mutual support.

CONCLUSION

Japanese Americans contributed to their city's transpacific urbanism in a number of ways. Their work added, in some cases, to San Francisco's actual Japanese connections and the iconography in the Japanese Cultural and Trade Center as well as to other cultural and economic institutions. This enriched the city's Pacific geography and deepened the transportation and commercial links connecting the metropolis to Japan and the Orient. Moreover, this Japanese American activity helped demonstrate their city's acceptance of them. This was not only a promotion of the city's racial liberalism; it also supported the city's claims to an embrace of all things Japanese.

Japanese Americans benefited from specific and contingent circumstances. Businesses and regular Americans were newly fascinated with the economy and culture of Japan, as Japanese industry boomed and its businesses expanded abroad (there were at least thirty-five branches or affiliates in San Francisco by 1960, according to one estimate).[151] At the same time, Americans perceived an extraordinary cultural gap between themselves and their Japanese counterparts: few Americans could speak Japanese, few had consistent commercial contact with Japanese companies, few had prior experience in Japanese relationships. Because popular thinking regarded Japanese Americans as possessing intimate cultural knowledge of Japan, a small but significant class of Japanese Americans used their self-conscious, hard-earned Japanese cultural knowledge to take advantage of US interest in and ignorance about Japan. To be sure, Japanese knowledge was never more than an additional attribute, supplementing professional or vocational skills that required significant development and refinement over years of training. But this was not unlike their Japanese knowledge, which also required education and hard work. Once honed, however, Japanese Americans could use these cultural skills to advance in mainstream employment or civic participation.

Japanese culture was one avenue to inclusion, but Japanese Americans did not feel entirely comfortable embracing it. They had been buffeted by the turmoil of international relations before and remained conscious of the dangers of betting inclusion on US-Japanese relations. Yet as Japanese Americans found opportunities for socioeconomic assimilation, they felt increasingly comfortable cultivating these ethnic associations.

Local Struggles: Japanese American and African American Protest and Cooperation after 1960

After almost eight years of delays, the Japanese Trade and Cultural Center opened in grand style on March 28, 1968. Well-dressed attendees—some in dark suits, some in kimonos—crowded into the courtyard between Minoru Yamasaki's stark concrete walls. Guests passed under the gaily waving Japanese and US flags and through the covered and pagoda-like walkway, neatly filing into the many rows of folding chairs; the overflow jostled for space on the walkway (fig. 6.1). The day was bright and sunny, and the occasional folded jacket attested to an early warmth.[1]

The event marked not just the opening of a long-delayed project, but also the progress of two decades of redevelopment. As Mayor Alioto contended, the project arose "out of what was an ugly ghetto" to become a testament to the city's will and ability.[2] Not only could attendees observe the modern buildings around them, they could peer over the courtyard's rear wall onto the widened, eight-lane Geary Expressway. But there were also signs of the disjointed, divisive nature of urban renewal. Beyond Geary lay the project area's southern portion, where the occasional vacant block testified to its slow-moving and disruptive process. In the courtyard itself, the Peace Pagoda centerpiece was still months from completion. Set in the shimmering, encircling reflective pool was a small model perched by the pagoda's flat, empty base. The ceremony, therefore, focused visitors' gaze on the international networks the center was meant to embody.

The celebration mirrored the ways that the transpacific relationship was defined in the Japanese Center: an expression of two distinct nations brought together symbolically, economically, politically, and socially. "Hostesses in kimono" escorted guests to their seats, their demure attentions offering a

FIGURE 6.1 "The finest view of Japan this side of the Pacific." The completed Peace Pagoda and courtyard of the Japanese Cultural and Trade Center, no date. San Francisco Convention and Visitors Bureau photograph, photo ID# AAB-9224, San Francisco Historical Photograph Collection, San Francisco History Center, San Francisco Public Library.

familiar and pliant symbol of Japan. The ceremony began with gestures of international exchange. A US Army band played both nations' anthems and "Japanese background music" throughout the ceremony. Next, multi-generational "'pioneer' Japanese-American families" solemnly strode into the courtyard followed by three Shinto priests.[3] San Francisco and Osaka officials came on their heels, with the Japanese ambassador, president of the Japanese Federation of Economic Organizations, and Kintetsu and Dream Entertainments executives filling the speakers' platform. Finally, Japan Air Lines flight attendants, one in a trim suit and two more in kimonos, bore a flame from the Todaiji Temple in Japan to the tune of a soft Japanese melody. Smoke filled the air as the women passed through the audience to pass the torch to the Japanese ambassador, who lit the eternal flame set in the reflective pool. The band then closed with "Auld Lang Syne." The ceremony's cultural markers transformed an official municipal ceremony into a lively civic event evidencing many markers of the city's Gateway to the Pacific identity.

Japanese American San Franciscans were key participants in the ceremonies, just as they were in their city's transpacific ambitions. The "pioneer" families narrated a beloved and generations-long connection between Japan and the city, while the designation situated them in the most verdant of US myths. The center's location at the former heart of the historical Japanese American enclave justified and intensified these stories. Its development had required the dispersal of thousands of Japanese Americans, but the surrounding discourse reinterpreted this dispersal as integration into city life. At the same time, the festivity accentuated their difference, since Japanese Americans were either proxies for or connections to Japan. They remained visibly and publicly different, reinforcing their city's connection to Japan. The center cemented the racial thinking around Japanese Americans that was cultivated in San Francisco.

The Japanese Center's grand opening also contrasted Japanese Americans with African Americans, the district's largest nonwhite group. Japanese Americans were celebrated as links to Japan; African Americans were absent. By the late 1960s, this configuration appeared natural for a Japanese institution placed "in the heart of Nihonmachi."[4] As officials celebrated the Japanese Center's opening, however, others decried the process that built it as "Negro removal." Pre-redevelopment "Nihonmachi" had been home to almost as many African Americans as Japanese Americans. Historically, the center sat in a decades-old Japanese enclave, but from 1942 on, it was also a part of black San Francisco. The Japanese Center represented a choice on the part of municipal officials, a filtering of who to celebrate and how. Redevelopment highlighted Japanese Americans as valued and cooperative contributors to city life and African Americans as excluded foreigners.

In actuality, the district's African Americans and Japanese Americans shared as much as they differed. The Western Addition's second redevelopment project, A-2, unfolded as the center was built, and its formal announcement in 1962 had brought to life for residents all the fears and lessons of the first project, A-1. The first had rebuilt twenty-eight blocks with the Japanese Center and other large projects; A-2 encompassed sixty-two blocks, housing about thirteen thousand residents, 63 percent of them nonwhite.[5] The two groups responded to these shared conditions as well as developments in other parts of the city and across the country. Not only was the civil rights freedom struggle electrifying people around the nation, but also urbanites were rising up against redevelopment. Throughout San Francisco, residents of color reconceived their place in city life, from Chinatown to the Mission district to the little-acknowledged Filipino enclave. In the Western Addition, African American and Japanese American residents articulated their own responses to A-2's threats. Here, too, the two groups shared tactics of both cooperation and protest. Quite different social and economic resources, of course, meant that their methods and tactics would at times differ substantially. Nonetheless, Japanese Americans and African Americans alike publicly and vocally protested the second redevelopment project in their district and partnered with city agencies on neighborhood redevelopment projects. The redevelopment process highlighted their differences, but the two communities shared many conditions and goals.

The progress of the second Western Addition redevelopment project from the 1960s through the 1970s supported diverging views of Japanese and African Americans. Both sought to protect their communities from the bulldozer, but their relations with the city were based on distinctive understandings of their place in city life and reflected the differing opportunities generated by the celebration of one group at the expense of the other. These divergences veiled their most fundamental similarity: both struggled toward the intimate, local goal of maintaining control over their homes, businesses, and neighborhood.

THE CHANGING WESTERN ADDITION

The Japanese Cultural and Trade Center itself clarified racialized and geographic distinctions. The widened Geary Boulevard, now Geary Expressway, became an arterial road that split the Western Addition in two. It acted, in the memory of one resident, as "a physical barrier between the Japanese community and the African American community."[6] The Japanese Center was poised on Geary's northern edge, and its windowless, concrete southern walls created an imposing barricade for those from the southern approach that contrasted with the welcoming walkways and courtyard on the north side (fig. 6.2). And while prewar Japanesetown had spread out in all directions from Geary, the expressway helped consolidate Japanese American residence to its north. This new residential pattern had the effect of distilling a new racial geography, reinforcing a divide between the predominantly white, wealthy Pacific Heights to the north and the black Fillmore to the south (fig. 6.3).

This racial geography reflected the diverging conditions of the district's Japanese and African Americans. First, in both percentages and absolute numbers, there were fewer Japanese Americans in the Western Addition than there had been in 1950. By 1970, their number in Japanesetown's three census tracts had shrunk by over half, even as the number living in the city

FIGURE 6.2 The southern view of the Japanese Cultural and Trade Center, with the Geary Expressway at the front. The building's design minimized entry from the south below Geary, and emphasized the barrier that the expressway posed between the southern and northern portions of the Western Addition. San Francisco Convention and Visitors Bureau photograph, photo ID# AAB-9223, San Francisco Historical Photograph Collection, San Francisco History Center, San Francisco Public Library.

had almost doubled, to 11,705.[7] The biggest decrease was in the two census tracts bordering the new Geary Expressway and almost entirely engulfed by redevelopment. These tracts dropped from almost 2,500 Japanese Americans to 400. The northern, third tract was largely untouched by renewal and declined less: from 1,500 Japanese Americans, when they had been 25 percent of the population, to 726, when they were 20 percent of a much smaller total. Many varied and personal reasons motivated this decline, of course, from the well-documented out-migration of the second generation to the anti-black racism of some Japanese Americans. Still, the redevelopment process—with its disruption, demolitions, and stigma—seemed to accelerate the emigration of Japanese Americans from the district.[8]

Those emigrants, along with Chinese Americans, faced fewer residential barriers. Japanese Americans lived all over the city, with the largest number in the Richmond district where Asian Americans had been only 1 percent of the population in 1950. By 1970, it was home to 4,471 Japanese Americans, compared to only 1,885 in the Western Addition. The district was firmly middle class, with an average family income higher than the citywide average, indicating the "very rapid change in occupational status of Nisei" after the war as well as lowered anti-Asian segregation.[9] In contrast, many African Americans were forced out of the city altogether by segregation and rising costs; median rents in the Western Addition rose 64 percent between 1950 and 1960 in the census tract that bore the brunt of the district's redevelopment.[10] Those who remained were restricted to just a few neighborhoods, including the long-time middle-class enclave of Ingleside and the Western Addition. Perhaps most striking was the extreme black segregation in Bayview–Hunters Point, a physically isolated margin of the city home to the naval shipyard, industry, and the Hunters Point housing project; African Americans composed almost 70 percent of the district's population, and in one tract, 96 percent.[11]

FIGURE 6.3 The Japanese Cultural and Trade Center lies barrier-like in the surrounding district. The primarily white, higher-income Pacific Heights is at the top, while the lower-income, nonwhite Fillmore neighborhood is in the bottom half. As this 1970 photograph demonstrates, much of the demolition and clearance of the Western Addition redevelopment area was in the latter portion. Photo by Gill, photo ID# AAC-1868, San Francisco Historical Photograph Collection, San Francisco History Center, San Francisco Public Library.

Rising black segregation in San Francisco starkly contrasted with gains made by Asian Americans. As African Americans moved from a scant 1 percent of the population to a much larger group associated with, as Herb Caen noted in 1947, the "Negro Problem" of unemployment and blighted neighborhoods, their opportunities contracted.[12] From about 3,000 black Western Addition residents in 1940, there were 14,631 twenty years later. This was only about 20 percent of black San Franciscans—most lived in Bayview–Hunters Point—but it represented the largest single nonwhite group in the central city. In 1940, only five Western Addition census tracts had African Americans as more than 5 percent of residents; the highest African American population in any tract was 14 percent. By 1960, there were six tracts with a majority of African American residents and almost all had over 20 percent.[13] This was partly about income. Aside from low-income white seniors, black families were the largest low-income group in the Western Addition. In the San Francisco metropolitan area in 1960, which also experienced the "last hired, first fired" pattern evident nationally, African Americans were twice as likely as whites to be out of work: 11 percent of African Americans were unemployed in contrast to 5 percent of white people.[14] But residence was not just about economic means. As a fair-housing group noted, "San Franciscans of Oriental ancestry have made some moves into neighborhoods in which they were not found in 1950; but last year's census showed that virtually no Negroes lived in the Mission, Sunset and downtown apartment house districts, which contain housing of the price level needed by many Negro families."[15] One 1961 housing survey found that while 67 percent of vacancies were open to "Orientals," only 33 percent were open to African Americans.[16] Black San Franciscans faced barriers that were falling for Asian Americans.

Of course, these interracial comparisons were messier on the ground. Black San Franciscans, like Japanese Americans, also lived throughout the city. Thousands lived in the Richmond, and certainly there was no neighborhood without hundreds of black residents. Nor did Japanese Americans experience complete integration. After all, Japanese Americans did not so much disperse as congregate anew. They found it difficult to buy into the mushrooming suburbs in the East Bay. By the time the Richmond district opened to Asian Americans, its housing stock was decades old and was briefly considered for its own renewal project. There was also evidence that reasons besides income kept many Japanese Americans in the Western Addition. According to a Redevelopment Agency study, over half of the 372 Japanese American families in the A-2 area had the ability to pay $100 or more a month in rent, well above the district average of $57; this was much higher than the 35 percent of all families in the district. Undoubtedly some preferred the familiarity. But others perhaps remained sequestered by limited residential opportunity. Sociologist Harry H. L. Kitano's survey found that of those living in the "old Japanese area," only 55 percent were satisfied with their neighborhood compared to 78 percent of those elsewhere.[17]

There was nonetheless a divergence between the many Japanese Americans who fled redevelopment for middle-class areas and the far larger proportion of African Americans who moved within the Western Addition. Of course, people move for myriad reasons, and it would be a stretch to the point of inaccuracy to claim that redevelopment caused the Japanese American departure. Nonetheless, the Japanese Cultural and Trade Center emphasized a celebratory story of their declining segregation by highlighting, even hastening, their withdrawal from their enclave.

NIHONMACHI AND JAPANESE AMERICAN REDEVELOPMENT

Japanese American contributions to Western Addition redevelopment and the city's Gateway identity laid the groundwork for the thwarted but long-desired project of Victor Abe's group of merchant-planners. The project was small and in many ways a compromise, but it was theirs. Now organized as the Nihonmachi Community Development Corporation, the group's victory was as much the product of changed conditions and new pressures on the Redevelopment Agency as it was of the group's persistent, adaptive efforts. The most important change was, perhaps, the Japanese Cultural and Trade Center itself. The center gave the group new opportunities, as long as they were willing to accommodate its Japanese parameters.

Abe and his group had suffered a crushing blow after Tokioka won the Japanese project bid, but they were dogged. When the Redevelopment Agency announced plans for the district's second redevelopment project in late 1961, the *Nichi Bei Times* grimly noted, "Much of what remains of this city's Nihonmachi is in this new project area."[18] And so a group of Japanese American merchants and property owners once again gathered to protect their businesses and neighborhood. This time, they were motivated by the cautionary tale of the district's first redevelopment project. Masao Ashizawa, who became the president of their revived group, recalled the "community uproar": "we were really up in arms" and determined not "to be bulldozed out" again.[19] In January 1962, much as they had in 1953, members of the business community and property owners organized "to be able to get back into Japantown after redevelopment."[20]

The merchant-planners were now much more established than in the uncertain years immediately after their wartime incarceration. Many, such as the Benkyodo manju pastry shop and the Uoki K. Sakai Grocery, had firm prewar foundations and long-standing patrons with shared institutions and a sense of place developed through the interwar decades, incarceration, and resettlement. Others, like plumber Sam Seiki, benefited from the city's fair-employment policies and supplemented ethnic business with new mainstream clients. Still more, such as Soko Hardware and Honnami stationary, capitalized on the new postwar popularity of Japan to attract new, white customers. Some, of course, had gone out of business or left the area entirely. But most of the remaining Japanese American proprietors were relatively stable.[21]

The Japanese American merchant-planners also had a stronger relationship with the city. Members had come and gone since 1953, but some veterans remained, including Ashizawa, Soko's owner; Honnami, the book and stationery shop owner; and Abe, the staunch promoter.[22] After a decade of accumulated correspondence, meetings, and even lawsuits, they were M. Justin Herman's reliable (if not necessarily representative) Japanese American contacts for announcements and input, and they had their own contacts for queries or ideas. They were also now quite fluent in the language, values, and practices of urban progress prized by city officials. Their plans, therefore, reflected Herman's modern taste and new federal and local provisions for rehabilitation, an incremental process suited to their limited resources. Furthermore, they incorporated the mixed residence and commerce now in vogue with planners, since years of residential dispersal had lowered the toxicity of Japanese American clustering. Last, what they did not know, they acquired from Noboru Nakamura, the Japanese Center's architect of record. Nakamura designed Nihonmachi's master plan and a number of its buildings, bringing with him his experience with the Redevelopment Agency.[23]

The merchant-planners' acquired knowledge was not automatically useful. While Herman agreed that another Japanese-themed project "adjacent to the new Japanese Cultural and Trade Center" would "be a good thing for the city," he made no promises: "The Japanese community ought to and deserves to be given every *opportunity* to have such an area."[24] However, the Redevelopment Agency desperately needed allies in the face of mounting criticism. The A-1 project had swept across seven hundred parcels of land, demolishing more than six hundred buildings.[25] This displaced 2,700 households, most of whom were poor or nonwhite. As Herman himself conceded, most of these residents simply moved to the surrounding blocks, from "blighted area to blighted area."[26] This indicated, in the US comptroller general's words, a "general pattern of unsatisfactory relocation."[27] As the earliest opponents to Western Addition redevelopment plans had feared, the agency's relocation plan had not taken into account the "limited availability of private housing to some groups."

Established safeguards had not protected the most vulnerable residents. Federal and state regulations mandated affordable, safe, and decent rehousing for displaced individuals. The Board of Supervisors and the Redevelopment Agency Commission passed their own anti-discriminatory policies for redevelopment projects. But, as critics across the country pointed out, such liberal measures were hardly relevant in the face of structural barriers. This was evident in the International Longshore and Warehouse Union-sponsored St. Francis Square cooperative, the agency's model of affordable Western Addition housing. Not only did its sponsors claim to offer the lowest monthly cost for redeveloped housing, but it also prioritized displaced families and was proudly interracial. Yet as the union unhappily noted a year after the cooperative opened in 1963, none of its occupants were displaced A-1 residents from the district's first project. The square was also unlikely to absorb many residents from the second project area. Its monthly cost of $125 to $140 for the smallest units was unaffordable to the agency-estimated 31 percent with average monthly incomes of less than $250 a month. As Tarea Hall Pittman, officer for the regional National Association for the Advancement of Colored People (NAACP), flatly summarized, all the measures instituted to protect residents "have not proven effective."[28]

By the eve of the second project in the Western Addition, redevelopment had an expansive opposition. The San Francisco NAACP called for renewal programs to be "curtailed" because while they "beautif[ied]" former slum areas," they only created "new ghettos" for the displaced.[29] The Japanese American Citizens League (JACL) made a point of vocal support for the Nihonmachi plan, but declared itself "opposed to the wholesale uprooting as indicated in the A-1 project."[30] The Council for Civic Unity, once an ardent supporter, critiqued projected low-income housing as "very small compared to what was there before."[31] Even members of the San Francisco Housing Authority censured the agency for "uprooting Negros" and "forcing them into inadequate public housing."[32]

Residents had their own criticisms. The earliest evictions led a handful of particularly vocal "hold-out" tenants to ignore the eviction notices and in some cases to refuse their rent to the Redevelopment Agency, now their landlord.[33] They also fumed at the owners of their homes who, they alleged, avoided maintenance work in light of impending demolition. Homeowners, too, were angry. They had "stinted and saved to buy their homes," only to lose them in an inflated real-estate market in which they could no longer afford to purchase.[34] Some also objected to what they saw as unfair valuations of their buildings, even as scandals erupted over city officials who received inflated prices for their property. The redevelopment process impacted lives in other ways, as well. Slow progress meant that people might live for extended periods among emptied buildings or lots. These vacancies attracted vandalism, theft, and squatting, while resident James Howard Pye Jr., who lived in the district during the second redevelopment project, saw arson and drug use in vacated structures. Few residents wanted to repeat their experiences with the first redevelopment project.[35]

Residents' anger joined an increasingly critical national and international debate over redevelopment. None of these tribulations were unique to San Francisco. By this point, there were 1,300 projects in approximately 650 US cities. Activists, scholars, and critics all over the country castigated redevelopment as a "failure" on social, economic, policy, and even aesthetic grounds.[36] The "most common charge" was that the process "displace[d] minority group families from slum areas and rebuilt cleared land with luxury apartments."[37] The highly visible uprooting of large African American populations led to the frequent claim of "urban redevelopment as 'Negro removal.'"[38] National debates thereby framed local ones in particularly resonant terms, even as they slightly misrepresented local conditions. In San Francisco, Japanese American households were also among the displaced. However, in stressing the consequences for the most numerous and poor in the Western Addition, local opponents drew from the language of their contemporaries to underscore Japanese and African American divergences.[39]

San Francisco officials responded with carefully demonstrated nonwhite community participation in the A-2 project. In a 1964 report, the agency stressed that "Negro leadership has . . . been consulted and kept informed of planning." However, this relay of information did not compare with the collaboration of a group of "Japanese civic and business representatives" who "cooperated with the Agency in developing suitable plans for a four-block area."[40] The Japanese American group substantiated the agency's claims that it "worked with residents, businessmen, and property owners" to "encourage and assist them in assuming the role of redevelopers in their own area."[41] The Nihonmachi group also highlighted their cooperative relationship with the city, although they emphasized their guiding role: the "unique renewal of an entire community . . . would not be possible without neighborhood participation."[42] The merchant-planners allowed the city to deflect criticism about its heavy-handed approach to redevelopment.

The merchant-planners incorporated as the Nihonmachi Community Development Corporation in 1964 to redevelop the four-block area immediately to the north of the Japanese Center (fig. 6.4), conceiving and carrying out plans in conjunction with agency officials. The corporation shared a name, local control, and a nonwhite identity with the Community Development Corporations (CDCs) that would become widespread by the 1970s, many with Office of Economic Opportunity support. However, the Nihonmachi group did not share a base in the "poor and disadvantaged" as most did, nor did they have any links to Black Power ideas or, quite pointedly, to the War on Poverty.[43] Still, like those of many other similarly named organizations, its members were fixed on preserving their presence in an embattled neighborhood, as they had been for the past decade. The corporation therefore allocated redevelopment parcels among members, oversaw rehabilitation when feasible, collectively financed and developed the area's shared spaces or facilities, and "coordinat[ed] community interests in negotiations with the Agency."[44] At long last, the group of Japanese American

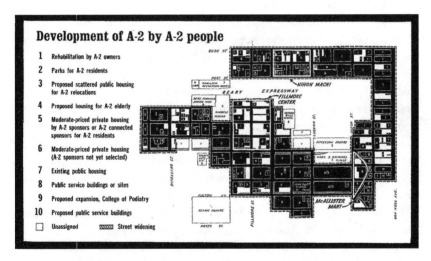

FIGURE 6.4 Map of the A-2 project area, speaking directly to redevelopment critics with an emphasis on community participation and accessible housing. Nihonmachi is highlighted. Image from San Francisco Redevelopment Agency, *Annual Report* (San Francisco: San Francisco Redevelopment Agency, 1971).

merchant-planners controlled the renewal process in their neighborhood, albeit in the small section the group could afford.

Their participation came with costs, however, especially their emphasis on property ownership. To some degree, this reflected Japanese Americans' high rates of ownership. In a survey of the A-2 project area, "Orientals" were 33 percent of resident-owned households even though they composed only 9 percent of the area's households; this rate exceeded the citywide average. This nonetheless left approximately 60 percent of Japanese American households, or 620 families and single people, as renters.[45] These tenants were largely unrepresented in the Nihonmachi development, although they could buy membership stock in the corporation. They could also rent in the mixed-use project. However, they could not be poor. Low-income tenants were completely excluded from the Nihonmachi development, and even moderate-income people might have struggled. Unlike other Western Addition developments, almost all the Nihonmachi housing was market rate. This selective program may have been less exclusionary in intent than a matter of financial viability or an assumption of which groups had the most to lose. Nonetheless, the Nihonmachi Corporation and its project was composed of those whom the city was most ready to work with: financially secure property owners.

The Nihonmachi Corporation also had to work within the Japanese Cultural and Trade Center's framework. The center was its commercial anchor, intended to "stimulate the economic development of the surrounding area, especially the informal Nihonmachi, its neighbor to the north."[46] The group's "collaborative planning" willingly accommodated this. Merchants decorated their shops with colorful banners for the center's opening and organized the Cherry Blossom Festival, with three days of festivities and a grand parade. Bilingual signs, "bright new Japanese shops," and restaurants made Nihonmachi the "community extension" of the center, giving the neighborhood "the reputation as the place to buy Japanese merchandise" (fig. 6.5).[47] However, businesses that would not attract outside visitors or did not have the capital to adapt, such as Yamato Auto Repair, found no place in the project area.[48] Such businesses had served the residents' needs, but Nihonmachi as a visitor attraction had no room for them.

The project's name, Nihonmachi, which translates as "Japan Town," reflected the neighborhood's reorientation. "Nihonmachi" was a recent moniker. "Uptown" had been the popular, and deracinated, name for the neighborhood after World War II. "Nihonmachi" was only sporadically used during the latter part of that decade, and interchanged with "Nihonjin-

FIGURE 6.5 Redeveloped Nihonmachi, as viewed from the Japanese Center opposite. The design, as planned by Noboru Nakamura, adopted a modern style similar to the Japanese Cultural and Trade Center and included Japanese features such as bilingual signs and windows that resembled shoji screens. Photo ID# AAZ-0965, San Francisco Historical Photograph Collection, San Francisco History Center, San Francisco Public Library.

machi" ("Japanese People Town" or "Japanese Town").[49] While the latter defined the place by the people in it, "Nihonmachi" defined it through Japan; perhaps not coincidentally, the use of "Nihonmachi" surged in 1959, largely replacing "Uptown," as interest in Japan rose throughout the city. By 1966, the merchant-planners' use of the term for their development simply paralleled the language of the ethnic press. Yet it perhaps remained uncommon in everyday use: the corporation's internal correspondence occasionally lapsed into the older term, "Japanese Town," to describe their project area.[50] But "Nihonmachi" had civic value by highlighting their contribution to the city's Pacific urbanism.

While Japanese Americans redeveloped their neighborhood in the Japanese Cultural and Trade Center's framework of Japanese culture, Nihonmachi businesses provided it a sense of authenticity. In July 1962, *Sunset Magazine* featured a color spread on the neighborhood's tempura houses, Japanese sweet shops, and yearly Obon festival. Although "those who know San Francisco's Japantown are considerably fewer" than those familiar with Chinatown, the journalist pointed to its similar appeal as a "primarily residential district that caters to the people who live here."[51] Japanese American stores and restaurants, it turned out, had an organic charm that the Japanese Center lacked. When *Sunset* covered the center in 1969, it underlined a sense of artifice: visitors heard "piped-in music" as they browsed items "similar to those sold on Chinatown's Grant Avenue." But while those stores "also suppl[y] the needs of a large Chinese population, the Japan Center is planned primarily for Caucasian visitors. Most of the Japanese community eats and shops in the Nihonmachi district across Post Street from the center."[52] Nihonmachi and its Japanese American businesses and patrons balanced the center's perceived lack of authenticity.

Nihonmachi both responded to and rearticulated the racial thinking about Japanese Americans during this period. The lessons and experience gained in the merchant-planners' failed attempts years before helped secure their role in the neighborhood's redevelopment. The merchant-planners centered Japan in their neighborhood's identity, producing a necessarily filtered and incomplete portrayal of Japanese American community that excluded those who did not conform to Nihonmachi's revised image or the Redevelopment Agency's priorities. But their project was collaborative and evocative of the city's desired transpacific identity.

The Nihonmachi project also echoed the model-minority image of Asian Americans that scholars have traced to this period. Mayor Joseph Alioto neatly articulated this in his praise of the Nihonmachi Community Development Corporation. In Alioto's telling, the pre-redevelopment Japanese American enclave had been "a slum that bred crime, violence, and unsanitary conditions." However, the corporation had created "a better environment for its people" by working "with the Redevelopment Agency in a constructive way." Alioto, literally making the group a model, urged "Negro leaders to . . . follow its example" rather than just "talk[ing] and moaning."[53] The Japanese American story of success was also a local rebuke. Redevelopment and San Francisco's Gateway to the Pacific identity calcified Japanese Americans as a model, just as it responded to the emergent black freedom struggle in early-1960s San Francisco.

THE "NEGRO REVOLT FINALLY COMES TO SAN FRANCISCO"

Black San Franciscans did not need Alioto to scold them about cooperation or hard work when they were well aware of the disparity between their opportunities and those of Japanese Americans. When the Japanese Cultural and Trade Center was announced, the *Sun-Reporter*'s editors pressured the state attorney general to decide on its legality: "we have misgivings about the authority of the Redevelopment Agency to take such land, which has been obtained through public condemnation and the right of eminent domain for redevelopment purposes, and specifically allocate it to any racial or ethnic group." This was a clear indication of "favored group prerogatives and special privileges": "if the Japanese community feels that such a center is essential to their welfare and best interests," they should build it, but *"without the assistance of Urban Redevelopment statutes."*[54] Japanese Americans had worked hard for years to participate in redevelopment programs. However, African Americans, even middle-class professionals such as the editors of the *Sun-Reporter*, realized that their Japanese American neighbors had prospects unavailable to them.

Just like the Japanese American merchant-planners, black Western Addition residents wanted to fight displacement, protect their community, and shape their neighborhood's future. But without a revalued racialized identity, African Americans turned to alternative methods to shape the redevelopment program. Their tactics reflected a different vision of their place in the city as well as the burgeoning black freedom struggle in San Francisco of the early 1960s. Similar to those elsewhere in the nation, activists struggled against displacement using direct-action protest and nonviolent, civil disobedience rather than negotiation or legal action. The tactics and values of this emerging movement in the Western Addition recalibrated anti-redevelopment grassroots activity as black protest and reshaped official response.[55]

Activists in San Francisco, as elsewhere, were electrified by southern civil rights activism and applied their tactics and models to local problems. On May 27, 1963, a confederation of groups organized a march and rally in solidarity with the Birmingham campaign, where the virulent opposition of Sheriff "Bull" Connor and Governor George Wallace to peaceful student protesters had shocked the nation. An estimated twenty to thirty thousand people attended the solidarity event, and high-profile speakers included "some of the community's most prominent religious, labor, political and community leaders."[56] After many speakers used the event to make clear that "we have common problems here," Mayor George Christopher called for a biracial committee to "study San Francisco's racial problems."[57] A cohort of black activists immediately denounced the measure. Wilfred T. Ussery of the Congress of Racial Equality (CORE) dismissed it as "tokenism, paternalism, and gradualism."[58] He and others instead formed the United San Francisco Freedom Movement to create a "unified front" of civil rights organizations and build an effective coalition against recalcitrant discrimination.[59]

The Freedom Movement's common set of ideals cohered organizations such as the local branches of CORE, NAACP, and the Student Nonviolent Coordinating Committee (SNCC), as well as the newly formed Ad Hoc Committee to End Discrimination, a radical, predominantly white, local student group originally formed at the University of California, Berkeley. Activists coalesced around two principles. The first was black leadership: African Americans themselves should determine "the timetable for confrontation of the white power structure and the order of business to be considered." Therefore, while the Freedom Movement's membership was decidedly mixed, black organizers led the work.[60] Second, as chair Dr. Thomas Burbridge, also president of the San Francisco NAACP, argued, activists wanted "freedom immediately, yesterday."[61] The Freedom Movement included "only groups who have direct action as an M.O." Moderate groups, activists argued, had a place at the "implementation" but "not the initiation stage" of the struggle.[62] Black leadership and direct-action tactics exemplified the San Francisco freedom struggle of the early 1960s and helped cast civil rights activism in the multiracial city as black protest.[63]

Unsuccessful negotiations quickly turned into protest. Meetings with city officers accomplished mixed results. For example, negotiations won a municipal Human Rights Commission and liaisons in the education department, but little in the prime demands for employment or civilian oversight of police.[64] The group consequently led a number of high-profile demonstrations in 1964 to direct San Franciscans' attention to hiring inequities. The past year had seen a series of protests that commanded attention and resulted in black employment at individual firms, largely organized by CORE. But the Freedom Movement's most significant coup was the "massive civil rights offensive" at the Sheraton Palace Hotel, one of the city's oldest and most revered institutions.[65] After negotiations over nondiscriminatory hiring broke down, demonstrators descended on the hotel in March. On the protest's biggest day, an estimated fifteen hundred protesters picketed outside while five hundred occupied the hotel. Locking arms amid the gilded interiors, the white and black protesters clogged the lavish lobby while thousands of spectators ogled outside. Although Mayor John Shelley originally castigated the protest's methods, he eventually stepped in to lead discussions between activists and management. The demonstrations finally ended with an agreement on minority hiring and, even more importantly, a nondiscrimination hiring pact with the San Francisco Hotel Employers Association, representative of all the city's major hotels.[66]

This protest ushered in a "new phase" of the "Negro revolt" in San Francisco, according to the San Francisco NAACP.[67] Demonstrations were nothing new in the Bay Area, but the direct-action protests of the early 1960s departed from past educational picket lines and boycotts. Protesters organized well-publicized demonstrations at a number of other businesses in the area, including Bank of America, retail outlets, supermarkets, and automobile dealerships. Together, these were a "serious fight against economic discrimination."[68] Protesters exacted employment agreements that were previously unattainable. Their disruptions of everyday activities brought attention to the discriminatory practices of individual businesses and wider conditions of unemployment, underemployment, poverty, and prejudice in the lives of many of the city's African Americans.[69]

The Freedom Movement's leadership, the targets of its protests, and black politics more broadly were complicated in a multiracial region struggling to assert its racial liberalism. While protested businesses were generally white owned and operated, a notable exception was the Berkeley realty office of Tad Nakamura. After CORE denounced his "token integration" for black clients, the Sun-Reporter's editorial expressed mock incredulity that a realtor of color would employ such practices—"Certainly Not a Japanese Realtor?"—and called for "a complete boycott."[70] Nakamura himself refuted the charges, arguing that his firm had "sold or rented more homes to Negroes and Orientals in previously all-white neighborhoods than any other East Bay realty firm."[71] Yet the Sun-Reporter's coverage hinted at the believability and irony of anti-black bias from a nonwhite business owner. The Nakamura firm was an exception for protesters, who rarely took aim at nonwhite employers and businesses, but their demands indicated that employment and residential discrimination fell most heavily on black San Franciscans.

Protest framed black politics in San Francisco, although in reality the politics were quite heterogeneous. African Americans served in formal politics on municipal commissions, the Board of Supervisors, the California assembly, and other offices. The black freedom struggle in San Francisco also took many forms. For example, the San Francisco NAACP concurrently attacked de facto school segregation with negotiations; the instigation of local-, state-, and national-agency investigations; and a lawsuit. Furthermore, many black leaders viewed direct-action protest and civil disobedience as "reprehensible," "endangering the civil rights movement here."[72] Although the San Francisco NAACP was a formative group in the Freedom Movement, direct-action tactics bitterly divided the branch. A cohort of officers ambivalent about "militant action" launched a vigorous campaign to moderate branch activities, although they were swept out of office by the voting membership.[73] Protest was by no means representative of the diversity of black politics.[74]

Black protest nonetheless framed African American activists and the black community as undeserving outsiders. Columnists deplored protests as "government by mob" even as they claimed to sympathize with abstract notions of civil rights. Citizens flooded City Hall with letters, especially in response to the Sheraton Palace Hotel demonstration. While some praised the protesters, many more decried them. A few people were openly racist, but most used more coded language to reject the radical tactics they associated with, as one constituent wrote, "the negro problem."[75] As anti-Asianists had in the past, observers saw activists as foreigners undeserving and incapable of inclusion. Self-proclaimed "native San Franciscan[s]" railed against the "southern negroes" who "think they can come in here and shove the natives out."[76] Such rhetoric built on media descriptions of black San Francisco. For example, a ten-part series on local African Americans in the *San Francisco News-Call Bulletin* described a "rising tide of immigrants" from "tiny villages and share-crop farms of the rural South." They were shown as different in so many ways: rural, southern, black, and poor.[77] As descriptions of Asian foreignness had done through World War II, such discourse made black exclusions appear reasonable. In contrast, the foreignness of Japanese Americans was domesticated and possessive: "our longtime Japanese colony" and "our Japanese citizenry." This allowed boosters to claim Japan—"San Francisco's bit of Japan," as the neighborhood was called in one front-page story—and dissemble racial inequities as a foreign importation.[78] San Franciscans, who most likely would have been horrified at the comparison, paralleled southern segregationists' rhetoric of black civil rights activism as the work of foreign subversion or "outside agitators." In the situational racial thinking of the Bay Area, however, "foreign" was both a frame for black exclusion and Japanese American inclusion.

Direct-action protest and civil disobedience burst into San Francisco politics in the early 1960s. These tactics were a potent tool for black activists and their allies that also reshaped public understandings of civil rights struggles in the city. However, they were also framed in contrast with Asian Americans, and especially Japanese Americans.

A "NEW FIGHT HAS BEEN DECLARED"
IN THE WESTERN ADDITION

Freedom Movement activists and allies brought this new wave of activism to the Western Addition, shaping the black community's response to redevelopment in ways that paralleled and contrasted with the highly publicized actions of the Nihonmachi Community Development Corporation. Like the corporation, black community-based groups struggled for local control in redevelopment. But without the same kind of civic valuation, they largely did so through protest. Community organizing helped poor, black residents build on freedom-struggle tactics in order to demand the forms of neighborhood and civic participation that Nihonmachi members could gain through cooperation. Through a very different route, these activists also achieved a new, hard-won, participatory relationship with city government.

Redevelopment Agency planners had in fact implemented changes to the A-2 project in response to critics. New measures included improved information dissemination, listings of preapproved housing, resident liaisons, homemaking classes, and an elderly-assistance program. Revised federal and state legislation further allowed for new forms of aid for the displaced. Legislation encouraged moderate-priced housing, which the agency used to propose 3,600 moderate-priced units for the A-2 project, a substantial increase from the less than 800 in A-1. These measures, in Herman's view, "solv[ed] housing problems of low-to-moderate income families" in the second Western Addition project.[79]

Critics still predicted damages, which motivated activists. The city's own data showed that less than 7 percent of the projected housing would be low income, yet 30 percent of the project area's families were. Critics highlighted the plight of the black majority of residents who would be priced out of redeveloped housing, discriminated against in most low-range and private-market rentals, and forced into either public or private housing in "new ghettos and new slums."[80] Residents, having witnessed the results of the district's first redevelopment project and the freedom struggle in the city and country, foresaw impending displacement and upheaval. They resolved to fight. An owner of a senior boardinghouse announced, "When the bulldozer comes . . . they'll find Serena Smith and all her tenants sitting on the doorstep."[81] Percy Jones declared that the "first bulldozer that comes into my neighborhood is going to have to run over me."[82] Residents predicted the same displacement that they had seen before, but were determined to create a different outcome for the district's second project.

Joye Goodwin, chair of the housing committees of both the San Francisco NAACP and the Freedom Movement, attempted negotiations with the city, but dismissed her meetings as an insincere "matter of form" for Redevelopment Agency officials.[83] Residents and activists therefore joined together, most visibly as Freedom House in 1964. The grassroots group aimed for "the people of Area 2 . . . to oppose and resist the Redevelopment Agency's attempt to evict them and build homes too expensive for them to afford." The group's activities were based on a critique of urban renewal as the "removal of low-income, working class people who aren't politically organized and have had the bad luck of living on 'prime real-estate.'" Freedom House activists therefore organized residents to "confront" the agency.[84] As historian Robert O. Self finds among their contemporaries in Oakland, Western Addition activists built on the "civil rights liberalism" of early 1960s San Francisco to create a "community-centered politics of place," one that was embedded in a demand for participation in government decisions.[85] Such place-based organizing not only voiced a faith in community solutions and approaches; it sought to foster political, economic, and cultural power within black urban neighborhoods.[86]

Freedom House borrowed from both the southern and local freedom struggle in its goal to empower local residents. It shared its name with a multitude of other localized civil rights headquarters and situated itself as "part of the fight for Freedom begun in the South."[87] Freedom Movement activists such as Joye Goodwin and Wilfred T. Ussery were joined by Mike Miller, an experienced field worker who had toiled with SNCC in Mississippi.[88] These origins shaped an organization based in and primarily led by black Western Addition residents. One example of this distinctive leadership was Pleasant Carson, the group's elected chief executive. Carson was a Western Addition resident, had attended San Francisco City College, and firmly held the belief that "only through mass protest can the power structure be forced to grant the Negro his political, social, and economic rights."[89] Still, he was very different from the college students and middle-class professionals of San Francisco's Freedom Movement: he had served time in jail, was frequently unemployed, and had been for a period a black nationalist. In contrast, Thomas Burbridge was a professor at the University of California, San Francisco; Goodwin was a nurse married to an attorney; and Ussery was an architectural draftsman, all at a time when less than 4 percent of Bay Area African Americans had professional occupations. The former two also lived in the Sunset, while Ussery lived on the Western Addition's northern, more prosperous edge outside of the renewal project areas.[90] Freedom House activists, in contrast, came from the project areas.

The group's primary strategy was resident organization. As organizer Miller's background with SNCC would suggest, the group sought a transformed community through resident empowerment rather than the Freedom Movement's spectacular but short-lived public mobilization. The group kept a walk-in storefront office on Fillmore Street, the black Western Addition's commercial heart, and distributed a biweekly newsletter, the *A-2 Stand*. Its block-by-block structure reached as many residents as possible. "Block workers" fanned out across the district, organizing block clubs by "walking door to door, talking and arguing, spreading the news of the danger and the fight, organizing meetings."[91] Each club elected a captain to represent it, who in turn elected a head from their number. Not all of the blocks claimed by Freedom House were organized, but the structure nonetheless gave it, as one contemporary political scientist argued, a very deep and organized base in the district's residents.[92]

This meant that residents could be quickly mobilized. This speed was key during the summer of 1964 when A-2 project plans were publicly assessed. The redevelopment process demanded a cascade of hearings: plans had to be introduced, passed by the Redevelopment Agency Commission, and approved by the Board of Supervisors before they could be sent to federal agencies for ultimate authorization, each step requiring several meetings. Activists and residents made these meetings the site of "furious public controversy."[93] For example, in July 1964, Freedom House activists learned with little advance notice of an unpublicized agency meeting to approve the A-2 plans. Block clubs and workers sprang into action. They "spent the weekend informing residents of the meeting and circulating a petition." Their efforts brought a crowd of one hundred people to the small hearing room meant for fifty, and a petition for delay with 707 residents' signatures.[94] Freedom House continued to coordinate with other groups; for instance, the San Francisco NAACP arranged free childcare to facilitate family attendance. However, Freedom House activists stressed that the vocal and animated crowd were not outside advocates but "the people in Area-2 themselves."[95]

A growing chorus of respected and expert critics augmented resident voices and pressured officials to respond. Nationally known developer Edward P. Eichler, who had a building in the district's first project, and Berkeley sociologist Nathan Glazer both spoke at hearings in opposition to the plans. The Council for Civic Unity, once a keen supporter of the Redevelopment Agency, had switched sides. Even Mayor Shelley, the first Democrat to hold the office for fifty years and who came to politics through the American Federation of Labor, argued that the "city should first worry about the human factors involved" and advocated for accommodations and subsidies for the displaced.[96] As a result, the City Planning Commission "acknowledged the roar of controversy" with a special investigative committee.[97] Even two members of the Redevelopment Agency Commission voted against the A-2 plans, arguing that they "fail[ed] to consider the 'little man.'"[98] Others dug in. Commission chair Everett Griffin declared he would "brook no delays regardless of what the hearings produced."[99] M. Justin Herman even unsuccessfully attempted to fast-track plans through municipal approvals. The project passed unmitigated through bureaucratic channels. Members were forced to pound gavels and shout to be heard, but the Board of Supervisors approved the plans in October 1964.[100]

This might appear disappointing to residents, but Freedom House goals were always more complex than simply ending redevelopment. The group's mission was to encourage "people living in Area-2" to "get together to develop plans to solve community problems," of which redevelopment was crucial, but not the only one.[101] Their "improvement movement" encompassed a capacious set of endeavors.[102] Consonant with its organization, much was done at the block level. One of the group's earliest activities was block cleanups, inexpensive but effective undertakings. Additionally, activists worked to alleviate larger, pressing problems. For instance, block captains Lawrence Custus and Gilbert Pope held a press conference to charge city agencies with lax enforcement of city codes, pressuring officials to investigate neighborhood conditions. Freedom House activists addressed a broad range of "community problems," including high rents, poor public education, unemployment, and police brutality.[103]

Seen from this perspective, Freedom House activists scored important victories. They helped make the neighborhood livable and attractive for current Western Addition residents in ways that redevelopment had not. This corresponded with activists' and residents' efforts to make the redevelopment program more responsive to residents' needs. The group engaged municipal government in other ways, as well. Members drew from all available resources, such as borrowing equipment from the Public Works Department for the cleanup program. The *A-2 Stand* published the names and phone numbers of the city-code inspectors handling sanitation, building maintenance, and pests as well as the Police Community Relations Bureau. These actions educated residents about city resources, connecting them to city bureaucracy and raising expectations for municipal receptivity. Freedom House helped change the relationship between residents and city government.[104]

"NO ONE IS GOING TO PUSH US AROUND ANY MORE": WACO AND COMMUNITY ORGANIZATION

Freedom House was short lived, but it was the foundation of a subsequent provocative and ultimately quite successful group, the Western Addition Community Organization (WACO). WACO built on the Freedom Movement's tactics and resident mobilization in order to successfully halt the renewal process and force the Redevelopment Agency to integrate resident input. This group used a number of strategies, but their methods of organizing, their foundation in black civil rights struggles, and their reception by the San Francisco public contrasted them with the cooperative efforts of the Nihonmachi group.

WACO formed in 1967 to bring the district's residents, businesses, preexisting groups, and institutions together to oppose the current renewal program. Redevelopment in California stalled immediately after the second Western Addition project was approved in 1964, when voters passed Proposition 14. The state measure allowed discrimination in housing in explicit defiance of federal policies, cutting off federal funding until the California Supreme Court ruled against the proposition in 1966. But when redevelopment began again, so did protest. WACO was instigated by a trio of progressive ministers concerned about future redevelopment, but fundamental to its viability were Freedom House groups, such as the Tenants Union, and its organizers, such as Miller. They brought leadership, structure, and determination. As Hannibal Williams, a resident who became a WACO leader, declared, they were "tired of being buffet[ed] around like chess pieces."[105] Eva Brown, another resident and leader, insisted that "WACO must have the right to guide the hand of the Redevelopment Agency, in order to be sure that the mistakes made in the A-1 area of the Western Addition Project are not repeated in A-2."[106]

Williams, Brown, and the other men and women of WACO had disparate pasts, but they shared residence in and commitment to their district. Williams joined the organization as an elder with Howard Presbyterian, one of the founding churches. Also, he had been active with the Black Student Union while at San Francisco State College; the politicized student union was a partner in United San Francisco Freedom Movement and Freedom House activities and would play a key role in the college's Third World Strike in 1968. Helen O. Little, another outspoken leader and a welfare mother, came to the group through work in welfare rights. Eva and Ken Brown led the group's Homeowners' Association, a holdover from Freedom House. Mary Rogers, another welfare mother and the leader of the Tenants Union, had moved to the Western Addition only in 1965, after all the protests and organizing of 1963 and 1964. But she shared the raised expectations from municipal government that Freedom House had kindled in others. She had lived in a "deplorable" third-floor apartment with holes in the walls and falling hazards for her children. So although she "didn't know much about government and how it operates," she tracked down a host of city inspectors, who pressured her landlord into repairs. Energized by her success and increasingly fluent in civic engagement, she joined others who refused to "be displaced by the Redevelopment Agency."[107] WACO coordinated hundreds of like-minded residents and dozens of groups. By spring of 1967, the organization had over twenty member groups; by the end of the year, over forty.[108]

Activists took the direct-action tactics of Freedom House a step further, in ways that city officials could not ignore. Williams, Little, and others did not wait for public hearings to make their demands known. In one of the group's earliest actions, sixty residents stormed the mayor's office, chanting, "We are going to stop redevelopment! We are going to see the Mayor!"[109] A month later, activists not only picketed a local event featuring Secretary of Housing and Urban Development Robert C. Weaver; they also bought tickets to fill the audience and barrage him with questions. WACO activists also mobilized hundreds for municipal hearings, turning some meetings into twelve-hour marathons. And when building began, members prostrated themselves in front of bulldozers to stop demolitions. These protests kept the Western Addition in the newspapers and provoked a flurry of correspondence between the mayor's office, the Redevelopment Agency, and other city bodies. These actions infuriated Herman. He believed that he had developed good-faith relationships with "leaders of the neighborhood," only for WACO to "accuse us of not having worked with the community."[110] Nor did he appreciate their tactics: "we can't solve problems with mass meetings."[111]

But WACO had a far more diverse toolkit than Herman acknowledged. City officials groused about the group's "technique of generating public anger and frustration," but activists were catholic in their tactics.[112] Some of Freedom House's most effective lessons had been in municipal engagement. Not only had the prior group alerted residents to city resources; its activists had also demonstrated how to use even the most recalcitrant of them. WACO built on those lessons. They initiated negotiations with city agencies, and effectively used the press to amplify their message. Reporters eagerly covered their provocative activities, flocked to the group's press conferences, and ate up their press releases. Additionally, WACO knew where and how to pressure city agencies. After the group boisterously occupied City Hall, members quietly sent Rogers for a private meeting. She acquiesced to a conference with a lower-level official—whose description of Rogers as "pleasant and intelligent" indicated not only condescension but apparent surprise at the decorum of WACO's representative—and asked "if they can expect any help from the Mayor's office" in the absence of a response from Herman.[113] Such activity, along with officials' acknowledgment of the "fairly large number of individuals" the group represented, compelled the mayor's staff to facilitate communication between WACO and the Redevelopment Agency for "constructive meetings."[114]

WACO's mix of direct-action protest, mass mobilization, formal politics, and lawsuits brought results. For example, Herman responded to the group's pressure with a "unique plan": "certificates of priority" for displaced residents to use in redeveloped housing.[115] This measure was limited—after all, they were useless if housing was unaffordable—but activists won a bigger victory when they forced city supervisors to reverse their project approvals. In June 1967, the Board of Supervisors had passed the Redevelopment Agency's annual outline of renewal activities even though the sole dissenting supervisor claimed it "leaves out relocation"; WACO members described it as "totally unacceptable."[116] By October, however, after an entire summer of sustained protest, the board rescinded their trust. In a special five-hour meeting, the supervisors called the current relocation plan "insufficient on its face" and halted all progress.[117] This was a momentous and unprecedented step that also suspended regional and federal approvals. WACO politicking had laid the groundwork for the resolution. Members testified and allied with liberal supervisors such as Jack Morrison, who the group endorsed for mayor. Indeed, the board specifically requested WACO's presence at the hearings, the only non-municipal body issued a special invitation. The raucous disruptions of residents and activists captured the public's attention and facilitated their lobbying in formal municipal politics. Williams, Rogers, the Browns, and other WACO members had worked on multiple fronts.[118]

WACO's victory did not, however, indicate a harmonious or lasting municipal relationship. Mayor Shelley quickly vetoed the legislation.[119] Herman held fast: he had "no plans to alter his Redevelopment Agency's plans for the controversial Western Addition A-2 Project."[120] If anything, he held a special grudge against the group who consistently called for his resignation. Activists made some progress through formal politics, but it was limited with a municipal government generally supportive of urban renewal.

PROTEST, COOPERATION, AND PARTICIPATION IN THE BLACK FILLMORE

WACO's protests and demands were not the only form of black response to redevelopment. There were also those who favored participation. From the first large-scale evictions in 1958, some black San Franciscans had criticized those who were "Yelling and Hollering for civil rights." Instead, they urged their contemporaries to adopt the "quiet" and "specific positive action . . . of our Japanese Neighbors" to create a better neighborhood.[121] A handful of black entrepreneurs and institutions took this to heart. Like the Nihonmachi Community Development Corporation, these developers sought community betterment and, perhaps, personal gain through municipal cooperation. This at times brought them into conflict with WACO, but in the end, WACO too established a lasting form of cooperation with the city. The line between protest and cooperation was blurry, and both could lead to measures of community control and disappointment.

This contrast can be seen in WACO's protest of the Martin Luther King Square residential development, named just after King's assassination and the target of one of the activists' most high-profile actions in 1968. Ulysses J. Montgomery was project coordinator and the "power behind" the sponsoring "non-profit, grass roots" Fillmore Community Development Corporation.[122] Montgomery's group shared a name with its Nihonmachi counterpart, the city's model of successful municipal partnership, and the goal of community oversight through resident ownership and development of land. His group shared far more with the other Black Power–inspired Community Development Corporations emerging around the country by this point. In a departure from the Nihonmachi organization, his project sought out resident protections and, as a middle-class and professional developer, attempted to connect state resources with the disadvantaged in the Western Addition. The square was intended to provide affordable housing and "help black people in the Western Addition achieve a sense of pride by participating in the rehabilitation of their own community" through on-site job training and the promotion of "the culture and heritage of the black people."[123] The group had first unsuccessfully floated the idea for a cooperative housing complex. After securing financing from business and society fixture Mortimer Fleishhacker Jr., who had given up his own project after being "attacked" as a "master of the plantation" by activists, the corporation settled on a rental project.[124] Montgomery's group had to sacrifice financial autonomy and residential ownership, but it secured participation.

Martin Luther King Square was the first residential project built in the second Western Addition project area and was applauded for its black developer. While most of its housing was for moderate incomes, 30 percent were designated for low-income families. This was not enough for WACO, however, who argued that it was "not housing for poor people" and would "set the rent pattern" for subsequent projects.[125] To make their point, Williams and other WACO members demonstrated at the groundbreaking ceremony, attended by the mayor, a federal judge, and two state representatives. Subsequent pickets brought work on the site "to a standstill" (fig. 6.6)[126] Herman grumbled that WACO "found a pressure point that is difficult for the City to accept but tremendously appealing."[127] Some staid groups indeed

FIGURE 6.6 WACO protesters halted construction at the Martin Luther King Square residential project. Protests such as these stalled the square, the first residential development in the Western Addition A-2, for months and resulted in additional low-income housing. BANC PIC 2006.029:140596.04.08-NEG, Fang Family San Francisco Examiner Photograph Archive Negative Files, circa 1930–2000, © The Regents of the University of California, The Bancroft Library, University of California, Berkeley.

supported WACO's opposition, such as the social-service Family Service Agency, who encouraged such projects to mitigate the "frustration in the black community."[128] Montgomery himself took the opportunity to register his own preferences for increased low-income housing. Another setback stemming from increased land costs contributed to another few months' delay, but when the project began, it had WACO's endorsement, more low-income housing, and a black contractor.[129]

This outcome was mixed for Montgomery and the Fillmore Community Development Corporation, both of whom garnered "lavish" praise from Herman at the square's completion a year later.[130] The modern townhouses were comfortable and well-appointed with landscaping and children's playgrounds. Subsidies and public housing meant that the units were available to a range of incomes. Personally, Montgomery used this success to enter a long career in urban development. However, the setbacks stacked the financial decks against the Fillmore Corporation. The group had to accept far higher costs and razor-thin margins in order to increase the number of low-income units. Without the Nihonmachi Community Development Corporation's ready pool of investors and property owners, the Fillmore group bore crippling burdens that led to their default just four years later.[131]

The Martin Luther King Square project was a clear victory for WACO, on the other hand. It demonstrated the group's ability, if not to direct redevelopment, then at least to reshape it in significant ways. Another action had even more consequence, eventually leading to WACO's formal absorption by the Redevelopment Agency as a citizen review board. With the assistance of the San Francisco Neighborhood Legal Assistance Foundation, WACO first filed an administrative complaint with the Department of Housing and Urban Development in 1967.[132] When that failed, the group filed a lawsuit, claiming that the project "will deprive the 15,000 Negroes, Japanese and whites of their right to freedom of association" and that the agency failed to relocate residents into "decent, safe and sanitary housing which they can afford." Herman dismissed the suit as "goofy," but in a "rare legal move," a federal judge ruled in WACO's favor.[133] The district court temporarily halted involuntary relocation and federal funding pending "a satisfactory relocation plan." WACO's lawyers claimed this was the first time a court had issued an injunction on an urban-renewal project. This was the action and the validation WACO leaders had sought: "someone has finally listened to our peaceful plea for justice."[134]

The judge dissolved the injunction four months later when the Department of Housing and Urban Development formally approved the Redevelopment Agency's relocation plans, but the temporary legal victory gave the group a newfound legitimacy in the district and with the agency. Former Western Addition–based critics such as the Urban League and the Baptist Ministers Union joined WACO to form the Western Addition Project Area Committee (WAPAC); WACO activists including Rogers and the Browns held leadership positions while Williams served for a period as chair of both organizations. Within months of the lawsuit in 1969, the agency was forced to accept it as the Western Addition's official Project Area Committee, a new designation for resident participation by federal mandates. Although Herman resented their new power, the designation gave members the ability to review "all plans and procedures . . . undertaken by the Redevelopment Agency in rebuilding the Western Addition."[135] Members now had a say in "the formulation and execution of plans," could "modify the project program" when appropriate, and could "give any project area–based groups preferential consideration." This kind of resident input had long been activists' central demand. As Williams argued, they had not been "trying to stop redevelopment."[136] Instead, as Eva Brown stated, they wanted to make sure it was "by and for the area's residents."[137]

WACO's success was partly timing. It coincided with new federal demands for resident participation, largely motivated by similar activism all over the country. Activists were also buttressed by the explosive nature of urban race relations in the mid- to late 1960s. In 1965, Watts shook with an uprising that rattled the entire country just months after the Voting Rights Act was signed. The following year, San Francisco saw its own Hunters Point burst into revolt. These events clearly haunted officials leery of violence, and activists and their allies obliquely referenced them in their demands. After the mayor vetoed the Board of Supervisors' halt of the redevelopment program, Ken Brown grimly observed, "The Mayor did not realize when he picked up his pen to sign this order that he may also be putting a torch to this city."[138] Williams warned that the redevelopment program could "increase tension and trigger violence."[139] This precipitated a long-delayed meeting between the mayor and Williams, out of fear for public safety. Like officials across the country, San Francisco officers saw such concessions as a form of "riot insurance."[140]

WAPAC's victory was notable but not clear-cut. Its work more than doubled the low-cost and subsidized housing planned for the A-2 project, encouraged community sponsorships of housing developments, and monitored displacement. Activists such as Williams joined the permanent staff in the Redevelopment Agency, bringing with them their concerns and values.

However, significant limitations remained. WAPAC's role in the redevelopment process meant members shouldered some of the agency's most contentious operations, such as affordable, "decent, safe and sanitary housing" for those displaced or "maintain[ing] in a habitable condition any occupied dwelling units owned by the Agency."[141] This allowed the Redevelopment Agency to shed some accountability, while WAPAC had limited power to enforce their decisions; one contemporary scholar viewed WACO's absorption into WAPAC as cooption. Not unlike the victories in Nihonmachi, Western Addition protest secured critical but uneven interventions.

WACO's complexity was also evident in its claimed and actual members. The group varied from presenting itself as an organization of the black, disenfranchised, and poor to one of "all faiths, the poverty people, business groups, homeowners. . . . We represent Caucasians and Negroes and Japanese."[142] Even though the Nihonmachi Community Development Corporation was celebrated at the expense of protest organizations like WACO, the latter in fact included Japanese Americans. One founding minister was David Hawbecker, with the Japanese American Christ United Presbyterian Church. Yori Wada, head of the Buchanan YMCA and formerly with the Booker T. Washington Center, was a member. Fred Y. Hoshiyama, an executive with the local YMCA, sat on the steering committee.[143] In fact, WAPAC included the Nihonmachi Corporation itself. However, the group's roots in the black freedom struggle led to an emphasis on black leaders and the concerns of the most vulnerable in the district. So even at a meeting organized for "all ethnic and civic associations in the area," WACO articulated its purpose as allowing "black people [to] build black housing" and "prevent the Negro ghetto from becoming a police state." This "predominantly Negro" frame was amplified by the media and city officials.[144] Municipal agents, the press, and the resident activists solidified an association between militant protest in the Western Addition with African Americans.

JAPANESE AMERICAN ANTI-EVICTION PROTEST

Black activists, however, were not the only ones in the Western Addition to use direct-action protest to challenge the course of redevelopment in their neighborhoods. Young Japanese American activists organized in the early 1970s to stop the "dispersal and destruction of the Japanese community . . . carried on by the San Francisco Redevelopment Agency."[145] These activists organized the Committee against Nihonmachi Evictions (CANE) to challenge redevelopment in Nihonmachi. Like WACO and Freedom House, the group sought community controls and protections for low-income and minority residents and small businesses.

CANE activists came out of the Asian American movement and were, like other Asian American radicals, nurtured by contemporary political ferment. Some had begun in the antiwar movement, while others developed an "identity of being an Asian" from the Black Power movement.[146] Other activists were politicized by the 1968–1969 Third World strikes by students at San Francisco State College and the University of California, Berkeley, which resulted in the nation's first ethnic studies departments.[147] Japanese American activists therefore came to Nihonmachi politicized. Many were the children of parents who had left Japanesetown for residential districts and the suburbs, but the Asian American movement connected "community activism and campus" for such students.[148] Carole Hayashino, for instance, learned about Japanesetown and its problems from her Asian American studies class. She and others organized in Nihonmachi because "our community, the Japanese community, should control the institutions that affect our lives."[149] Borrowing from the activities of the Black Panthers, activists initially focused on health initiatives, social services, policing, and other concerns.[150] Their attention quickly shifted to redevelopment, which seemed a clear threat to the community they imagined as both ethnic and spatial and which echoed past injustice. During the war, "we were rounded up [en masse] and were 'relocated' into America's prisoner of war camps. In 1958, once again San Francisco Japanese were faced with relocation . . . at the hands of the San Francisco Redevelopment Agency." Protest was an opportunity to "plot our own destiny": "we intend to fight and defend J-Town."[151] Activists drew from the same ideals of self-determination and racial pride that had animated WACO.

These young Japanese American activists "joined together with residents and small businesspeople" in early 1973 as CANE, an "intergenerational and multiracial formation to oppose redevelopment" and "preserve the character and spirit of Nihonmachi."[152] This translated into two goals: (1) "Stop the destruction and dispersal of the Japanese Community" and (2) "Uphold the rights of residents and small businesses."[153] Like their Western Addition predecessors, they relied on militant, direct-action tactics (fig. 6.7). For instance, nine activists chained themselves together and conducted a two-hour sit-in at the Redevelopment Agency office in the Western Addition while "50 Japanese Americans, mostly young," picketed outside.[154] Padlocking doors and marching, CANE members prevented potential developers from viewing possible sites. Activists gathered thousands of petition signa-

FIGURE 6.7 CANE protesters confronting police officers in the activists' attempt to stop the eviction of a Nihonmachi tenant. Young Japanese American activists and local residents focused on halting demolitions of specific buildings and evictions of individuals and families. BANC PIC 2006.029-PIC, folder San Francisco—Japanese—Japan Town (2 of 2), carton L086, Fang Family San Francisco Examiner Photograph Archive Photographic Print Files, circa 1874–2000, © The Regents of the University of California, The Bancroft Library, University of California, Berkeley.

tures to halt particular developments and staged a rally at a demolition site to protest declining affordable housing. And like Western Addition residents before them, they disrupted municipal meetings and drowned out city representatives.[155]

Such activities brought CANE into opposition with co-ethnic developers, as happened with WACO years earlier. Activists' demands to maintain the Nihonmachi community meant that they theoretically supported the efforts of Japanese American developers as part of "the just struggle of the surviving owners and small developers to remain within Nihonmachi."[156] For example, activists supported the Japanese American Religious Federation's low-income housing development.[157] Relations with the Nihonmachi Community Development Corporation, however, were more contentious. According to activists, the corporation represented "only a handful of businessmen" and was a "dependent accomplice" of the Redevelopment Agency, supporting its "plan for tourism in Nihonmachi."[158] Furthermore, the corporation counted Kintetsu, an "outside developer from Japan," among its dues-paying, property-owning members. This allowed the company to become "the dominant financial interest in San Francisco Japantown," with a bowling alley in the Nihonmachi project area and, supported by the corporation's advocacy, a second hotel immediately adjacent to the project's boundaries.[159]

As Kintetsu's growing presence suggests, CANE could not stop evictions outright. In practice the demands of activists and residents "to stay" moderated into limited and achievable goals. This was partly due to timetables. The student activists were young and recently politicized, so CANE emerged years after the second Western Addition project had passed local and federal approvals. Protests took place not at Board of Supervisors or Redevelopment Agency hearings but at the Board of Permit Appeals; the focus of struggle shifted from overarching plans to the permits, evictions, and demolition of specific buildings. Still, CANE could be a useful last resort. As one small business owner recalled, "It seemed they were the only ones that really wanted to help us back then."[160] Particularly painful for residents and enraging to activists was the juxtaposition of rapid evictions and slow development, which moved residents off property that might stand vacant for years. CANE insisted that "tenants should never be moved out of their homes until completed blueprints and plans for immediate rebuilding is definite," nor should they be "forced to move out of the community."[161] These moderated demands allowed them to claim successes, and indeed to shift the impact of eviction for residents. One resident remembered that CANE's advocacy "prolonged our stay [in their apartment] for another year or two" before his family "was forced to move."[162] This was limited, to be sure, but translated to a savings of time and resources that could be significant for a working-class family. This redefined "success" into actions that took residents into account and softened the blow of redevelopment for those unwilling or unable to participate in redevelopment themselves.

CANE's activism thereby challenged the Nihonmachi Community Development Corporation's portrayal of Japanese American cooperation. In doing so, it fought the now-established stereotype of Asian Americans as, in the words of one activist, "quiet, passive, hard-working, polite, etc.," a framing that prevented Japanese Americans from "defining and controlling ourselves and our community."[163] Their struggle was therefore about more than a handful of buildings: they also sought to redefine Japanese American identity more broadly. First, activists self-consciously adopted the militant protest tactics associated with African American activists, an engagement that had been formative to the Asian American movement as a whole. They also supported cross-racial alliances and a broad definition of Nihonmachi that included non-Japanese membership. While activists saw Nihonmachi, as the name suggests, as definitely a "Japanese community," officers included African American resident Helen Jones and grocery-store owner Freddie Powell, as well as the Wongs, Chinese American bait-shop owners; the group likewise worked on behalf of any displaced resident or business that sought their assistance. Furthermore, while their numbers were small, two hundred at CANE's height, activists pointedly included renters, small non-touristic businesses, and the poor in their group: those largely excluded from the Nihonmachi project and ignored by the prevalent model-minority stereotype. For example, working-class renters, such as longshoreman Charles Toyooka and barber Mr. Abbey, were active members of CANE. Finally, they fought Nihonmachi's conflation between Japanese Americans and Japan. They argued that Kintetsu and other Japanese corporations were an "outside" and destructive force that corroded the integrity and self-determination of the enclave. At the same time, though, activists could not entirely escape Nihonmachi's racialized formulations. Not only did CANE incorporate "Nihonmachi" into its name; it participated in activities like the Cherry Blossom Festival, which activists referred to by its Japanese name, "Sakura Matsuri." Activists opposed racial formulations expressed by the Nihonmachi Corporation, but they sustained an identity built through years of association between Japanese Americans and Japan.[164]

CANE was tightly focused on Nihonmachi, but activists' critique encompassed a broader terrain. The group built not just on predecessors in the Western Addition, but also on groups such as the Mission Council on Redevelopment, which in 1966 prevented renewal of the increasingly Latino Mission district. They also allied with anti-redevelopment groups across the city: "In Chinatown, in South of Market, all over the city . . . people are rising up against the powerful forces trying to drive them out through the redevelopment process."[165] CANE activists participated in the demonstrations at International Hotel, a Chinatown residential hotel slated for demolition, occupied mostly by elderly and poor Chinese and Filipino bachelors. Peter Mendelsohn, the former merchant seaman and labor organizer who led Tenants and Owners in Opposition to Redevelopment in the South of Market, spoke in CANE's support at municipal hearings. CANE activists focused like a laser on the most localized of sites—lots, buildings, apartments, and individual homes—but this connected them to the "broader struggle" going on around the city and, indeed, "across the nation."[166]

CHINATOWN AND THE POSTWAR
GATEWAY TO THE PACIFIC

Chinatown was among the many San Francisco neighborhoods slated for a redevelopment project in the 1960s and 1970s. The neighborhood's long-standing symbolism for the city's cosmopolitan, Pacific-oriented identity had made it a touchstone for the Japanese Center and for Nihonmachi, and also marked it as another potential manifestation of the city's Gateway to the Pacific ambitions. However, transformations abroad and in Chinatown produced a cramped and compromised Chinese Culture Center. The project's trajectory highlighted Chinatown's limited utility to the city's Gateway to the Pacific identity and Nihonmachi and Japanese Americans' relative advantages.

The Japanese Cultural and Trade Center was viewed as an inspiration and a "challenge" by some in Chinatown. Dai-Ming Lee, the editor of the bilingual *Chinese World*, saw it as possible competition but also an "excellent example of imaginative planning" that Chinese Americans "might apply to their own circumstances."[167] A group of residents and business owners took this to heart in the early 1960s. J. K. Choy, Chinatown bank branch manager and officer as well as a long-time ally of Lee, led the effort. Born in 1892 in Hawai'i, Choy had served with the nationalist Kuomintang but, disillusioned with its leader, Chiang Kai-shek, moved to San Francisco where he lobbied US officials in support of a "Third Force" alternative to both the Communists and the Nationalists. Choy's politics distanced him from Chinatown's established institutions such as the ardently National-ist Chinese Six Companies, but his wealth and influence in the statewide Democratic Party made him a community leader. By 1964, his group had raised money toward a proposed Chinese cultural and trade center adjacent to Chinatown.[168]

The Chinese center project was viewed as "similar to the Japanese Cultural Center" by the Board of Supervisors, and in many ways it was.[169] Choy's group "[took] a leaf from the Japanese cultural center" to mix commerce with culture in a building with "contemporary Chinese character."[170] Choy even considered the Japanese Center's financing models.[171] Finally, Choy argued that "San Francisco's large Chinese community should be given at least as much opportunity and cooperation to help develop its plans as the Japanese community."[172]

But while the Chinese center was accepted as a "fitting and desirable addition to the cultural resources of San Francisco," it had crucial and hobbling differences from its Japanese counterpart.[173] The project was embedded in troubled transpacific Cold War politics. A Taiwanese architect designed part of the Chinese development, while the newly designated sister-city of Taipei donated a cache of artifacts that promised to, with the Brundage collection, "make San Francisco rank first outside of Asia in Oriental art."[174] Still, organizers were compelled to reject the Republic of China's request to lease an information office, in order to leave "no doubt that the organization is not the agent of any foreign government" and ensure that "cultural groups, both from Taiwan and the Chinese mainland will be welcome."[175] Economics were at least as problematic as politics. Choy had argued that a renewed Chinatown would be a target for "new capital . . . from wealthy Chinese businessmen abroad."[176] This was a far cry from Japan's booming industries, and even that fell through. The Chinese Culture Center instead was forced to rely on domestic donations and loans.[177] As a result, Choy's original vision for a Chinese cultural and trade center shrank to a cultural center occupying one floor of a privately held hotel. Chinese Americans, one local reporter summarized, had never before been "screwed on so grand an architectural scale": "the new 'Chinese Cultural [sic] Center' . . . s' help me — is really a Holiday Inn."[178]

Organizers focused on their project's "truly Chinese-American" origins in "the largest Chinese community outside of Asia," rather than troublesome overseas connections, but even these were complicated.[179] Changing perceptions of and conditions among Chinese Americans caused the organizers and city officials to view the Chinese Culture Center less as a contribution to the municipality than as vital community assistance. In part, it was intended to provide "a feeling of identity" for Chinese Americans "right within their own neighborhood."[180] More urgently, it was a necessary "shot of assistance" for the neighborhood's "housing, social and economic problems."[181] Even Choy argued that the center was a necessary intervention into the "social and physical jungle" of Chinatown.[182] The rosy, successful view of Chinatown had clearly dimmed by the late 1960s.

Increasingly dark portrayals of Chinatown, ironically, reflected the rising fortunes of many Chinese Americans as well as rising post-1965 migration. Second-generation Chinese Americans, like Japanese Americans, were able to leave Chinatown for new neighborhoods. Furthermore, while thousands of new Chinese migrants arrived in the 1950s and 1960s, they came through student-exchange or refugee programs that prioritized skills and likely self-sufficiency and so largely bypassed the enclave. This distilled Chinatown's population to the "the elderly, the timid, the very poor, and that largest segment, the non-English speaking."[183] The provisions of the

1965 Immigration Act and its enlarged quota added to this population with more low-skilled, poor entrants who were "virtually forced to fit [themselves] into the already overcrowded confines of Chinatown."[184]

The poverty and limited English of these new arrivals encouraged San Franciscan observers to resurrect stories of Chinatown as a problem. Journalists and officials once again began to describe the neighborhood as "America's oldest continuous ghetto," whose long-standing problems were "compounded by increased immigration."[185] By the Planning Department's estimate, over 9,000 Chinese migrants arrived in San Francisco in the past decade and officials predicted another 1,100–1,400 every future year.[186] This migration dwarfed that of other sending countries and included "many" migrants with "low or non-existent income."[187] While the planning for the Chinese Culture Center was under way, Mayor Alioto "carried a special plea for federal assistance for Chinatown directly to the White House," arguing that "the City's resources are already strained" without "the needs of the thousands of immigrants who arrive annually in San Francisco from Hong Kong and Taiwan."[188] Certainly, his meetings with a number of federal officials indicated a familiar urban lobby for federal funds, especially as the newly inaugurated president Richard Nixon threatened the largess of Great Society programs. Yet Alioto's plea suggested the salience of the nationally known neighborhood's needs—the *Atlantic* described Chinatown as a "refugee camp"—for federal monies.[189] Compared to Nihonmachi, Chinatown perhaps had more transpacific connections by the late twentieth century, but they were less useful for the city's transpacific urbanism.

Japanese Americans were an easy group to celebrate in comparison to other Asian Americans. San Francisco's second-largest Asian American group, Filipinos, had also suggested a "Philippine Cultural and Trade Center," but their project did not even make it to the proposal stage. Filipinos were even less of a "model" minority than Chinese and represented a nation perceived as having little to offer the city in either economic or geopolitical terms. Furthermore, postwar migration from Japan was the smallest of any Asian country. The most numerous and visible were Japanese women who arrived as the wives of, most frequently, white US servicemen stationed in Japan. These women tended to settle with their husbands outside of Japanesetown, and there were thus few new arrivals to refresh Nihonmachi's population or even to add to the Japanese American population as a whole. The Nihonmachi project thus memorialized a declining ethnic enclave, one less troubling to San Francisco's professed racial liberalism and without relation to troublesome geopolitical conflicts.[190]

CONCLUSION

This story of transpacific San Francisco ends with one of the most frequently told stories of postwar space and race: that of urban redevelopment and its critics. African and Japanese Americans in San Francisco were
not the only ones to debate redevelopment or to view it as a crucial forum
for community development and control. Chinese Americans also tried to
use redevelopment to reshape Chinatown in ways that reflected community formations, while Latinos and others in the Mission district had managed to prevent it for the same reason. And of course, all over the country,
redevelopment programs had similarly unintended, scarring consequences
that residents and their allies protested or participated in on terms similar to those seen here. This commonality with national patterns, however,
helps clarify the singularity of San Francisco's transpacific aims. Japanese
and African Americans at times used divergent methods, but both communities struggled toward the intimate, local goals of preserving homes and
businesses and communities. These battles were fought building by building, block by block. Yet the war was waged for a much larger field. Both
Japanese and African American representatives, activists, and organizers
challenged redevelopment and their neighborhoods' futures, but they were
more fundamentally seeking a place in a city that was in turn struggling to
redefine itself in a transpacific economy and politics.

CONCLUSION

In the years following the Pacific War, a variety of San Franciscans transformed their city's Gateway to the Pacific moniker into a transpacific urbanity that reshaped the city. Japanese Americans and the twin institutions
of the Japanese Cultural and Trade Center and Nihonmachi served as the
most desirable evidence of San Francisco's postwar transpacific links and its
Pacific identity. But these were embedded in a much larger environment of
Japanese connections, institutions, artifacts, and ideas. This postwar prominence of Japan reflected transformations in the city's racial terrain as well as
shifts in possibilities and problems across the Pacific. New transpacific dynamics and new postwar tools at the local, state, federal, and international
level provided both innovative means of connection and revised conceptions of how business, government, and people could or should engage phe-
nomena outside national borders. These connections and ideas helped turn
San Francisco's longtime identity as a Pacific metropolis into a transpacific
urbanity expressed in the city's built environment, politics, economy, civic
life, and racial terrain.

San Francisco's distinct Cold War–era urbanity took an expansive form.
Of course, it guided the development of the Japanese Cultural and Trade
Center, the preeminent institutionalization of the city's transpacific ties.
But it also oriented the municipal government, guiding the travels of mayors and shaping the collections in the city's art museums. City government
hosted visiting officials and executives, for which they learned the structures of Japanese municipalities and some Japanese formalities for polite,
effective encounters. Furthermore, San Francisco's identity relied on more

than just bureaucrats and officials. The flourishing of popular and well-regarded Japanese restaurants shifted attention from the European-derived, often French slant of San Francisco eating and made the city a US center of Japanese cuisine. The breadth of these Japanese links and the wide participation of a range of San Franciscans breathed life into the city's Gateway to the Pacific identity.

The development of jet-age travel amplified the city's status as a transpacific hub and carved deeper ruts in the many routes crossing the Pacific Ocean. San Francisco officials, in their attempt to solidify their own city's status, leveraged opportunities and resources in other parts of the Pacific region. This meant, of course, Japan, but Hawai'i was another site that tethered boosters' ambitions. M. Justin Herman and the Redevelopment Agency took advantage of Masayuki Tokioka's financial and social resources, fostered in the territory's unique conditions, in order to build the Japanese Center. This partnership traversed well-worn paths between the city and the islands, and it created new economic and political networks between the new state and the mainland. The "Pacific" in the Gateway to the Pacific formulation was a reformulation that evolved along with needs, resources, and opportunities.

San Francisco's transpacific urbanity also reshaped the city's racial terrain, framing Japanese Americans as a "model" for other people of color in San Francisco. To be sure, this Japanese American racial rearticulation did not rely entirely on the transpacific for expression. Historians have created a rich body of work in recent years that has documented the many domestic factors informing newfound socioeconomic opportunity for the group. Yet this pattern could not have emerged without their associations with Japan, particularly in San Francisco. There, the city's fascination with Japan and the many links developed with the nation built on and furthered Japanese American status. These commercial, cultural, and social developments opened doors for the once-maligned minority and cast a newly positive light on their association with a foreign nation.

Japanese Americans therefore assumed a hitherto unimaginable role in civic life by furthering San Francisco's transpacific urbanity. But they were not alone in their task. City boosters, local politicians, large and small business owners, neighborhood residents, activists, and others fought and disagreed with each other, as well as allied in shifting configurations. But their actions and ideas collectively helped recreate San Francisco's urbanism. This framework was by no means consistent in either intent or in outcome, as goalposts moved and intentions changed. At times, Chinese Americans served as vibrant models of success and assimilation, but by the early 1970s, the successful professional living in the suburbs was overshadowed by the figure of the poor, non-English-speaking, unskilled immigrant huddled in Chinatown. In contrast, Japanese Americans moved from a distinct second place—disliked but little considered, overshadowed by the image of the noxious Chinese—to valued civic participants. This move not coincidentally occurred as another racialized group moved to the center of civic fears, not only of poverty and social degradation but also of violence and unrest. African Americans were not explicitly a part of the city's transpacific urbanism, which focused most steadily on real or imagined connections to the Orient. But they were a part of it nonetheless, as their freedom struggle and everyday practices helped redefine its stakes and consequences. And racialized groupings were just one of the many ways in which people identified and worked. Amateur and budding developers worked in similar terms, while business executives shared ideas about urban growth and activists fought in numerous ways for community control.

The transpacific urbanism that took root in San Francisco was not limited to the city. San Francisco boosters had taken advantage of the tools available to all cities at the time—People-to-People affiliations, redevelopment programs, private investment, and self-promotion—but they did so early on and with novel intent. They affiliated with Osaka in the first year of the sister-city program, incorporated a transpacific center into the city's first redevelopment project, sourced international capital in the earliest years of postwar Japanese overseas direct investment. Their efforts shaded into the surrounding Bay Area, as well. As we have seen, Oakland cultivated its own sister-city a few years after San Francisco and rehabilitated its port for greater Pacific trade, while advancing select Japanese Americans into visible seats of municipal power. To the south, Silicon Valley would develop its own transpacific urbanism, as urbanist Shenglin Chang has shown, connected more to Taiwan than to Japan, but with a similarly intimate scale of transpacific shared ideas and uses of space that transformed homes and neighborhoods on both sides of the Pacific.[1] The work of San Francisco boosters to remake their city along transpacific lines was part of a larger shift across the entire Bay Area.

Transpacific urbanism reshaped all of the West Coast in varying ways. Seattle, San Diego, Los Angeles, and other cities formed their own sister-city affiliations with Japanese municipalities. The Japan-American Conference of Mayors and Chamber of Commerce Presidents, which contained mostly West Coast cities, regularly brought together municipal officials from across the Pacific in scattered cities in the United States and Japan. While these developments were indigenous, Los Angeles, San Francisco's most heated rival, borrowed directly from its northern peer at times. Japanese American merchants in Little Tokyo observed their Northern California counterparts and initiated their own redevelopment project, also with Japanese capital, modern design, and tourists in mind; Asian American movement activists in turn convened with Berkeley and San Francisco peers in movement building and community work.[2] Transpacific urbanism was perhaps more pervasive in San Francisco than in other cities along the coast, but it was one example among many.

Indeed, all over the country, localities and businesses sought their own paths toward participation with Japan's high-speed growth. In this, San Francisco was years ahead of the United States as a whole. The city put Japan at the front of its international relations, recognizing the nation as an equal, desirable, and formidable partner well before the Kennedy administration did. And it was not until 1966 that Lyndon Johnson declared a "Pacific Era," albeit a late articulation of a policy already evidenced by ongoing and recent US wars. Relations with and ideas about Asia might have been distinctly unequal in the first postwar decades, but they increasingly shaped the contours of US power in the second half of the twentieth century.

San Francisco's transpacific urbanity helps us better understand the postwar US urban West, but it also provides ways of historicizing current urban conditions. Now cities regularly have foreign-relations agendas, and mayors travel the world to trade ideas about municipal programs and to promote their cities. As I write, mayors and city councils across the country are asserting their own position in international relations, by denying or welcoming Syrian refugees and by crafting "sanctuary" policies of cooperation with or independence from federal immigration policies and agents. Many of the tools they use for such statements were honed decades before, and they declared positions not only in support of federal policies but in contradistinction to them. As scholars of the United States in the world will tell us, such efforts are, of course, not new. The constitutional diction that international relations are the domain only of the federal government has always been historical fiction. As long as people and goods have moved across national borders, localities and individuals have crafted, if not intentionally or explicitly, their own set of international relations. Cities, including San Francisco, created their own immigration policies in relation to Japan in the early twentieth century. And of course, migrants crafted their own foreign relations that reshaped both municipal mechanisms and national policy.

The urban sphere, which historians are only just beginning to move into the transnational frame, has much to tell us about how the international works in our everyday lives. It is significant, I would argue, that the most vocal and politicized expression of support for refugees and denials of federal immigration policy have occurred not in state assemblies, county seats, or among cross-state associations, but in city halls. This is where the rubber hits the road for those of us in the United States. These are our communities, the most lived experience of citizenship or residence, where even in the smallest cities we rub elbows with people from around the world, and obtain goods from all over, and where people, things, and ideas move in and out of nations.

The creation of San Francisco and the West Coast's transpacific urbanism, of course, does not analytically bring us all the way to early-twenty-first-century sanctuary cities. But it can help us better understand how municipal governments and urban dwellers—in an age of more rapid and affordable transoceanic travel, of Cold War emphases on international connections and stakes, and of rising international direct investment—came to view themselves as direct international actors in new ways. The choices made in San Francisco were perhaps early in their conscious and expansive transpacific intent, but they joined places like New York City and London, whose boosters also envisioned their city's identity as global. While they may not have been consciously trying to do so, entrepreneurs, restaurant patrons, festivalgoers, and school children all helped give life to San Francisco's Gateway to the Pacific identity. As African American and Japanese American resident activists in the Western Addition showed, not everyone valued their city's international identity, and indeed, some struggled to retain sovereignty in the workplaces, homes, and identities they saw as threatened. The local, the neighborhood, the block: this was the scale at which grassroots organizations and activists could work, but it also gained new meaning in the distant yet looming presence of overseas interests. Still, even their oppositional work shows the extent to which people engaged their communities in increasingly worldly terms. Events and decisions abroad reshaped the most intimate and localized spheres of life, not just in terms of abstract Cold War threats or wars abroad, but the very streets people walked on, the buildings they slept in, and the neighbors they encountered on a daily basis.

Notes

INTRODUCTION

1. "Far Eastern Treasures," *San Francisco Chronicle*, July 19, 1959.

2. Biographical information of the Takahashis is drawn from Tomoye Takahashi, interview with the author and Judy Hamaguchi, July 16, 2004, San Francisco, CA, transcript, "Back in the Day" Collection, National Japanese American Historical Society, San Francisco, CA; Tomoye Takahashi, interview with Sandra Taylor, August 29, 1989, San Francisco, CA, transcript, Topaz Oral History Project Collection, Accn 1002, Special Collections and Archives, J. Williard Marriott Library, University of Utah, Salt Lake City; Sandra C. Taylor, *Jewel of the Desert: Japanese American Internment at Topaz* (Berkeley: University of California Press, 1993), 41, 235–36, 276, 277; "Philanthropist Martha Suzuki Dies at 90," *Rafu.com*, February 29, 2012, http://www.rafu.com/2012/02/philanthropist-martha-suzuki-dies-at-90/; Henri Hiroyuki Takahashi obituary, *SFGate.com*, April 18, 2002, http://www.sfgate.com/news/article/TAKAHASHI-Henri-Hiroyuki-2850141.php.

3. George Christopher to the Board of Supervisors of the City and County of San Francisco, October 6, 1958, folder 28, box 7, series 5, George Christopher Papers (SFH 7), San Francisco History Center, San Francisco Public Library (hereafter SFPL).

4. "Chamber Group Reports on 'Unequalled' Japanese Boom," *Bay Region Business*, August 3, 1956, 2; Aaron Forsberg, *America and the Japanese Miracle: The Cold War Context of Japan's Postwar Economic Revival, 1950–1960* (Chapel Hill: University of North Carolina Press, 2000), 199; Walter LaFeber, *The Clash: U.S.-Japanese Relations throughout History* (New York: W. W. Norton, 1997), 327.

5. I thank Becky Nicolaides, who suggested the framing term "transpacific urbanity." Any failings in its interpretive utility, of course, lie solely with me.

6. Louis Wirth, "Urbanism as a Way of Life," *American Journal of Sociology* 44 (1938): 7.

7. David Eltis, "Atlantic History in Global Perspective," *Itinerario* 23 (1999): 141.

8. For work through the early twentieth century, see Kornel Chang, *Pacific Connections: The Making of the U.S. Canadian Borderlands* (Berkeley: University of California Press, 2012); Epeli Hau'ofa, *We Are the Ocean: Selected Works* (Honolulu: University of Hawai'i Press, 2008); Jean Heffer, *The United States and the Pacific: History of a Frontier*, trans. W. Donald Wilson (Notre Dame, IN: University of Notre Dame Press, 2002); Madeline Y. Hsu, *Dreaming of Gold, Dreaming of Home: Transnationalism and Migration between the United States and South China, 1882–1943* (Stanford, CA: Stanford University Press, 2000); David Igler, *The Great Ocean: Pacific Worlds from*

Captain Cook to the Gold Rush (New York: Oxford University Press, 2013); Marilyn Lake and Henry Reynolds, *Drawing the Global Colour Line: White Men's Countries and the International Challenge of Racial Equality* (New York: Cambridge University Press, 2008); Shelley Sang-Hee Lee, *Claiming the Oriental Gateway: Prewar Seattle and Japanese America* (Philadelphia, PA: Temple University Press, 2011); Gary Okihiro, *Island World: A History of Hawai'i and the United States* (Berkeley: University of California Press, 2008); Elizabeth Sinn, *Pacific Crossing: California Gold, Chinese Migration, and the Making of Hong Kong* (Hong Kong: Hong Kong University Press, 2013). For work that looks at the latter half of the twentieth century, much of it from nonhistorical fields, see Shenglin Chang, *The Global Silicon Valley Home: Lives and Landscapes within Taiwanese American Trans-Pacific Culture* (Stanford, CA: Stanford University Press, 2006); Bruce Cumings, *Dominion from Sea to Sea: Pacific Ascendency and American Power* (New Haven, CT: Yale University Press, 2010); Arif Dirlik, ed., *What Is in a Rim? Critical Perspectives on the Pacific Region Idea* (1993; repr., Lanham, MD: Rowman & Littlefield, 1998); Matt K. Mastuda, *Pacific Worlds: A History of Seas, Peoples, and Cultures* (New York: Cambridge University Press, 2012); Rob Wilson, *Reimagining the American Pacific: From* South Pacific *to* Bamboo Ridge *and Beyond* (Durham, NC: Duke University Press, 2000); Henry Yu, "Los Angeles and American Studies in a Pacific World of Migrations," *American Quarterly* 56 (2004): 531-43.

9. LaFeber, *Clash*, 256.

10. Igler, *Great Ocean*, 11.

11. Cumings, *Dominion from Sea to Sea*, xiii. An earlier essay by Cumings also influenced my thinking of the Pacific as an idea: Bruce Cumings, "Rimspeak; or, The Discourse of the 'Pacific Rim,'" in Dirlik, *What Is in a Rim?*, 29-47. Igler also pointed out the need to clarify "whose Pacific?" for analytical precision and to avoid "elid[ing] native histories and reif[ying] imperial agendas." Igler, *Great Ocean*, 11.

12. Lyndon B. Johnson, "Speech on U.S. Foreign Policy in Asia," July 12, 1966, Miller Center, http://millercenter.org/president/lbjohnson/speeches/speech-4038.

13. E. Guy Talbott, "The Pacific Era," *Overland Monthly and Out West Magazine*, September 1933, 117.

14. For the mutual relationship between US policy and Japan's economic growth, see, for example, William S. Borden, *The Pacific Alliance: United States Foreign Economic Policy and Japanese Trade Recovery, 1947-1955* (Madison: University of Wisconsin Press, 1984); Forsberg, *America and the Japanese Miracle*; Chalmers Johnson, *MITI and the Japanese Miracle: The Growth of Industrial Policy, 1925-1975* (Stanford, CA: Stanford University Press, 1982); Michael Schaller, *The American Occupation of Japan: The Origins of the Cold War in Asia* (New York: Oxford University Press, 1985); Howard Schonberger, *Aftermath of War: Americans and the Remaking of Japan, 1945-1952* (Kent, OH: Kent State University Press, 1989); Sayuri Shimizu, *Creating People of Plenty: The United States and Japan's Economic Alternatives, 1950-1960* (Kent, OH: Kent State University Press, 2001); Yoneyuki Sugita, *Pitfall or Panacea: The Irony of US Power in Occupied Japan* (New York: Routledge, 2003).

15. Rob Wilson, *Reimagining the American Pacific*, 29.

16. The few transnational urban or local histories include S. Chang, *Global Silicon Valley Home*; Jeffrey A. Engel, ed., *Local Consequences of the Global Cold War* (Stanford, CA: Stanford University Press, 2008); Jesse Hoffnung-Garskof, *A Tale of Two Cities: Santo Domingo and New York after 1950* (Princeton, NJ: Princeton University Press, 2010); Nancy H. Kwak, *A World of Homeowners: American Power and the Politics of Housing Aid* (Chicago: University of Chicago Press, 2015); A. K. Sandoval-Strausz, "Latino Landscapes: Postwar Cities and the Transnational Origins of a New Urban America," *Journal of American History* 101 (2014): 804-31; and Sinn, *Pacific Crossing*.

17. Charles Tilly, "What Good Is Urban History?," *Journal of Urban History* 22 (1996): 702-19.

18. US Bureau of the Census, *16th Census of the Population, 1940: Population: Characteristics of the Population, Part I* (Washington, DC: GPO, 1942), 542; US Bureau of the Census, *17th Cen-*

sus of the Population, 1950: Characteristics of the Population, California (Washington, DC: GPO, 1952), 103.

19. The canonical works on urban renewal and racial change include Arnold R. Hirsch, *Making the Second Ghetto: Race and Housing in Chicago, 1940–1960* (Chicago: University of Chicago Press, 1983); Robert O. Self, *American Babylon: Race and the Struggle for Postwar Oakland* (Princeton, NJ: Princeton University Press, 2003); Thomas J. Sugrue, *The Origins of the Urban Crisis: Race and Inequality in Postwar Detroit* (Princeton, NJ: Princeton University Press, 1996); and June Manning Thomas, *Redevelopment and Race: Planning a Finer City in Postwar Detroit* (Baltimore: Johns Hopkins University Press, 1997). Other interpretations include Andrew R. Highsmith, *Demolition Means Progress: Flint, Michigan, and the Fate of the American Metropolis* (Chicago: University of Chicago Press, 2015); David Schuyler, *A City Transformed: Redevelopment, Race, and Suburbanization in Lancaster, Pennsylvania, 1940–1980* (University Park, PA: Penn State University Press, 2002); and Jon C. Teaford, *The Rough Road to Renaissance: Urban Revitalization in America, 1940–1985* (Baltimore: Johns Hopkins University Press, 1990). For a fresh interpretation of renewal in light of Cold War ideology, see Samuel Zipp, *Manhattan Projects: The Rise and Fall of Urban Renewal in Cold War New York* (New York: Oxford University Press, 2010). For a compelling transnational view of renewal's ideas and aesthetics, see Christopher Klemek, *The Transatlantic Collapse of Urban Renewal: Postwar Urbanism from New York to Berlin* (Chicago: University of Chicago Press, 2011). For a larger discussion of the intertwined dynamics of urban and racial change, see, for example, Kevin M. Kruse, *White Flight: Atlanta and the Making of Modern Conservatism* (Princeton, NJ: Princeton University Press, 2007); Matthew Lassiter, *The Silent Majority: Suburban Politics in the Sunbelt South* (Princeton, NJ: Princeton University Press, 2007); Beryl Satter, *Family Properties: How the Struggle over Race and Real Estate Transformed Chicago and Urban America* (New York: Metropolitan Books, 2009); and Amanda I. Seligman, *Block by Block: Neighborhoods and Public Policy on Chicago's West Side* (Chicago: University of Chicago Press, 2005).

20. See, for example, Mark Brilliant, *The Color of America Has Changed: How Racial Diversity Shaped Civil Rights Reform in California, 1941–1978* (New York: Oxford University Press, 2010); Wendy Cheng, *The Changs Next Door to the Díazes: Remapping Race in Suburban California* (Minneapolis: University of Minnesota Press, 2013); Scott Kurashige, *The Shifting Grounds of Race: Black and Japanese Americans in the Making of Multiethnic Los Angeles* (Princeton, NJ: Princeton University Press, 2008); Natalia Molina, *Fit to Be Citizens? Public Health and Race in Los Angeles, 1879–1939* (Berkeley: University of California Press, 2006); Quintard Taylor Jr., *In Search of the Racial Frontier: African Americans in the American West, 1528–1990* (New York: W. W. Norton, 1998); and Allison Varzally, *Making a Non-White America: Californians Coloring outside Ethnic Lines, 1925–1955* (Berkeley: University of California Press, 2008).

21. With the exception of the literature on historical preservation, Asian American history is only just beginning to engage the built environment. See, for example, Gail Lee Dubrow with Donna Graves, *Sento at Sixth and Main: Preserving Landmarks of Japanese American Heritage* (Seattle: Seattle Arts Commission, 2002); Hillary Jenks, "The Politics of Preservation: Power, Memory, and Identity in Los Angeles's Little Tokyo," in *Cultural Landscapes: Balancing Nature and Heritage in Preservation Practice*, ed. Richard W. Longstreth (Minneapolis: University of Minneapolis Press, 2008), 35–54; Willow Lung-Amam, *Trespassers?: Asian Americans and the Battle for Suburbia* (Oakland: University of California Press, 2017); and Becky Nicolaides and James Zarsadiaz, "Design Assimilation in Suburbia: Asian Americans, Built Landscapes, and Suburban Advantage in Los Angeles's San Gabriel Valley since 1970," *Journal of Urban History* 43 (2017): 332–71.

22. Sandoval-Strausz, "Latino Landscapes."

23. Ibid., 807. Notable exceptions are Charlotte Brooks and Scott Kurashige, although the built environment plays a minor role in his story. Charlotte Brooks, *Alien Neighbors, Foreign Friends: Asian Americans, Housing, and the Transformation of Urban California* (Chicago: University of Chicago Press, 2009; Scott Kurashige, *Shifting Grounds of Race*.

24. Ellen D. Wu, *The Color of Success: Asian Americans and the Origins of the Model Minority* (Princeton, NJ: Princeton University Press, 2014).

25. Mae Ngai, *Impossible Subjects: Illegal Aliens and the Making of Modern America* (Princeton, NJ: Princeton University Press, 2004), 8.

26. Eiichiro Azuma, "Race, Citizenship, and the 'Science of Chick Sexing': The Politics of Racial Identity among Japanese Americans," *Pacific Historical Review* 78 (2009): 242–75; Brooks, *Alien Neighbors, Foreign Friends*; Cindy I-Fen Cheng, *Citizens of Asian America: Democracy and Race during the Cold War* (New York: New York University Press, 2013); Madeline Y. Hsu, *The Good Immigrants: How the Yellow Peril Became the Model Minority* (Princeton, NJ: Princeton University Press, 2015); Michael K. Masatsugu, "'Beyond This World of Transiency and Impermanence': Japanese Americans, Dharma Bums, and the Making of American Buddhism during the Early Cold War Years," *Pacific Historical Review* 77 (2008): 423–51; Arissa H. Oh, *To Save the Children of Korea: The Cold War Origins of International Adoption* (Stanford, CA: Stanford University Press, 2015); Wu, *Color of Success*; Chiou-Ling Yeh, *Making an American Festival: Chinese New Year in San Francisco's Chinatown* (Berkeley: University of California Press, 2008); Mary L. Dudziak, *Cold War Civil Rights: Race and the Image of American Democracy* (Princeton, NJ: Princeton University Press, 2000); Christina Klein, *Cold War Orientalism: Asia in the Middlebrow Imagination, 1945–1961* (Berkeley: University of California Press, 2003).

27. Michi Weglyn, *Years of Infamy: The Untold Story of America's Concentration Camps* (New York: Morrow Quill Paperbacks, 1976), 201.

CHAPTER ONE

1. Panama Pacific International Exposition (PPIE) Company, *Panama-Pacific International Exposition, San Francisco 1915* (San Francisco: PPIE, 1914), 2.

2. North American Press Association, *Standard Guide to San Francisco and the Panama-Pacific International Exposition* (San Francisco: North American Press Association, 1913), 21; PPIE Company, *Panama-Pacific International Exposition*, 6.

3. "World's Fair for This City," *San Francisco Chronicle*, November 25, 1906.

4. Ibid.; "Panama-Pacific International Exposition," February 9, 1911, 61st Cong., 3rd sess., S. Rept. 1133, 2–3.

5. Abigail M. Markwyn, *Empress San Francisco: The Pacific Rim, the Great West, and California at the Panama-Pacific International Exposition* (Lincoln: University of Nebraska, 2014), 32; Sarah J. Moore, *Empire on Display: San Francisco's Panama-Pacific International Exposition of 1915* (Norman: University of Oklahoma Press, 2013), 84–92.

6. Markwyn, *Empress San Francisco*, 139.

7. PPIE Company, *Panama-Pacific International Exposition*, 8; Markwyn, *Empress San Francisco*, 107.

8. Many Chinese San Franciscans criticized China's contributions, arguing that its labor conditions, delayed openings, and disinterest in authenticity displayed more corruption and incompetence than national pride. Yong Chen, *Chinese San Francisco, 1850–1943: A Transpacific Community* (Stanford, CA: Stanford University Press, 2000), 205–6.

9. Quote from Anthony W. Lee, *Picturing Chinatown: Art and Orientalism in San Francisco* (Berkeley: University of California Press, 2001), 171.

10. PPIE Company, *The Blue Book: A Comprehensive Official Souvenir View Book of the Panama-Pacific International Exposition at San Francisco, 1915* (San Francisco: Robert A. Reid, 1915), 218–20, 321; Lee, *Picturing Chinatown*, 170–72; Markwyn, *Empress San Francisco*, 105–7, 110–14.

11. Eiichiro Azuma, *Between Two Empires: Race, History, and Transnationalism in Japanese America* (New York: Oxford University Press, 2005), 62–66; Andrea Geiger, *Subverting Exclusion:*

Transpacific Encounters with Race, Caste, and Borders, 1885–1928 (New Haven, CT: Yale University Press, 2011), 154–55; Paul A. Kramer, *Blood of Government: Race, Empire, the United States and the Philippines* (Chapel Hill: University of North Carolina Press, 2006), 349; Lon Kurashige, *Two Faces of Exclusion: The Untold History of Anti-Asian Racism in the United States* (Chapel Hill: University of North Carolina Press, 2016), 99–101.

12. PPIE Company, *Blue Book*, 196–99, 310.

13. Kramer, *Blood of Government*, 348.

14. "San Francisco: The Metropolis of the Pacific," *Harper's Weekly*, June 28, 1902, 820.

15. J. P. Munro-Fraser, *Doxey's Guide to San Francisco and Vicinity: The Big Trees, Yo Semite Valley, the Geysers, China, Japan, and Sandwich Islands* (San Francisco: Doxey, 1881).

16. "Preparing for an Early Movement," *San Francisco Chronicle*, May 12, 1898.

17. "California Likely to Profit by War in China," *San Francisco Chronicle*, July 11, 1900.

18. H. W. Postlethwaite, "The Carthage of the West," *Overland Monthly*, August 1903, 168; "Metropolis of the Pacific," 823.

19. "Metropolis of the Pacific," 823.

20. Quote from Palace Hotel News Stand, *SF Guide & Souvenir* (San Francisco: Palace Hotel News Stand, 1903), 43. Frank Morton Todd, *The Chamber of Commerce Handbook for San Francisco: A Guide for Visitors* (San Francisco: San Francisco Chamber of Commerce, 1914), 6; Gray Brechin, *Imperial San Francisco: Urban Power, Earthly Ruin* (Berkeley: University of California Press, 1999); Dorothy B. Fujita-Rony, *American Workers, Colonial Power: Philippine Seattle and the Transpacific West, 1919–1941* (Berkeley: University of California Press, 2003), 34–36; Jean Heffer, *The United States and the Pacific: History of a Frontier*, trans. W. Donald Wilson (Notre Dame, IN: University of Notre Dame Press, 2002), 44–46; David Igler, *The Great Ocean: Pacific Worlds from Captain Cook to the Gold Rush* (New York: Oxford University Press, 2013), 110–11; Roger W. Lotchin, *Fortress California, 1910–1961: From Warfare to Welfare* (New York: Oxford University Press, 1992), 1, 42; Kramer, *Blood of Government*, 91, 93, 347–49; Mel Scott, *The San Francisco Bay Area: A Metropolis in Perspective* (Berkeley: University of California Press, 1959, 1985), 13–14.

21. Todd, *Handbook for San Francisco*, 5.

22. Allan J. Dunn, *Care-Free San Francisco* (San Francisco: A. M. Robertson, 1913), 31–32.

23. Robert Newton Lynch, "San Francisco's Growing Pains," *Overland Monthly and Out West Magazine*, September 1923, 37; Todd, *Handbook for San Francisco*, 5.

24. Writers Project of the Works Progress Administration, *California in the 1930s: The WPA Guide to the Golden State* (Berkeley: University of California Press, 2013), 268.

25. "The Chinese in San Francisco," *Friend*, January 15, 1876, 169.

26. Todd, *Handbook for San Francisco*, 67.

27. Munro-Fraser, *Doxey's Guide*, 66.

28. Hamilton Basso, "San Francisco," *Holiday*, September 1953, 39.

29. Thomas B. Wilson, "Old Chinatown," *Overland Monthly*, September 1911, 233; Charlotte Brooks, *Alien Neighbors, Foreign Friends: Asian Americans, Housing, and the Transformation of Urban California* (Chicago: University of Chicago Press, 2009), 33; Lee, *Picturing Chinatown*, 151–52, 174; Mae Ngai, *The Lucky Ones: One Family and the Extraordinary Invention of Chinese America* (Boston: Houghton Mifflin Harcourt, 2010), 127; Raymond Rast, "The Cultural Politics of Tourism in San Francisco's Chinatown, 1882–1917," *Pacific Historical Review* 76 (2007): 50–54.

30. Lillian Symes, "San Francisco—Our Other Metropolis," *Harper's Magazine*, April 1932, 625.

31. Frank J. Taylor, "San Francisco's New Chinese City," *Travel*, March 1929.

32. Basso, "San Francisco," 39; Sydney Herschel Small, "San Francisco's Chinatown," *Holiday*, August 1954, 101, 102, 107.

33. John Gerrity, "Racial Prejudice—How San Francisco Squelched It," *Collier's*, February 14, 1953, 36–37, 40–41.

34. Christopher Lowen Agee, *The Streets of San Francisco: Policing and the Creation of a Cosmo-

politan Liberal Politics, 1950–1972 (Chicago: University of Chicago Press, 2014), 22–27, 29–40; Charlotte Brooks, *Between Mao and McCarthy: Chinese American Politics in the Cold War Years* (Chicago: University of Chicago Press, 2015), 30–32, 64–69, 71–74; Brooks, *Alien Neighbors, Foreign Friends*, 135–58, 194–236; Nayan Shah, *Contagious Divides: Epidemics and Race in San Francisco's Chinatown* (Berkeley: University of California Press, 2001), 204–50; Ellen D. Wu, *The Color of Success: Asian Americans and the Origins of the Model Minority* (Princeton, NJ: Princeton University Press, 2014), 181–209. See also Brooks, *Alien Neighbors, Foreign Friends*, chapter 1; Chen, *Chinese San Francisco*; Lee, *Picturing Chinatown*; and Guenter B. Risse, *Plague, Fear, and Politics in San Francisco's Chinatown* (Baltimore: Johns Hopkins University Press, 2012).

35. Herb Caen, *The San Francisco Book* (Boston: Houghton Mifflin, 1948), 61.

36. Francis Bruguiere, *San Francisco* (San Francisco: H. S. Crocker, 1918), 58; Eleanor B. Caldwell, "The Midwinter Fair," *New Peterson Magazine*, May 1894, 404.

37. Quote from George Spaulding & Company, *Official Guide to the California Midwinter Exposition* (San Francisco: George Spaulding, 1894), 115.

38. Quote from Ngai, *Lucky Ones*, 127. Clay Lancaster, *The Japanese Influence in America* (New York: Walton H. Rawls, 1963), 98, 103; Kendall H. Brown, "Rashômon: The Multiple Histories of the Japanese Tea Garden at Golden Gate Park," *Studies in the History of Gardens & Designed Landscapes* 18, no. 2 (1998): 93–119.

39. Kendall H. Brown, "Rashômon," 96.

40. California Legislature, Senate Fact-Finding Committee on San Francisco Bay Ports, *Ports of the San Francisco Bay Area, Their Commerce, Facilities, Problems, and Progress* (Sacramento: California State Printing Office, 1951), 192.

41. Japanese Chamber of Commerce, *An Economic Analysis of United States–Japanese Trade* (San Francisco: Japanese Chamber of Commerce, 1940), 7; John Kuo Wei Tchen, *New York before Chinatown: Orientalism and the Shaping of American Culture, 1776–1882* (Baltimore: Johns Hopkins University Press, 1999), 40–59; Arrell Morgan Gibson and John S. Whitehead, *Yankees in Paradise: The Pacific Basin Frontier* (Albuquerque: University of New Mexico Press, 1993), 172–73; State Harbor Commissioners, *World Trade Center*, 222, 79–80; Board of State Harbor Commissioners, *Foreign Trade through San Francisco Customs District, 1938* (San Francisco: Board of State Harbor Commissioners, 1938), 6–7, 32–33. These figures for the Customs District include all the harbors of Northern California, which for the most part comprise a handful of ports on the San Francisco Bay. Until the 1950s, San Francisco's port held a large majority of the foreign traffic with the exception of a handful of commodities. According to harbor reporter Jack Fosie, even in 1954, San Francisco did 65 percent of all foreign trade through the Customs District. "Bay Area Cargo— Everything from Myrrh to Motor Cars," *San Francisco Chronicle*, May 18, 1954.

42. "San Francisco Again Leads in Foreign Trade," *San Francisco Today*, March 4, 1939; Mira Wilkins, "Japanese Multinationals in the United States: Continuity and Change, 1879–1990," *Business History Review* 64, no. 4 (1990): 585–629.

43. Roger Daniels, *The Politics of Prejudice: The Anti-Japanese Movement in California and the Struggle for Japanese Exclusion* (Berkeley: University of California Press, 1962), 21.

44. Ibid., 20; US Bureau of the Census, *Twelfth Census of the United States, 1900, Vol. I, Part I* (Washington, DC: GPO, 1901), vol. 1, part 1, section 9, 487.

45. Quoted in Sandra C. Taylor, *Jewel of the Desert: Japanese American Internment at Topaz* (Berkeley: University of California Press, 1993), 27.

46. Roger Daniels, *Politics of Prejudice*, 28; Ocean Howell, *Making the Mission: Planning and Ethnicity in San Francisco* (Chicago: University of Chicago Press, 2015), 103–9; A. E. Yoell, "Oriental vs. American Labor," *Annals of the American Academy of Political and Social Science* 34, no. 2 (1909): 27–36.

47. Frank F. Chuman, *The Bamboo People: The Law and Japanese-Americans* (Chicago: Japanese American Citizens League, 1981), 19–37; Roger Daniels, *Politics of Prejudice*, 24–29, 33, 91; Gei-

ger, *Subverting Exclusion*, 90-96; Donald Teruo Hata Jr., *"Undesirables": Early Immigrants and the Anti-Japanese Movement in San Francisco, 1892-1893; Prelude to Exclusion* (New York: Arno Press, 1978), 140-44; Izumi Hirobe, *Japanese Pride, American Prejudice: Modifying the Exclusion Clause of the 1924 Immigration Act* (Stanford, CA: Stanford University Press, 2001); Yuji Ichioka, *The Issei: The World of the First Generation Japanese Immigrants, 1885-1924* (New York: Free Press, 1988), 68-72; Lon Kurashige, *Two Faces of Exclusion*, 111-12, 132-38; Lotchin, *Fortress California*, 6, 59-63.

48. *National Defense Migration: Hearings Pursuant to H. Res. 113, Part 29, before the Select Comm. Investigating National Defense Migration*, 77th Cong., 2nd sess., 11231, 11232 (1942) (statement of James M. Omura, editor and publisher, *Current Life*).

49. Geiger, *Subverting Exclusion*, 90; Azuma, *Between Two Empires*, 41-42; Scott Harvey Tang, "Pushing at the Golden Gate: Race Relations and Racial Politics in San Francisco, 1940-1955" (PhD diss., University of California, Berkeley, 2002), 106-7.

50. For the use of "Nihonjinmachi" or "Japanesetown," see Lynne Horiuchi, "Coming and Going in the Western Addition of San Francisco: Japanese Immigrants and African Americans in Search of Housing," *Annals of Scholarship* 17, no. 3 (2008): 45. For examples of other descriptors and monikers, see "Emperor's Death Causes Deep Grief," *San Francisco Chronicle*, July 30, 1912; "Japtown Housing," *San Francisco Chronicle*, July 1, 1943; Workers of the Writers' Program of the Works Progress Administration for Northern California, *San Francisco: The Bay and Its Cities* (New York: Hastings House, 1947), 220, 224, 284; and Caen, *San Francisco Book*, 112; "San Francisco," *Sunset*, November 1957, 51-61. The last article contained the one map that I have seen marking the Japanese American neighborhood before 1960. US Bureau of the Census, *16th Census of the United States, 1940: Population: Vol. II: Characteristics of the Population—Part I* (Washington, DC: GPO, 1943), 567.

51. Anne Vernez Moudon, *Built for Change: Neighborhood Architecture in San Francisco* (Cambridge, MA: MIT Press, 1986), 1, 32-35; Suzie Kobuchi Okazaki, *Nihonmachi: A Story of San Francisco's Japantown* (San Francisco: SKO Studios, 1985), 35-43; Scott, *San Francisco Bay Area*, 111, 118.

52. "Wants Site of Book Concern," *San Francisco Chronicle*, June 16, 1906.

53. Quote from "Last Zoning Hearing Will Be Held Today," *San Francisco Chronicle*, July 12, 1921. "Plans to Zone City Approved by Committee," *San Francisco Chronicle*, July 20, 1921; "Anti-Japanese Win Fight for Bush Street," *San Francisco Chronicle*, September 1, 1921.

54. "Move Made by City Planning Commission," *San Francisco Chronicle*, June 22, 1921.

55. "The Art of Zoning Cities," *San Francisco Chronicle*, June 23, 1921.

56. Brooks, *Alien Neighbors, Foreign Friends*, 26; Horiuchi, "Coming and Going in the Western Addition," 36-45.

57. Sam Mihara, interview with Sam Redman, November 8, 2012, Berkeley, CA, "Japanese American Incarceration / WWII American Home Front Oral History Project," 9, Regional Oral History Office, Bancroft Library, University of California, Berkeley (hereafter BANC).

58. George Matsumoto, interview with David Dunham and Candice Fukumoto, November 16, 2013, Oakland, CA, "Japanese American Incarceration / WWII American Home Front Oral History Project," Regional Oral History Office, BANC, 6.

59. Japanesetown was spread over the census tracts J-2, J-6, and J-8. In these tracts, 4,551 people were categorized as "other" (as opposed to "native white," "foreign white," or "Negro") while another 400 were in two surrounding tracts. "Other" in this case can be assumed to be primarily Japanese American, although it also included several hundred Chinese or Filipino residents known to live in the district. US Bureau of the Census, *16th Census of the United States, 1940: Population and Housing, Statistics for Census Tracts: San Francisco* (Washington, DC: GPO, 1943), 11-12; Select Committee Investigating National Defense Migration, *National Defense Migration*, H. Rep. No. 77-1911, at 13 (1942); US Bureau of the Census, *16th Census of the United States, 1940: Population: Vol. II, Part I*, 567; US Bureau of the Census, *16th Census of the United States, 1940:*

Characteristics of the Nonwhite Population by Race (Washington, DC: GPO 1943), 109; Brooks, *Alien Neighbors, Foreign Friends*, 39; Dorothy Swaine Thomas and Richard Nishimoto, *The Salvage* (Berkeley: University of California Press, 1952), 32.

60. Edna Bonacich and John Modell, *The Economic Basis of Ethnic Solidarity: Small Business in the Japanese American Community* (Berkeley: University of California Press, 1980), 38–41, 43–45; Charlotte Brooks, "The War on Grant Avenue: Business Competition and Ethnic Rivalry in San Francisco's Chinatown, 1937-1942," *Journal of Urban History* 37 (2011): 314–16; Stephen S. Fugita and David J. O'Brien, *Japanese American Ethnicity: The Persistence of Community* (Seattle: University of Washington Press, 1991), 50–62; Yamato Ichihashi, *Japanese in the United States: A Critical Study of the Problems of the Japanese Immigrants and Their Children* (Stanford, CA: Stanford University Press, 1932), 129–33; Thomas and Nishimoto, *The Salvage*, 28–32; Tamotsu Shibutani, "The Initial Impact of the War on the Japanese Communities in the San Francisco Bay Region: A Preliminary Report," unpublished paper, 1942, 43-44, reel 10, frame 0515, Japanese American Evacuation and Resettlement Records, BANC MSS 67/14 c, BANC; Edward K. Strong, *The Second-Generation Japanese Problem* (Stanford, CA: Stanford University Press, 1934), 1-10; Sandra C. Taylor, *Jewel of the Desert*, 15–23, 32–37; US Department of the Interior, War Agency Liquidation Unit (formerly the WRA; hereafter WRA), *People in Motion: The Postwar Adjustment of the Evacuated Japanese Americans* (Washington, DC: GPO, 1947), 111-12.

61. Fujita-Rony, *American Workers*, 92–95; Rick Baldoz, *The Third Asiatic Invasion: Migration and Empire in Filipino America, 1898-1946* (New York: New York University Press, 2011), 62–69; Michel S. Laguerre, *The Global Ethnopolis: Chinatown, Japantown and Manilatown in American Society* (New York: St. Martin's Press, 2000), 76–109. Other examples of interethnic tensions can be found in Azuma, *Between Two Empires*, 187-207.

62. Joseph James, "Profiles: San Francisco," *Journal of Educational Sociology* 19 (November 1945): 166–78.

63. Quotes from Franzy Lea Ritcharrdson, interview with Albert S. Broussard and Linda Burham, January 31, 1978, transcript, 2, "Afro-Americans in San Francisco prior to World War II" Oral History Project (hereafter "Afro-Americans in San Francisco" project), Friends of the San Francisco Public Library and San Francisco African-American Historical and Cultural Society, San Francisco, CA; Albert Broussard, *Black San Francisco: The Struggle for Racial Equality in the West, 1900-1954* (Lawrence: University Press of Kansas, 1993), 29–35; Daniel Crowe, *Prophets of Rage: The Black Freedom Struggle in San Francisco, 1945-1969* (New York: Garland, 2000) 16–20; Douglas Henry Daniels, *Pioneer Urbanites: A Social and Cultural History of Black San Francisco* (Berkeley: University of California Press, 1990), 98-101, 104; Brian J. Godfrey, *Neighborhoods in Transition: The Making of San Francisco's Ethnic and Nonconformist Communities* (Berkeley: University of California Press, 1988), 71-72, 97; Arthur E. Hippler, *Hunter's Point: A Black Ghetto* (New York: Basic Books, 1974), 13-14; Tang, "Pushing at the Golden Gate," 120–21; David McEntire, "Postwar Status of Negro Workers in San Francisco Area," *Monthly Labor Review* 70 (1950): 614.

64. Quote from Kenneth Finis, interview with Jesse J. Warr III, June 7, 1978, transcript, 1, "Afro-Americans in San Francisco" project.

65. Brooks, *Alien Neighbors, Foreign Friends*, 4; Broussard, *Black San Francisco*, 38–52; Henry Daniels, *Pioneer Urbanites*, 31–43.

66. Charles Kikuchi, *The Kikuchi Diary: Chronicle from an American Concentration Camp*, ed. John Modell (Chicago: University of Illinois Press, 1993), 45.

67. Ibid.

68. Shibutani, "Preliminary Report," 59, 86; "S.F. Joins the Nation in Rounding Up Suspicious Characters and Business Men," *San Francisco Chronicle*, December 8, 1941; Tang, "Pushing at the Golden Gate," 78–79.

69. Shibutani, "Preliminary Report," 57.

70. *National Defense Migration*, 10968 (1942) (statement of Hon. Angelo J. Rossi, mayor of the City of San Francisco, CA).

71. Karl G. Yoneda, *Ganbatte: Sixty-Year Struggle of a Kibei Worker* (Los Angeles: Asian American Studies Center of the University of California, Los Angeles, 1983), 120.

72. "FBI Opens New Offense on Jap Aliens," *San Francisco Chronicle*, March 7, 1942.

73. Leonard Broom and John I. Kitsuse, *The Managed Casualty: The Japanese American Family in World War II* (Berkeley: University of California, 1956), 12–15; Roger Daniels, *Concentration Camps: North America; Japanese in the United States and Canada during World War II* (Malabar, FL: Krieger, 1993), 42–44, 83–88; Greg Robinson, *By Order of the President: FDR and the Internment of Japanese Americans* (Cambridge, MA: Harvard University Press, 2003), 115–24; Paul Spickard, *Japanese Americans: The Formation and Transformations of an Ethnic Group*, rev. ed. (New Brunswick, NJ: Rutgers University Press, 2009), 104–9; Sandra C. Taylor, *Jewel of the Desert*, 44–47; Shibutani, "Preliminary Report," 65–76; Tang, "Pushing at the Golden Gate," 112–13.

74. Shibutani, "Preliminary Report," 79; Greg Robinson *A Tragedy of Democracy: Japanese Confinement in North America* (New York: Columbia University Press, 2009), 109.

75. Quote from Miné Okubo, *Citizen 13660* (Seattle: University of Washington Press, 2014), 17; Tang, "Pushing at the Golden Gate," 81.

76. Quote from "S.F. Clear of All but 6 Sick Japs," *San Francisco Chronicle*, May 21, 1942. "8,000 More Japanese Are Ordered Out," *San Francisco Chronicle*, May 10, 1942; "Last Japs Are Moved out of S.F.," *San Francisco Chronicle*, May 11, 1942; Sandra C. Taylor, *Jewel of the Desert*, 61; Tang, "Pushing at the Golden Gate," 110.

77. Estelle Ishigo, *Lone Heart Mountain* (Santa Clara, CA: Communicart, 1972), 9. Ishigo, a white artist married to a Japanese man, was interned from Los Angeles, and so her description was of the Pomona detention facility. Like other "assembly centers," it was surrounded by barbed wire and so likely provoked a common response.

78. Yoshiko Uchida, *Desert Exile: The Uprooting of a Japanese American Family* (Seattle: University of Washington Press, 2015), 70.

79. Sandra C. Taylor, *Jewel of the Desert*, 63–67, 94–97; Uchida, *Desert Exile*, 69–77; Robinson, *Tragedy of Democracy*, 124–32, 155–63; Spickard, *Japanese Americans*, 115–21; Commission on Wartime Relocation and Internment of Civilians, *Personal Justice Denied: Report of the Commission on Wartime Relocation and Internment of Civilians* (Seattle: University of Washington Press, 1997), 135–44.

80. WRA, *Final Report: Japanese Evacuation from the West Coast, 1942* (Washington, DC: GPO, 1943), 77–78.

81. Dave Tatsuno, interview by Wendy Hanamura, May 17, 2005, San Jose, CA, denshovhtdave-03-0014, transcript and video, Japanese American Film Preservation Project Collection, Densho Digital Archive, densho.org.

82. Quote from John Tateishi, *And Justice for All: An Oral History of the Japanese American Detention Camps* (Seattle: University of Washington Press, 1984), 182. Sandra C. Taylor, *Jewel of the Desert*, 110; Uchida, *Desert Exile*, 86, 124; Robinson, *Tragedy of Democracy*, 143; Broom and Kitsuse, *Managed Casualty*, 37; Naomi Hirahara and Gwenn M. Jensen, *Silent Scars of Healing Hands: Oral Histories of Japanese American Doctors in World War II Detention Camps* (Fullerton: Center for Oral and Public History at California State, Fullerton), 40, 173; John Howard, *Concentration Camps on the Home Front: Japanese Americans in the House of Jim Crow* (Chicago: University of Chicago Press, 2008), 162–63; Stephen S. Fugita and Marilyn Fernandez, *Altered Lives, Enduring Community: Japanese Americans Remember Their World War II Incarceration* (Seattle: University of Washington Press, 2011), 11, 176, 194–96, 199.

83. As sociologists Stephen S. Fugita and Marilyn Fernandez argue, internment was undeniably a turning point that transformed the lives of those incarcerated. However, they also argue that ethnic community—in their argument, not necessarily spatial—also provided continuity for many even though it was fundamentally changed by the internment and subsequent events. Fugita and Fernandez, *Altered Lives, Enduring Community*.

84. "Jap Town Sells Out, Packs Its Bags and Awaits the Order to Depart," *San Francisco Chronicle*, April 9, 1942; Kikuchi, *Kikuchi Diary*, 51.

85. Herb Caen, "Eyes, Ears, Nose and Throat," *San Francisco Chronicle*, July 23, 1942.

86. Quote from Maya Angelou, *I Know Why the Caged Bird Sings* (New York: Bantam Books, 1969), 177-78. US Bureau of the Census, *16th Census of the Population, 1940: Population: Vol. II, Part I*, 542; US Bureau of the Census, *Special Reports*, series CA-3, no. 3, *Characteristics of the Population, Labor Force, Families, and Housing: San Francisco Bay Congested Production Area* (Washington, DC: GPO, 1944), 7, 8; US Bureau of the Census, *17th Census of the Population, 1950: Vol. II, Part 5, Characteristics of the Population, California* (Washington, DC: GPO, 1952), 90, 103; Charles S. Johnson, Herman H. Long, and Grace Jones, *The Negro War Worker in San Francisco: A Local Self-Survey* (San Francisco: YWCA, 1944), 2; Gerald D. Nash, *The American West Transformed: The Impact of the Second World War* (Lincoln: University of Nebraska Press, 1985), 66-68; Tang, "Pushing at the Golden Gate," 130-32.

87. James, "Profiles: San Francisco," 176; Brooks, *Alien Neighbors, Foreign Friends*, 142.

88. Wilson Record, "Willie Stokes at the Golden Gate," *Crisis*, June 1949, 176.

89. Marilynn S. Johnson, *The Second Gold Rush: Oakland and the East Bay in World War II* (Berkeley: University of California Press, 1993), 52-55; Tang, "Pushing at the Golden Gate," 122; Charles S. Johnson et al., *Negro War Worker*, 4; Gretchen Lemke-Santangelo, *Abiding Courage: African American Migrant Women and the East Bay Community* (Chapel Hill: University of North Carolina Press, 1996), 51-52; Donna Jean Murch, *Living for the City: Migration, Education, and the Rise of the Black Panther Party in Oakland, California* (Chapel Hill: University of North Carolina Press, 2010), 17-19, 24.

90. "Slum Danger in 'Jap Town' under Study," *San Francisco News*, April 13, 1942; Tang, "Pushing at the Golden Gate," 115-16; Floyd A. Cave, "Cultural Conflicts of Minority Groups in San Francisco," *Social Science* 22 (1947): 24.

91. Carlton B. Goodlett, "Statement of SF Branch, NAACP to Joint Congressional Hearing on Housing Held in San Francisco, November 13, 14, 1947," November 13, 1947, folder 3, carton 93, series 4.1, National Association for the Advancement of Colored People, Region 1, Records, BANC MSS 78/180c, BANC; Charles S. Johnson et al., *Negro War Worker*, 20-21; Broussard, *Black San Francisco*, 172-75; Roger W. Lotchin, *The Bad City in the Good War: San Francisco, Los Angeles, Oakland, and San Diego* (Bloomington: Indiana University Press, 2003), 128-29.

92. "Home Front," *San Francisco Chronicle*, July 3, 1943.

93. Marilynn S. Johnson, *Second Gold Rush*, 55; Tang, "Pushing at the Golden Gate," 133; Broussard, *Black San Francisco*, 143-46, 165, 172-76.

94. Cave, "Cultural Conflicts," 24; "Home Front"; "Japtown Housing."

95. "Japtown Tenement Owner Is Held," *San Francisco Chronicle*, November 26, 1943.

96. Charles S. Johnson et al., *Negro War Worker*, 3.

97. Tang, "Pushing at the Golden Gate," 122.

98. Charles S. Johnson et al., *Negro War Worker*, 29; Cave, "Cultural Conflicts," 25.

99. Deirdre L. Sullivan, "'Letting Down the Bars': Race, Space, and Democracy in San Francisco, 1936-1964" (PhD diss., University of Pennsylvania, 2003), 53-55, 60-63.

100. "Citizen League for Japanese Is Reopened," *San Francisco Chronicle*, January 4, 1945.

101. "Two Held in Attack on U.S.-Japanese," *San Francisco Chronicle*, September 22, 1945; "Anti-Japanese Terror—in Watsonville," *San Francisco Chronicle*, September 25, 1945; Mark Brilliant, *The Color of America Has Changed: How Racial Diversity Shaped Civil Rights Reform in California, 1941-1978* (New York: Oxford University Press, 2010), 37-41; Greg Robinson, *After Camp: Portraits in Midcentury Japanese American Life and Politics* (Berkeley: University of California Press, 2012), 198-200.

102. About 30,000 people had relocated from the detention centers by the end of 1944, when the West Coast was still closed to them but while they could procure leaves for jobs or education

elsewhere. A total of 54,254 listed destinations other than California, Washington, and Oregon upon leaving the camps, regardless of their date of exit. WRA, *People in Motion*, 9–10, 111; Roger Daniels, *Asian America: Chinese and Japanese in the United States since 1850* (Seattle: University of Washington Press, 1988), 283–95; Robinson, *Tragedy of Democracy*, 258–59.

103. Most of the half dozen or so hostels for resettlers in San Francisco had less than fifteen beds; the Buddhist hostel had two hundred. Charles F. Miller to Duncan Mills, September 17, 1945, enclosure, frame 381, reel 64, Japanese American Evacuation and Resettlement Records (hereafter JAERR), BANC. For more on these institutions, see Jeffrey C. Copeland, "Stay for a Dollar a Day: California's Church Hostels and Support during the Japanese American Eviction and Resettlement, 1942–1947" (MA thesis, University of Nevada, Reno, 2014).

104. WRA, "Northern California Area Final Report: Area and All Districts," May 15, 1946, 5, frame 0179, reel 64, JAERR, BANC; WRA, "Final Report—San Francisco District," 1946, 5, frame 0262, reel 64, JAERR, BANC.

105. WRA, *People in Motion*, 111, 113; US Department of the Interior, WRA, *The Relocation Program* (Washington, DC: GPO, 1946), 73; "Anti-Nisei Terrorism," *San Francisco Chronicle*, September 21, 1945; "New Local Business Houses Open," *Nichi Bei Times*, May 22, 1946; Robert O'Brien, "San Francisco," *San Francisco Chronicle*, August 30, 1945; WRA, "Final Report—San Francisco," 5, 9; "Most of S.F. Japanese Have Returned," *San Francisco Chronicle*, November 23, 1947; Sandra C. Taylor, *Jewel of the Desert*, 278–80. For similar conditions elsewhere, see Hilary Jenks, "'Home Is Little Tokyo': Race, Community, and Memory in Twentieth-Century Los Angeles" (PhD diss., University of Southern California, 2008), 170–72; Robinson, *Tragedy of Democracy*, 255–59; and Spickard, *Japanese Americans*, 144–51.

106. "Japanese-Americans Are Advised by Ickes to Settle Away from Coast," *San Francisco Chronicle*, March 18, 1945.

107. Mitziko Sawada, "After the Camps: Seabrook Farms, New Jersey, and the Resettlement of Japanese Americans, 1944–1947," *Amerasia* 13 (1986–1987): 121.

108. US Department of the Interior, *WRA: A Story of Human Conservation* (Washington, DC: GPO, 1946), 192.

109. Michi Weglyn, *Years of Infamy: The Untold Story of America's Concentration Camps* (New York: William Morrow, 1976), 196–97.

110. Quote from Allan W. Austin, "Eastward Pioneers: Japanese American Resettlement during World War II and the Contested Meaning of Exile and Incarceration," *Journal of American Ethnic History* 26 (2007): 64.

111. Commission on Wartime Relocation and Internment of Civilians, *Personal Justice Denied*, 229; Austin, "Eastward Pioneers," 58–84; Brian Masaru Hayashi, *Democratizing the Enemy: The Japanese American Internment* (Princeton, NJ: Princeton University Press, 2004), 107; Scott Kurashige, *The Shifting Grounds of Race: Black and Japanese Americans in the Making of Multiethnic Los Angeles* (Princeton, NJ: Princeton University Press, 2008), 177–93; Thomas M. Linehan, "Japanese American Resettlement in Cleveland during and after World War II," *Journal of Urban History* 20 (1993): 54–55; Mae Ngai, *Impossible Subjects: Illegal Aliens and the Making of Modern America* (Princeton, NJ: Princeton University Press, 2004), 175–202; Robinson, *After Camp*, 15–42; Orin Starn, "Engineering Internment: Anthropologists and the War Relocation Authority," *American Ethnologist* 13, no. 4 (November 1986): 700–721.

112. "The Little Tokyos," *Pacific Citizen*, August 31, 1946.

113. WRA, *People in Motion*, 14.

114. "Little Tokyos."

115. Austin, "Eastward Pioneers," 65.

116. Scott Kurashige, *Shifting Grounds of Race*, 183; Leonard Broom and John I. Kitsuse, "Current Leadership Problems among Japanese Americans," *Sociological and Social Research* 57 (1953): 161.

117. WRA, "Final Report—San Francisco," 2; WRA, *People in Motion*, 9-10; "Mayor's Message," *San Francisco Chronicle*, January 3, 1945; "Anti-Nisei Terrorism," *San Francisco Chronicle*, September 21, 1945; "Windows Broken in S.F. House Bought by Nisei," *San Francisco Chronicle*, October 16, 1945; "Two Suspects Questioned in Nisei Attack," *San Francisco Chronicle*, March 8, 1945; "State Is near Showdown on Nisei Question," *San Francisco Chronicle*, April 18, 1945; "'Limited Aid' for Returning Nisei," *People's World*, January 2, 1945; "Japanese-Americans: Returning Children Pledged Aid by State School Heads," *People's World*, January 6, 1945; Tang, "Pushing at the Golden Gate," 189, 195-98, 205-9. For aid in other cities, see Scott Kurashige, *Shifting Grounds of Race*, 167; Linehan, "Japanese American Resettlement"; Setsuko Matsunaga Nishi, "Japanese American Achievement in Chicago: A Cultural Response to Degradation" (PhD diss., University of Chicago, 1963); Robinson, *Tragedy of Democracy*, 256-57; and Robinson, *After Camp*, 50-53, 58, 62-63.

118. Thelma Thurston Gorham, "Negroes and Japanese Evacuees," *Crisis*, November 1945, 314.

119. James, "Profiles: San Francisco," 178.

120. Tang, "Pushing at the Golden Gate," 197; Miller to Mills, enclosure, September 17, 1945; Yori Wada, "Working for Youth and Social Justice: The YMCA, the University of California, and the Stulsaft Foundation," interview with Frances Linsley and Gabrielle Morris, 1983 and 1990, 41-44, Regional Oral History Office, BANC.

121. Quoted in Tang, "Pushing at the Golden Gate," 201.

122. Alice Setsuko Sekino Hirai, interview with Megan Asaka, June 3, 2008, Salt Lake City, UT, video, denshovh-halice-01-0009, Topaz Museum Collection, Densho Digital Archive.

123. See, for example, Charlotte Brooks, "In the Twilight Zone between Black and White: Japanese American Resettlement and Community in Chicago," *Journal of American History* 86, no. 4 (March 2000): 1655-1688; Linehan, "Japanese American Resettlement"; Robinson, *After Camp*, 48-52; Matthew M. Briones, *Jim and Jap Crow: A Cultural History of 1940s Interracial America* (Princeton, NJ: Princeton University Press, 2012), 162-91.

124. Masao Ashizawa interview, interview by author, Ken Yamada, and Clement Lai, San Francisco, CA, July 15, 2004, transcript, NJAHS, 14.

125. Thurston Gorham, "Negroes and Japanese Evacuees," 315.

126. At the census-tract level, the census counts only black, white, and "other." "Other" resident owners therefore included Chinese or Filipino owners, as well.

127. "Plans Told for Western Addition," *San Francisco Chronicle*, August 20, 1952.

128. Jenks, "Home Is Little Tokyo," 176.

129. Willie K. Ito, interview with Kristen Kuetkemeier, December 5, 2013, Los Angeles, CA, video, denshovh-iwillie-01-0018, Manzanar Historic Site Collection, Densho Digital Archive.

130. Quote from Briones, *Jim and Jap Crow*, 184.

131. Quote from Wada interview, 42. The Japanese American to African American neighborhood transition was common in cities such as Los Angeles and Seattle, and resettlers in cities such as Detroit or Chicago, cities in which there had been few Japanese Americans prior to the war, also found themselves in segregated neighborhoods with black neighbors. Therefore, harmony and tension, and the wide spectrum between them, were found in all these cities. For instance, see Robinson, *After Camp*, 51, 64-65; Brooks, "In the Twilight Zone"; Jenks, "Home Is Little Tokyo," 170-92; Scott Kurashige, *Shifting Grounds of Race*, 169; Kariann Yokota, "From Little Tokyo to Bronzeville and Back" (MA thesis, University of California, Los Angeles, 1994); and Valerie J. Matsuomoto, *City Girls: The Nisei Social World in Los Angeles, 1920-1950* (New York: Oxford University Press, 2014), 186-90.

132. Benh Nakajo, interview with Reth Meas and Sarah Lew, May 11, 2004, San Francisco, CA, transcript, NJAHS, 4.

133. US Bureau of the Census, *17th Census of the United States, 1950: Census of the Population, Vol. III, Part IV: Census Tracts Statistics, San Francisco-Oakland California and Adjacent Area* (Wash-

ington, DC: GPO, 1952), 29. Tang, "Pushing at the Golden Gate," 189–91; James Richardson, *Willie Brown: A Biography* (Berkeley: University of California Press, 1996), 48–49; R. L. Polk & Co., *Polk's San Francisco City Directory 1953* (San Francisco: R. L. Polk, 1953), 1869; Medea Isphording Bern, *San Francisco Jazz* (Charleston, SC: Arcadia Press, 2014), 18; Elizabeth Pepin and Lewis Watts, *Harlem of the West: The San Francisco Fillmore Jazz Era* (San Francisco: Chronicle Books, 2006), 135, 169.

134. "Tea Garden to Be Dismantled," *San Francisco Chronicle*, May 16, 1942; Caen, *San Francisco Book*, 63.

135. Herb Caen, "Paragraph Parade," *San Francisco Chronicle*, March 6, 1942; "Tea Garden Will Stay in Park under New Name," *San Francisco Chronicle*, April 25, 1942. For examples of debated possibilities, see Chingwah Lee, letter to the editor, *San Francisco Chronicle*, March 9, 1942; Herb Caen, "If I Had a Million," *San Francisco Chronicle*, March 10, 1942; Mrs. Shirley Walker, letter to the editor, *San Francisco Chronicle*, March 11, 1942; and Mrs. Alice Winenow, letter to the editor, *San Francisco Chronicle*, March 20, 1942.

136. Herb Caen, "Tales of the Town," *San Francisco Chronicle*, October 29, 1947; "Local Issei to Operate Japanese Tea Garden," *Nichi Bei Times*, February 8, 1958.

137. "San Francisco Again Leads in Foreign Trade," *San Francisco Today*, March 4, 1939.

138. California Board of State Harbor Commissioners for San Francisco Harbor, *World Trade Center in San Francisco* (San Francisco: Board of State Harbor Commissioners for San Francisco Harbor, 1947), 16, 141–44.

CHAPTER TWO

1. "Authorizing and Requesting the President to Issue a Proclamation with Respect to the 1959 Pacific Festival, and for Other Purposes," S. Rep. No. 86-816 at 1 (1959).

2. "'Pacific Festival Days' Proclaimed by Mayor in Honor of States and Nations of Ocean Area," *Bay Region Business*, August 15, 1958, 1.

3. Board of State Harbor Commissioners for San Francisco, *World Trade Center*, 16.

4. "Pacific Festival Calendar," *Bay Region Business*, September 12, 1958, 2.

5. "Land of the Rising Export," *Time*, October 8, 1956, 95.

6. See, for example, Laura A. Belmonte, *Selling the American Way: U.S. Propaganda and the Cold War* (Philadelphia: University of Pennsylvania Press, 2008); Nicholas J. Cull, *The Cold War and the United States Information Agency: American Propaganda and Public Diplomacy, 1945–1989* (New York: Cambridge University Press, 2009); Brian C. Etheridge and Kenneth Osgood, eds., *The United States and Public Diplomacy: New Directions in Cultural and International History* (Boston: Brill, 2010); Justin Hart, *Empire of Ideas: The Origins of Public Diplomacy and the Transformation of U.S. Foreign Policy* (New York: Oxford University Press, 2013); Frank A. Ninkovich, *The Diplomacy of Ideas: U.S. Foreign Policy and Cultural Relations, 1938–1950* (New York: Cambridge University Press, 1981); and Kenneth Osgood, *Total Cold War: Eisenhower's Secret Propaganda Battle at Home and Abroad* (Lawrence: University of Kansas Press, 2006). This rich literature on public diplomacy focuses on the federal or national level, and this pattern extends to the small, primarily nonhistorical literature on sister-city affiliations. See, for example, Brian C. Etheridge, "The Sister City Network in the 1970s: American Municipal Internationalism and Public Diplomacy in a Decade of Change," in *Reasserting America in the 1970s: U.S. Public Diplomacy and the Rebuilding of America's Image Abroad*, ed. Halvard Notaker, Giles Scott-Smith, and David Snyder (Manchester, UK: Manchester University Press, 2016), B. Ramasamy and R. D. Cremer, "Cities, Commerce and Culture: The Economic Role of International Sister-City Relationships between New Zealand and Asia," *Journal of the Asia Pacific Economy* 3 (1998): 446–61; and Wilbur Zelinsky, "The Twinning of the World: Sister Cities in Geographic and Historical Perspective," *Annals of the Association of American Geographers* 81 (1991): 1–31. An exception is Daniel Alan Bush, "Seattle's Cold War

Foreign Policy, 1957–1990: Citizen Diplomats and Grass Roots Diplomacy, Sister Cities and International Exchange" (PhD diss., University of Washington, 1998).

7. "Queen of the Pacific—or Just an Aging Matron?," *San Francisco Chronicle*, September 12, 1958.

8. San Francisco Department of City Planning, *The Population of San Francisco: A Half Century of Change* (San Francisco: San Francisco Department of City Planning, 1954), 7; William Issel and Robert W. Cherny, *San Francisco 1865–1932: Politics, Power, and Urban Development* (Berkeley: University of California Press, 1986), 14–17, 22–25.

9. Quoted from Roger W. Lotchin, "The Darwinian City: The Politics of Urbanization in San Francisco between the World Wars," *Pacific Historical Review* 48, no. 3 (1979): 361.

10. Board of Harbor Commissioners Port of Los Angeles, *Los Angeles Harbor: Premier Port of the Pacific* (Los Angeles: Board of Harbor Commissioners Port of Los Angeles, no date), 2.

11. Japanese Chamber of Commerce, *American-Japanese Trade and Treaty Abrogation: An Economic Analysis of the Possible Effects of the Abrogation of the American-Japanese Treaty of 1911 upon American-Japanese Trade* (San Francisco: Japanese Chamber of Commerce, 1939), 14–15.

12. Quote from Roger W. Lotchin, *Fortress California, 1910–1961: From Warfare to Welfare* (New York: Oxford University Press, 1992), 13.

13. Quote from San Francisco Chamber of Commerce, *San Francisco: Financial, Commercial and Industrial Metropolis of the Pacific Coast* (San Francisco: H. S. Crocker, 1915); Citizens National Bank, *Los Angeles: Industrial Focal Point of the West* (Los Angeles: Citizens National Bank, [circa 1959]), Lotchin, *Fortress California*, especially 5–6, 11–18, 42–46, 145–48; Valerie J. Matsuomoto, *City Girls: The Nisei Social World in Los Angeles, 1920–1950* (New York: Oxford University Press, 2014), 185; Joseph A. Rodriguez, "Planning and Urban Rivalry in the San Francisco Bay Area in the 1930s," *Journal of Planning Education and Research* 20 (2000): 67–68.

14. See, for example, "Insured Units, Employment and Payroll," in Research Department of the San Francisco Chamber of Commerce (hereafter Research Department), *San Francisco and the Bay Area* (San Francisco: San Francisco Chamber of Commerce, 1950), 12; "San Francisco— Reporting Units, Insured and Payroll Employment—a 1955 Preliminary," in Research Department, *San Francisco and the Bay Area* (San Francisco: San Francisco Chamber of Commerce, 1956), 12; "Employment and Payrolls, by Industry, City and County of San Francisco for the Year Ending June 30, 1959," in Research Department, *San Francisco and the Bay Area* (San Francisco: San Francisco Chamber of Commerce, 1960), 12.

15. "1955 Action Program for Progress," *Bay Region Business*, February 18, 1955, 1.

16. Research Department, *San Francisco and the Bay Area: An Economic Survey and Yearly Review, 1955* (San Francisco: San Francisco Chamber of Commerce, 1955), 4; Susan S. Fainstein, Norman I. Fainstein, and P. Jefferson Armistead, "San Francisco: Urban Transformation and the Local State," in *Restructuring the City: The Political Economy of Urban Redevelopment*, ed. Susan S. Fainstein, Norman I. Fainstein, Richard C. Hill, Dennis R. Judd, and Michael P. Smith (New York: Longman, 1983), 204–5; Issel and Cherny, *San Francisco*, 50; Mel Scott, *The San Francisco Bay Area: A Metropolis in Perspective* (Berkeley: University of California Press, 1959, 1987), 169–201, 223–24.

17. "Chamber Announces Priority Goals for 1956," *Bay Region Business*, April 27, 1956, 1.

18. Trade measured in tons. "The New Port of San Francisco," *San Francisco Chronicle*, May 24, 1950.

19. California Legislature, Senate Fact-Finding Committee on San Francisco Bay Ports, *Ports of the San Francisco Bay Area, Their Commerce, Facilities, Problems, and Progress* (Sacramento: California State Printing Office, 1951), 281, 288, 193; Board of State Harbor Commissioners, *Foreign Trade through San Francisco Customs District, 1938*, (San Francisco: Board of State Harbor Commissioners, 1939), 6; Board of State Harbor Commissioners, *Foreign Trade through San Francisco Customs District, 1956* (San Francisco: Board of State Harbor Commissioners, 1957), 11; "New Port of San Fran-

cisco"; *Pacific Trade Patterns: Hearings, Day 4, before the Senate Committee on Commerce*, 88th Cong. 168 (1963) (statement of Goodman, president, San Francisco World Trade Association, and chairman, Getz Bros. & Co., Inc.), 127–45, 151–73; "Veg. Fiber Semimfgs x . . .' and World Trade," *San Francisco Chronicle*, May 22, 1952; "Gain for Japan," *San Francisco Chronicle*, September 17, 1957.

20. "Gain for Japan."

21. Quoted in John W. Dower, *Empire and Aftermath: Yoshida Shigeru and the Japanese Experience, 1878–1954* (Cambridge, MA: Harvard University Press, 1979), 419.

22. Walter LaFeber, *The Clash: U.S.-Japanese Relations throughout History* (New York: W. W. Norton, 1997), 270–83; William S. Borden, *The Pacific Alliance: United States Foreign Economic Policy and Japanese Trade Recovery, 1947–1955* (Madison: University of Wisconsin Press, 1984), 82–98; John W. Dower, *Embracing Defeat: Japan in the Wake of World War II* (New York: W. W. Norton, 1999), 87–120; Aaron Forsberg, *America and the Japanese Miracle: The Cold War Context of Japan's Postwar Economic Revival, 1950–1960* (Chapel Hill: University of North Carolina Press, 2000), 67–76.

23. US Department of State, *Foreign Relations of the United States, 1951*, NSC 48/5 (Washington, DC: GPO, 1979), 6:43, 54–55.

24. LaFeber, *Clash*, 257–59, 270–75; Michael Schaller, *Altered States: The United States and Japan since the Occupation* (New York: Oxford University Press, 1997). As historians have recently emphasized, US-Japanese relations were far more contested and mutual than US ambitions acknowledged. See, for instance, Dower, *Embracing Defeat*; Jennifer Miller, "Fractured Alliance: Anti-base Protests and Postwar U.S.-Japanese Relations," *Diplomatic History* 38 (2014): 953–86; Yoneyuki Sugita, *Pitfall or Panacea: The Irony of U.S. Power in Occupied Japan, 1946–1952* (New York: Routledge, 2003).

25. Board of State Harbor Commissioners, *Foreign Trade through San Francisco Customs District, 1951* (San Francisco: Board of State Harbor Commissioners, 1952), 62–63.

26. "Japan Returns to the Pacific Trade," *San Francisco Chronicle*, May 22, 1952.

27. "Chamber Group Reports on 'Unequalled' Japanese Boom," *Bay Region Business*, August 3, 1956, 3.

28. For comparative business activity, see, for example, the "General Business Activity" column by the Research Department in the chamber publication *Bay Region Business*; "C of C Takes Exception to ICC Examiner Report," *Bay Region Business*, May 14, 1954, 2; "I.C.C. Petition Filed," *Bay Region Business*, June 11, 1954, 3; "Chamber Protecting S.F. Interest in ICC and Air Line Cases," *Bay Region Business*, July 9, 1954; and "BOAC Begins First One-Plane European Service after Two Years' Effort by City, Chamber," *Bay Region Business*, March 15, 1957, 5.

29. "SF Port to Expand Orient Trade Program," *San Francisco Chronicle*, May 19, 1957.

30. "Japan Society Officers Elected," *San Francisco Chronicle*, June 28, 1953; "Lester Goodman Dies in Tokyo," *San Francisco Chronicle*, August 3, 1965; "'Unequalled' Japanese Boom."

31. "San Francisco Adopts Osaka as Sister City," *Japan Times*, clipping, no date, folder "C1-2, Community Participation, San Francisco," box 18, finding aid A1, entry 56 (Subject Files, 1953–1967), Records of the US Information Agency (RG 306), National Archives and Records Administration, College Park, MD (hereafter NARA).

32. "Osaka International Trade Fair," *Osaka Chamber of Commerce & Industry Overseas Bulletin*, January 1958, 1; "Chamber Sponsors Air Tour of Japan, Philippines, Hong Kong, May 5-28," *Bay Region Business*, February 15, 1957, 1.

33. San Francisco Chamber of Commerce, press release, March 28, 1958, folder 16, box 6, series 3, George Christopher Papers (SFH 7), San Francisco History Center, San Francisco Public Library (hereafter SFPL); "Business Tour: Japan, Hong Kong, Philippines," *International Bulletin*, March 15, 1957, 1.

34. "Chamber Sponsors Air Tour," 1.

35. San Francisco Chamber of Commerce, press release, March 28, 1958.

36. Telegram from the US Information Agency to USIS Tokyo, May 8, 1957, folder "C1–2, Community Participation, San Francisco," box 18, finding aid A1, entry 56, RG 306, NARA.

37. Robert McLaughlin, "US Cities Have Twins in Other Countries in Community People-to-People Programs," *American City*, February 1960, 163.

38. William Harlan Hale, "Every Man an Ambassador," *Reporter*, March 21, 1957, 18.

39. Dwight D. Eisenhower, "An Epidemic of Friendship," *Reader's Digest*, November 1963, 133.

40. Osgood, *Total Cold War*, 215; John E. Juergensmeyer, *The President, the Foundations, and the People-to-People Program* (Indianapolis, IN: Bobbs-Merrill, 1965); Christina Klein, *Cold War Orientalism: Asia in the Middlebrow Imagination, 1945–1961* (Berkeley: University of California Press, 2003), 24–29, 41–56.

41. Eisenhower, "Epidemic of Friendship," 136.

42. American Municipal Association, "Your Community in World Affairs: The People-to-People Program," no date, folder 16, box 6, series 3, SFH 7, SFPL; "U.S. and Foreign Cities Affiliate in 'People-to-People' Program," *American City*, July 1959, 33.

43. Charles C. Deil to J. J. Tornes, May 8, 1957, folder "C1–2, Community Participation, San Diego," box 18, finding aid A1, entry 56, RG 306, NARA.

44. Maurice E. Lee to Muriel Tolle, October 3, 1957, ibid.

45. "S.F., Osaka May Join in Affiliation," *Nichi Bei Times*, April 13, 1957; "Osaka Greeting Slated as Part of Chamber's Far East Business Tour," *Bay Region Business*, April 26, 1957, 2; telegram from the USIA to USIS Tokyo, March 6, 1957, folder "C1–2, Community Participation, San Francisco," box 18, finding aid A1, entry 56, RG 306, NARA.

46. Marianne Besser and Joseph Alvarez, "The Two-Way Rewards of City-to-Foreign-City Exchanges," *Reader's Digest*, March 1960, 254.

47. Bush, "Seattle's Cold War Foreign Policy," 14–18.

48. "San Francisco Adopts Osaka."

49. San Francisco Chamber of Commerce, "Fact Sheet: San Francisco–Osaka Town Affiliation," November 8, 1957, folder 16, box 6, series 3, SFH 7, SFPL; "S.F. Delegate Will Present Sister Cities Resolution to Mayor of Osaka Soon," *Mainichi Daily News*, no date, folder "C1–2, Community Participation, San Francisco," box 18, finding aid A1, entry 56, RG 306, NARA.

50. Telegram from the USIA to USIS Tokyo, June 5, 1957; "San Francisco Desires Strong Ties with Osaka," *Mainichi Daily News*; both in folder "C1–2, Community Participation, San Francisco," box 18, finding aid A1, entry 56, RG 306, NARA.

51. "US, Russia to Compete in All Exhibits in Osaka," *Mainichi Daily News*, January 24, 1958, folder 500.7 Osaka Int'l Trade Fair, 1957, box 12, General Records, 1936–1961, US Consulate Kobe, Japan, Records of the Foreign Service Posts of the Department of State (RG 84), NARA; "Osaka Trade Fair, Completed Project Report," attachment to Edgar S. Clark to George Christopher, no date (1962), folder 1, box 7, series 3, SFH 7, SFPL; "Large San Francisco Good Will Mission Leaving Sunday for Osaka and Expo '70," press release, March 25, 1970, 2, folder 38, box 13, subseries E, series 2, Joseph L. Alioto Papers (SFH 5), SFPL.

52. Jeffrey Hanes, *City as Subject: Seki Hajime and the Reinvention of Modern Osaka* (Berkeley: University of California Press, 2002), 195.

53. US Consulate Kobe, "Kansai Economic Review, 1956," folder 500 (Economic Notes, 1956), box 11, Classified General Records, 1952–1963, Kobe Consulate, Japan, RG 84, NARA.

54. "The Kansai and the Nation: Year-End Economic Review, 1956," 1, folder 500 (Economic Notes, 1956), box 11, General Records, 1936–1961, US Consulate Kobe, Japan, RG 84, NARA; dispatch 129 from George M. Emory to the Department of State, January 3, 1958, folder 500, box 11, General Records, 1936–1961, US Consulate Kobe, Japan, RG 84, NARA; "General Economic Matters, 1956–1958," folder 500, box 11, General Records, 1936–1961, US Consulate Kobe, Japan, RG 84, NARA; "History of Osaka Fair," *Osaka Chamber of Commerce & Industry Overseas Bulletin*, January 1958, 2; Dower, *Empire and Aftermath*, 418–27; Forsberg, *America and the Japanese Miracle*,

11-17; Hirohisa Kohama, *Industrial Development in Postwar Japan* (New York: Routledge, 2007), 32; Michael Schaller, *The American Occupation of Japan: The Origins of the Cold War in Asia* (New York: Oxford University Press, 1985); Sayuri Shimizu, *Creating People of Plenty: The United States and Japan's Economic Alternatives, 1950–1960* (Kent, OH: Kent State University Press, 2001).

55. DeLesseps S. Morrison to E. Snowden Chambers, February 13, 1957, folder "C1-2, Community Participation, New Orleans," box 18, finding aid A1, entry 56, RG 306, NARA.

56. Snowden Chambers to Glen Douthit, November 25, 1957, ibid.

57. Office of the Mayor, press release, circa January 1958, folder 16, box 6, series 3, SFH 7, SFPL.

58. Robert Taylor to George Christopher, January 8, 1958, ibid.

59. Phillips S. Davies to George Christopher, January 16, 1959, folder 17, box 6 series 3, SFH 7, SFPL.

60. Office of the Mayor, press release, circa January 1958; Davies to Christopher, July 29, 1958; and Charles von Loewenfeldt to John D. Sullivan, September 26, 1958, all in folder 16, box 6, series 3, SFH 7, SFPL; "Hirai, Shinmachi Fill OSTAC Posts," *Osaka–San Francisco Sister City Bulletin*, January 1962; "Kitazawa Succeeds Nagai as Chairman of OSTAC," *Osaka–San Francisco Sister City Bulletin*, May 1962.

61. Schaller, *Altered States*, 130.

62. Quote from "Japan: The Cherished Myths," *Holiday*, October 1961, 9. Klein, *Cold War Orientalism*; Sheila K. Johnson, *The Japanese through American Eyes* (Stanford, CA: Stanford University Press, 1988), 72–83; Gina Marchetti, *Romance and the "Yellow Peril": Race, Sex, and Discursive Strategies in Hollywood Fiction* (Berkeley: University of California Press, 1994), 126–39; Naoko Shibusawa, *America's Geisha Ally: Reimagining the Japanese Enemy* (Cambridge, MA: Harvard University Press, 2010).

63. "Yamato Was Imported in 7000 Tiny Pieces," *San Francisco Chronicle*, December 6, 1959.

64. "Japanese Cuisine Is Much [*sic*] Than Barbaric Bits of Seaweed," *San Francisco Chronicle*, June 9, 1958; Tokyo Sukiyaki advertisement, *San Francisco Chronicle*, July 17, 1951, 12.

65. "New Tokyo Sukiyaki Is the Perfect Rendezvous," *San Francisco Chronicle*, October 20, 1954.

66. "Delights for Delegates," *Life*, August 20, 1956, 49; Marjorie Trumbull, "Exclusively Yours," *San Francisco Chronicle*, November 21, 1955; "Like Sukiyaki? You Can Make It at Home in Japanese Style," *San Francisco Chronicle*, April 24, 1952.

67. See, for example, Herb Caen, *Herb Caen's Guide to San Francisco* (New York: Doubleday, 1957), 99; John Wesley Noble, "San Francisco: A Holiday Thrift Tour," *Holiday*, December 1956, 21; "San Francisco," *Sunset*, November 1957, 61.

68. "Delegates Vote to Limit Debate," *San Francisco Chronicle*, September 6, 1951; "Japanese Arrive, Talk with Acheson," *San Francisco Chronicle*, September 3, 1951.

69. Quotes from "Yoshida Accepts, Urges Peace," *San Francisco Chronicle*, September 7, 1951; and Joseph M. Henning, *Outposts of Civilization: Race, Religion, and the Formative Years of American-Japanese Relations* (New York: New York University Press, 2000), 22. John W. Dower, *War without Mercy: Race and Power in the Pacific War* (New York: Pantheon, 1987), 94–111.

70. "The Visitors from Tokyo," *San Francisco Chronicle*, January 18, 1950; "Japanese Visitors," *San Francisco Chronicle*, January 17, 1950.

71. "Great Japanese Chef Visits S.F.," *San Francisco Chronicle*, December 7, 1952.

72. "Japanese Judges," *San Francisco Chronicle*, April 21, 1950; "Buddhist Head Here," *San Francisco Chronicle*, February 4, 1952; "Japan Seamen Stop Here on Way to Sub Training," *San Francisco Chronicle*, January 25, 1955; "Japanese Farmers to Study at UC," *San Francisco Chronicle*, March 25, 1955.

73. "Too Many Japanese Officials Now Seek to Visit Washington," *Nichi Bei Times*, July 1, 1958.

74. "Japan Poses a Dilemma," *San Francisco Chronicle*, February 2, 1953.

75. Meghan Warner Mettler, "Gimcracks, Dollar Blouses, and Transistors: American Reactions to Imported Japanese Products, 1945-1964," *Pacific Historical Review* 79 (2010): 202-30; Michael R. Auslin, *Pacific Cosmopolitans: A Cultural History of U.S.-Japanese Relations* (Cambridge, MA: Harvard University Press, 2011), 180-81.

76. "Christopher Endorses 'Adopt' Osaka Plan," *San Francisco Chronicle*, April 24, 1957.

77. Charles von Loewenfeldt to Patricia Connich, January 9, 1958; Christopher to Nakai, January 20, 1958; both in folder 16, box 6, series 3, SFH 7, SFPL.

78. Quote from Clifton B. Forster, "Osaka Affiliation with San Francisco," no date, 2, folder "C1-2 Community Participation—San Francisco," box 18, finding aid A1, entry 56, RG 306, NARA; Akira Nishiyama to George Christopher, September 19, 1957; Christopher to Akira Nishiyama, September 25, 1957; Nakai Mitsuji to Christopher, telegram, October 3, 1957; Richard Abbot to J. Allen, September 27, 1957; and John Bolles to Christopher, September 30, 1957; all in folder 15, box 6, series 3, SFH 7, SFPL.

79. "大阪と姉妹になりましょう：サンフランシスコ都市縁組取決め [Let's Become Sisters with Osaka: Establishing San Francisco as a Sister City Decided]," *Osaka Shinbun*, April 27, 1957.

80. Quotes from San Francisco Chamber of Commerce, press release, October 28, 1957, folder 15, box 6, series 3, SFH 7, SFPL; and San Francisco Chamber of Commerce, "San Francisco and Osaka . . . Sister Cities in Trade," invitation, circa November 1957, folder 15, box 6, series 3, SFH 7, SFPL. James Mak folder "C1-2, Community Participation, San Francisco," box 18, finding aid A1, entry 56, RG 306, NARA *Tourism and the Economy: Understanding the Economics of Tourism* (Honolulu: University of Hawai'i Press, 2004), 103; "San Jose, California Salute to Okayama, Japan," memorandum, May 6, 1957, folder "C1-2, Community Participation, San Jose," box 18, finding aid A1, entry 56, RG 306, NARA; "San Jose-Okayama Tie to Become Official in Ceremonies on May 25," *Nichi Bei Times*, May 10, 1957; Osaka Mayor's Reception, guest list, November 1, 1957, folder 15, box 6, series 3, SFH 7, SFPL; Chalmers Johnson, *MITI and the Japanese Miracle: The Growth of Industrial Policy, 1925-1975* (Stanford, CA: Stanford University Press, 1982), 231-32.

81. "New Developments Japan, East Asia Trade," *International Bulletin*, November 15, 1957, 1.

82. Osaka Mayor's Reception, guest list, no date; San Francisco Chamber, "Sister Cities in Trade."

83. V. Fusco to John D. Sullivan, August 7, 1961, folder 21, box 6, series 3, SFH 7, SFPL.

84. John D. Sullivan to John Shelley, July 23, 1964, folder 9, box 6, series 3, John F. "Jack" Shelley Papers (SFH 10), SFPL.

85. Lester L. Goodman to George Christopher, January 23, 1961, folder 21, box 6, series 3, SFH 7, SFPL.

86. Conger Reynolds to Lester L. Goodman, August 21, 1961, folder 20, box 6, series 3, SFH 7, SFPL.

87. John D. Sullivan to George Christopher, December 4, 1963, folder 2, box 7, series 3, SFH 7, SFPL.

88. Sven Beckert, *Empire of Cotton: A Global History* (New York: Alfred A. Knopf, 2014), 231.

89. Von Loewenfeldt does not seem to have the other Japanese accounts by 1959; no reference was made to them in his introduction to Osakan affiliates, although his Japan Air Lines account was provided in evidence of his suitability. Phillips S. Davies to Christopher, January 16, 1959, folder 17, box 6, series 3, SFH 7, SFPL; Goodman to Mark Bortman, March 26, 1965, folder 8, box 6, series 3, SFH 10, SFPL; Pat Choate, "Political Advantage: Japan's Campaign for America," *Harvard Business Review* 68 (October 1990): 102.

90. *Pacific Trade Patterns*, 152 (statement of Lester L. Goodman, president, San Francisco World Trade Association and chairman, Getz Bros. & Co., Inc.).

91. John D. Sullivan to George Christopher, no date, circa February 1959, folder 17, folder 15, box 6, series 3, SFH 7, SFPL.

92. San Francisco Youth Association, "Report of the SFYA Cooperation with the Mayor's Com-

mittee on San Francisco–Osaka Town Affiliation," no date, folder 16, box 6, series 3, SFH 7, SFPL; "Greetings to Students in Osaka," *San Francisco Call-Bulletin*, April 18, 1958.

93. Herb Caen, *The San Francisco Book* (Boston: Houghton Mifflin, 1948), 63.

94. Akira Iriye, *Across the Pacific: An Inner History of American–East Asian Relations*, rev. ed. (Chicago: Imprint Publications, 1992), 3–7; Henning, *Outposts of Civilization*; Mari Yoshihara, *Embracing the East: White Women and American Orientalism* (New York: Oxford University Press, 2002).

95. Kanzo Matsumoto to Lester L. Goodman, March 30, 1961, folder 21, box 6, series 3, SFH 7, SFPL.

96. Dower, *War without Mercy*, 22, 111.

97. "Festival Sightseers Board Six Ships," *San Francisco Chronicle*, September 14, 1958; "6 Japanese Warships Due Here Friday," *San Francisco Chronicle*, September 7, 1958.

98. Shibusawa, *America's Geisha Ally*, 180–212.

99. "6 Japanese Warships Due Here Friday"; "S.F. Welcomes Six Japanese Warships," *San Francisco Chronicle*, September 13, 1958.

100. "Festival Sightseers Board Six Ships."

101. "Queen of the Pacific?"

102. "Pacific Festival Week Set for Sept. 12 to 19 Here," *Nichi Bei Times*, August 13, 1958.

103. "First Japanese Warships since War Tie Up in S.F.," *San Francisco Chronicle*, February 3, 1955; "Japanese Warships on Goodwill Visit," *San Francisco Chronicle*, October 27, 1955; "Five Japanese Training Ships Due Tuesday," *San Francisco Chronicle*, June 19, 1960; "Bay Visit Starts for Japan Ships," *San Francisco Chronicle*, August 12, 1964.

104. "Japan Centennial," *San Francisco Chronicle*, May 17, 1960.

105. "「咸臨丸入港記念碑」贈る：大阪姉妹都市のサンフランシスコへ ['Monument for the Kanrin Maru Arrival' Presented: From Osaka to the Sister City of San Francisco]," *Osaka Shinbun*, March 30, 1960.

106. Yukichi Fukuzawa, *The Autobiography of Yukichi Fukuzawa*, trans. Eiichi Kiyooka (Tokyo: Hokuseido Press, 1934), 113; Dana B. Young, "The Voyage of the Kanrin Maru to San Francisco, 1860," *California History* 61 (1983): 264–75.

107. Nakai Mitsuji to George Christopher, March 17, 1960, folder 3, box 7, series 3, SFH 7, SFPL; "Yoshida Says U.S., Japan Share Common Destiny," *Japan Times*, May 11, 1960; "Japan-U.S. Celebrations Mark Month," *Japan Times*, May 17, 1960; W. G. Beasley, *The Japanese Experience: A Short History of Japan* (Berkeley: University of California Press, 1999), 192–98.

108. "Parade Winds Up Centennial," *San Francisco Chronicle*, May 23, 1960.

109. "Japanese Pageant Today," *San Francisco Examiner*, May 22, 1960.

110. "Parade Winds Up Centennial."

111. "Kimonos, Samurai Delight Market Street," *San Francisco Examiner*, May 23, 1960; "Parade Winds Up Centennial."

112. "Parade Winds Up Centennial."

113. Auslin, *Pacific Cosmopolitans*, 182–83.

114. "Japan Trade Centennial Set for Sunday, May 22," *International Bulletin*, May 1960, 2; "Parade Ends World Trade Week Today," *San Francisco Chronicle*, May 22, 1960.

115. "Mayor Christopher Greeted at Osaka, Sister City to S.F.," *Nichi Bei Times*, February 20, 1959; "Mayor Christopher Given Keys to Tokyo, to Visit Osaka Next," *Nichi Bei Times*, February 19, 1959.

116. Japan-America Society of Osaka, newsletter, April 1959, folder 10, box 6, series 3, SFH 7, SFPL.

117. G. Joseph Hummel, "The Sister City and Citizen Diplomacy," *International Educational Cultural Exchange*, Fall 1970, 29.

118. Sara Fieldston, "Little Cold Warriors: Child Sponsorship and International Affairs," *Diplomatic History* 38, no. 2 (2014): 246; Joanne Meyerowitz, "'How Common Culture Shapes the

Separate Lives': Sexuality, Race, and Mid-Twentieth-Century Social Constructionist Thought," *Journal of American History* 96, no. 4 (2010): 1057-84.

119. San Francisco Youth Association, "Informative Program Highlights," July 1, 1957-June 30, 1958, folder 16, box 6, series 3, SFH 7, SFPL; Thomas Rowe to Lester L. Goodman, September 8, 1961, folder 21, box 6, series 3, SFH 7, SFPL.

120. Harold Spears, "San Francisco Unified School District Participation: Osaka-San Francisco Affiliation Program," no date (circa 1961), folder 21, box 6, series 3, SFH 7, SFPL.

121. San Francisco Youth Association, newsletter, June 8, 1964, folder 32, carton 106, National Association for the Advancement of Colored People, Region 1, Records, BANC MSS 78/180c (hereafter NAACP), Bancroft Library, University of California, Berkeley (hereafter BANC); "Report of Activities of the SF-Osaka Sister-City Affiliation," January 1, 1961-October 1, 1961," no date, 3, folder 21, box 6, series 3, SFH 7, SFPL; "Osaka-San Francisco Town Affiliation Program Narrative Report," 1957-1962, Draft," November 29, 1961, 11, folder 21, box 6, series 3, SFH 7, SFPL.

122. Lester L. Goodman to John D. Sullivan, July 26, 1961, folder 21, box 6, series 3, SFH 7, SFPL.

123. "San Francisco Seen thru the Eyes of an Osaka School Girl," no date (circa 1961), folder 21, box 6, series 3, SFH 7, SFPL.

124. Quote from Toshio Kudoh to John T. Buckley, February 8, 1962, folder 6, box 7, series 3, SFH 7, SFPL; "Personal History," no date (circa 1962), folder 6, box 7, series 3, SFH 7, SFPL.

125. "A Guest from Japan," *San Francisco Examiner*, April 2, 1962.

126. "Shopping with Miss Osaka," *San Francisco Chronicle*, April 6, 1962.

127. Goodman to Sullivan, July 26, 1961.

128. Roger Daniels, *The Politics of Prejudice: The Anti-Japanese Movement in California and the Struggle for Japanese Exclusion* (Berkeley: University of California Press, 1962), 39-40.

129. Quotes from "Osaka's Sister," *San Francisco Chronicle*, April 25, 1963; and "San Francisco to Conduct Contest to Select 'Miss Sister City' for Osaka Visit," *Nichi Bei Times*, March 26, 1963. "Pretty Envoy with Brains," undated newspaper clipping, folder 7, box 17, series 5, SFH 7, SFPL; "'Sukiyaki' Song at Airport to Send Off 'Miss Sister City' to Osaka Friday P.M.," *Hokubei Mainichi*, June 14, 1963, folder 7, box 17, series 5, SFH 7, SFPL; John D. Sullivan to George Christopher, March 15, 1963, 2, folder 8, box 7, series 3, SFH 7, SFPL.

130. San Francisco Chamber of Commerce World Trade Department, *Overseas Buyers' Guide to the San Francisco Area, U.S.A.* (San Francisco: San Francisco Chamber of Commerce, no date); "World Trade," *Bay Region Business*, January 30, 1959, 8.

131. "Report of Activities of the San Francisco-Osaka Sister-City Affiliation"; "Japanese-Language Guide to S.F. Issued," *Bay Region Business*, December 16, 1960, 4.

132. *Pacific Trade Patterns*, 159 (statement of James P. Wilson, Manager, World Trade Department, San Francisco Chamber of Commerce, and Executive Secretary of the San Francisco Area World Trade Association).

133. Quotes from George Christopher to the Board of Supervisors, October 6, 1958, folder 28, box 7, series 5, SFH 7, SFPL; John H. Mollenkopf, *The Contested City* (Princeton, NJ: Princeton University Press, 1983), 151; Frederick M. Wirt, *Power in the City: Decision Making in San Francisco* (Berkeley: University of California Press, 1974), 114-22.

134. "Mayor Seeks Closer U.S.-Japan Ties," *Nichi Bei Times*, February 25, 1959; "Head Table List," October 22, 1965, folder 9, box 7, series 3, SFH 7, SFPL.

135. Christopher to the Board of Supervisors, January 21, 1959, folder 17, box 6, series 3, SFH 7, SFPL; John D. Sullivan to John Shelley, February 3, 1965, folder 8, box 6, series 3, SFH 10, SFPL.

136. "Osaka-San Francisco Town Affiliation Program Narrative Report, 1957-1962, Draft," November 29, 1961, 11, folder 21, box 6, series 3, SFH 7, SFPL; President Eisenhower's Civic Committee, People-to-People, *Newsletter*, November 1959, 1.

CHAPTER THREE

1. San Francisco Redevelopment Agency (SFRA), *Report to Mayor Elmer E. Robinson for the Fiscal Year 1952-1953* (San Francisco: SFRA, 1952), 2; "Western Addition Slum Clearance Hearing," *San Francisco Chronicle*, October 15, 1952.

2. "First Public Hearings Speed Slum Clearance," *San Francisco Examiner*, October 15, 1952.

3. "Where Will You Live?," *Sun-Reporter*, October 4, 1952; "Plans Told for Western Addition," *San Francisco Chronicle*, August 20, 1952; "S.F. Slum Clearance Plan Goes Forward," *San Francisco Call-Bulletin*, October 15, 1952; "No Bias in Redevelopment Western Addition," *Sun-Reporter*, October 18, 1952.

4. "Western Addition Slum Clearance Hearing."

5. SFRA, *Narrative Report of Work Progress* (San Francisco: SFRA, 1952), 2; "Hearing on Western Addition Set," *San Francisco Chronicle*, September 3, 1952.

6. Mel Scott for the San Francisco City Planning Commission, *New City: San Francisco Redeveloped* (San Francisco: City Planning Commission, 1947).

7. Roger W. Lotchin, *Fortress California, 1910-1961: From Warfare to Welfare* (New York: Oxford University Press, 1992), 162.

8. SFRA, *Report to Mayor Robinson*, 2.

9. Stewart Black, "Redevelopment in California: Its Past, Present, and Possible Future," *California Journal of Politics and Policy* 6 (2014): 472.

10. San Francisco City Planning Commission, *The Master Plan of San Francisco: The Redevelopment of Blighted Areas: Report on Conditions Indicative of Blight and Redevelopment Policies* (San Francisco: San Francisco City Planning Commission, 1945), 8.

11. Mark I. Gelfand, *A Nation of Cities: The Federal Government and Urban America, 1933-1965* (New York: Oxford University Press, 1975), 105-13; Amy L. Howard, *More Than Shelter: Activism and Community in San Francisco Public Housing* (Minneapolis: University of Minnesota Press, 2014), 2-9; Allan A. Twichell, "Measuring the Quality of Housing in Planning for Urban Redevelopment," in *Urban Redevelopment: Problems and Practices*, ed. Coleman Woodbury (Chicago: University of Chicago Press, 1953), 16-19; Alexander von Hoffman, "A Study in Contradictions: The Origins and Legacy of the Housing Act of 1949," *Housing Policy Debate* 11 (2000): 299-326; Alexander von Hoffman, "Housing and Planning: A Century of Social Reform and Local Power," *Journal of the American Planning Association* 75 (2009): 232-35, 237-38; San Francisco City Planning Commission, *Report on Progress of Work Done to Date to Achieve a San Francisco Master Plan* (San Francisco: City Planning Commission, 1942); Roger W. Lotchin, "World War II and Urban California: City Planning and the Transformation Hypothesis," *Pacific Historical Review* 61, no. 2 (May 1993): 150, 155-58; Theresa J. Mah, "Buying into the Middle Class: Residential Segregation and Racial Formation in the United States, 1920-1964" (PhD diss., University of Chicago, 1999), 85-119; Jon C. Teaford, *The Rough Road to Renaissance: Urban Revitalization in America, 1940-1985* (Baltimore: Johns Hopkins University Press, 1990), 10-33.

12. Samuel Zipp, *Manhattan Projects: The Rise and Fall of Urban Renewal in Cold War New York* (New York: Oxford University Press, 2010), 5.

13. San Francisco City Planning Commission, *The Master Plan of the City and County of San Francisco: A Brief Summary* (San Francisco: City Planning Commission, 1946), 3.

14. City Planning Commission, *Redevelopment of Blighted Areas*, xi.

15. For similar uses of metaphors, see the widely influential Urbanism Committee of the National Resources Committee, *Our Cities: Their Role in the National Economy* (Washington, DC: GPO, 1937); and Coleman Woodbury and Frederick A. Gutheim, *Rethinking Urban Redevelopment* (Chicago: Public Administration Service, 1949).

16. Scott, *New City*, 4-5.

17. San Francisco City Planning Commission, *The Next Step in Urban Redevelopment: A Report*

to the Board of Supervisors on the Selection of the First Redevelopment Area (San Francisco: City Planning Commission, 1947), 5-6.

18. US Congress, House, Subcommittee on Naval Affairs, *Investigation of Congested Areas, Part 3: Hearings*, 78th Cong., 1st sess. (1943), 899.

19. Twichell, "Measuring the Quality of Housing,"11.

20. San Francisco Department of City Planning in Cooperation with the Redevelopment Agency, *Replanning the Geary Area in the Western Addition* (San Francisco: Department of City Planning, 1952), A-5. For more on the fraught relationship between segregation, profit, and real estate, see Thomas J. Sugrue, *The Origins of the Urban Crisis: Race and Inequality in Postwar Detroit* (Princeton, NJ: Princeton University Press, 1996), 47-51; and N. D. B. Connolly, *A World More Concrete: Real Estate and the Remaking of Jim Crow South Florida* (Chicago: University of Chicago Press, 2014). For a nuanced view of ethnic whites' struggle against deterioration before racial transformation, see Amanda I. Seligman, *Block by Block: Neighborhoods and Public Policy on Chicago's West Side* (Chicago: University of Chicago Press, 2005).

21. City Planning Commission, *Redevelopment of Blighted Areas*, 11.

22. Redevelopment Agency of the City and County of San Francisco in Cooperation with the Department of City Planning, *The Tentative Plan: For Redevelopment of Western Addition Project Area Number One as Submitted to the San Francisco Board of Supervisors, November 13, 1952* (San Francisco: SFRA, 1952), 5, 9.

23. Ibid., 9.

24. Mel Scott for the San Francisco City Planning Commission, *Western Addition District: An Exploration of the Possibilities of Replanning and Rebuilding one of SF's Largest Blighted Districts under the California Community Redevelopment Act of 1945* (San Francisco: City Planning Commission, 1947), 57, 4; Department of City Planning, *Replanning the Geary Area*, preface, 37; Twichell, "Measuring the Quality of Housing," 13. For similar ideas, see Redevelopment Agency, *Tentative Plan . . . as Submitted*; Redevelopment Agency of the City and County of San Francisco in Cooperation with the Department of City Planning, *The Tentative Plan: For Redevelopment of Western Addition Project Area Number One and Related Documents* (San Francisco: SFRA, 1952); Gelfand, *Nation of Cities*, 106-9; Jeanne R. Lowe, *Cities in a Race with Time: Progress and Poverty in America's Renewing Cities* (New York: Random House, 1967); Teaford, *Rough Road to Renaissance*, 10-11, 16.

25. Housing Authority of the City and County of San Francisco, *Real Property Survey, San Francisco, California* (San Francisco: City and County of San Francisco, 1941), 28, 24.

26. Quoted in Scott Tang, "Pushing at the Golden Gate: Race Relations and Racial Politics in San Francisco, 1940-1955" (PhD diss., University of California, Berkeley, 2002), 154.

27. For Chinatown's rehabilitation during the war, see, for example, Charlotte Brooks, *Between Mao and McCarthy: Chinese American Politics in the Cold War Years* (Chicago: University of Chicago Press, 2015), 57, 64-69; Kevin Scott Wong, *Americans First: Chinese Americans and the Second World War* (Cambridge, MA: Harvard University Press, 2005); Ellen D. Wu, *The Color of Success: Asian Americans and the Origins of the Model Minority* (Princeton, NJ: Princeton University Press, 2014), 54-55; Judy Yung, *Unbound Feet: A Social History of Chinese Women in San Francisco* (Berkeley: University of California Press, 1995), 249-77.

28. Redevelopment Agency, *Tentative Plan . . . as Submitted*, A-4.

29. City Planning Commission, *Western Addition District*, 3.

30. Scott, *New City*, 2-3.

31. City Planning Commission, *Redevelopment of Blighted Areas*, 1.

32. Scott, *Western Addition District*, 58, 11; US Bureau of the Census, *17th Census of the United States, 1950: Census of the Population, Vol. III, Part IV: Census Tracts Statistics, San Francisco* (Washington, DC: GPO, 1953), 46.

33. Wilson Record, "Minority Groups and Intergroup Relations in the San Francisco Bay Area," in *The San Francisco Bay Area: Its Problems and Future*, ed. Stanley Scott (Berkeley: University of California, Berkeley, Institute for Governmental Studies, 1966), 22; Scott, *Western Ad-*

dition District, 57, 58, 11; San Francisco Planning and Housing Association, *Blight and Taxes* (San Francisco: San Francisco Planning and Housing Association, 1947); Department of City Planning, *Replanning the Geary Area;* Redevelopment Agency, *Tentative Plan . . . and Related Documents.*

34. Department of City Planning, *Replanning the Geary Area,* letter of transmittal.

35. Redevelopment Agency, *Tentative Plan . . . and Related Documents,* 9.

36. Ibid., B-4; Scott, *New City,* 23; Board of Supervisors of the City and County of San Francisco, *Public Hearing on Redevelopment of the Western Addition* (San Francisco: City and County of San Francisco, June 3, 1948), 8.

37. Board of Supervisors, *Public Hearing,* 26.

38. Ibid., 39, 32, 55, 51.

39. Ibid., 15, 13.

40. Black, "Redevelopment in California," 471.

41. SFRA Commission, minutes, December 13, 1948, San Francisco Redevelopment Agency, San Francisco, CA [Please note: the minutes were held in the SFRA headquarters in San Francisco, now closed. The San Francisco History Center at the San Francisco Public Library has since acquired all its records.]; SFRA Commission, minutes, December 14, 1948; Council for Civic Unity of San Francisco, "Accomplishment," *Among These Rights . . . ,* newsletter, January–February 1955, 2, folder 10, box 4, subseries C, series 1, SFH 10, SFPL. For a close analysis of the entirety of the hearings, see William Issel, *Church and State in the City: Catholics and Politics in Twentieth-Century San Francisco* (Philadelphia, PA: Temple University Press, 2013), 150–60.

42. Board of Supervisors, *Public Hearing,* 26–27.

43. R. L. Polk & Co., *Polk's Crocker-Langley San Francisco City Directory, 1948–1949* (San Francisco: R. L. Polk, 1949) 43; Ulma A. Abels, Electronic Army Serial Number Merged File, circa 1938–1946 (Enlistment Records), World War II Enlistment Records, Record Group 64, NARA.

44. "Plans Told for Western Addition."

45. Board of Supervisors, *Public Hearing,* 23, 24.

46. The famed contemporary study *Black Metropolis* makes this connection clearly, and quotes segregated black residents to show their awareness of the problem's cause. St. Clair Drake and Horace R. Cayton, *Black Metropolis: A Study of Negro Life in a Northern City* (1945; repr., Chicago: University of Chicago Press, 1993), 199, 207.

47. "The City's Housing," *San Francisco Chronicle,* June 4, 1948.

48. "Supervisors Start Western Addition Action," *San Francisco Chronicle,* July 20, 1948.

49. Board of Supervisors, *Public Hearing,* 67.

50. "What Price Slum Clearance?," *Pacific Citizen,* July 31, 1948. Many thanks to Greg Robinson for bringing the article to my attention and sharing his transcribed copy.

51. "SF Redevelopment Project to Be Started in October," *Nichi Bei Times,* January 19, 1953.

52. Victor S. Abe to Dewey Mead, December 4, 1952, folder Block 700 #1, box D/RE 99, Project WA-A1, San Francisco Redevelopment Agency Central Records, San Francisco, CA (hereafter SFRA [Please note: All archival records of the SFRA were deposited with the San Francisco History Center at the San Francisco Public Library after the SFRA's dissolution in 2012]); "Western Addition Slum Clearance Hearing," *San Francisco Chronicle,* October 15, 1952.

53. "Western Addition Slum Clearance Hearing"; SFRA Commission, minutes, January 20, 1953.

54. SFRA Commission, minutes, January 20, 1953.

55. Redevelopment Agency, *Tentative Plan . . . and Related Documents,* 9.

56. SFRA Commission, minutes, January 20, 1953.

57. Oyama v. State of California, 332 U.S. 633 (1948); Torao Takahashi v. Fish and Game Commission, 334 U.S. 401 (1948); Sei Fuji v. State 38 Cal.2d 718 (1952).

58. Mark Brilliant, *The Color of America Has Changed: How Racial Diversity Shaped Civil Rights Reform in California, 1941-1978* (New York: Oxford University Press, 2010), 42–43; Cindy I-Fen Cheng, *Citizens of Asian America,* 176–77; Frank F. Chuman, *The Bamboo People: The Law and*

Japanese-Americans (Chicago: Japanese American Citizens League, 1981), 201–21; Roger Daniels, *Asian America: Chinese and Japanese in the United States since 1850* (Seattle: University of Washington Press, 1850), 296–99; Greg Robinson, *After Camp: Portraits in Midcentury Japanese American Life and Politics* (Berkeley: University of California Press, 2012), 200–211; Wu, *Color of Success*, 97–100.

59. Victor S. Abe to Joseph Alioto, December 2, 1958, folder Block 700 #1, box D/RE 99, Project WA-A1, SFRA.

60. Sam Seiki, interview by author, July 13, 2004, San Francisco, CA, transcript, National Japanese American Historical Society, San Francisco, CA (hereafter NJAHS); Uta Hirota, interview by author, July 8, 2004, San Francisco, CA, transcript, NJAHS; Sumi Honnami, interview with Ken Yamada, July 21, 2000, San Francisco, CA, transcript, NJAHS; Japantown Task Force, *San Francisco's Japantown* (San Francisco: Arcadia, 2005), 77; R. L. Polk & Co., *Polk's San Francisco City Directory 1953* (San Francisco: R. L. Polk, 1953), 578, 888, 1158; Nichi Bei Times, *Evacuation-Resettlement Report: 1948 Directory* (San Francisco: Nichi Bei Times, 1948), 3, 14.

61. Aaron Levine, *The Urban Renewal of San Francisco* (San Francisco: San Francisco Planning and Housing Association of the Blyth-Zellerbach Committee, 1959), 8.

62. Hirota interview.

63. Victor S. Abe to Dewey Mead, December 4, 1952.

64. US Bureau of the Census, *16th Census of the Population, 1940: Characteristics of the Nonwhite Population by Race* (Washington, DC: GPO, 1953), 79; US Bureau of the Census, *17th Census of the United States, 1950: Census of the Population, Vol. III, Part IV: Census Tract Statistics, San Francisco–Oakland and Adjacent Area* (Washington, DC: GPO, 1952), 7; Thomas Modell, *The Economics and Politics of Racial Accommodation: The Japanese of Los Angeles, 1900–1942* (Chicago: University of Illinois Press, 1977), chapters 6 and 7; Dorothy Swaine Thomas and Richard Nishimoto, *The Salvage* (Berkeley: University of California Press, 1952), 47; Edward K. Strong, *The Second Generation Japanese Problem* (Stanford, CA: Stanford University Press, 1934); Alexander Yoshikazu Yamato, "Socioeconomic Change among Japanese Americans in the San Francisco Bay Area" (PhD diss., University of California, Berkeley, 1986), 316.

65. Planning Committee of the Japanese Chamber of Commerce of Northern California (JCCNC), "A Report on the Redevelopment of the Western Addition," circa 1952, 2, file "Redevelopment," box 6, series 1, Japanese American Citizens League History Collection, Japanese American National Library, San Francisco, CA (hereafter JANL); Minako Kurokawa, "Occupational Mobility among Japanese Businessmen in San Francisco" (MA thesis, University of California, Berkeley, 1962), 18.

66. Planning Committee of the JCCNC, "Report," 2.

67. Ibid.; "City Agency Gets Plans for Japanese Shopping Center," *Nichi Bei Times*, July 9, 1953; SFRA Commission, minutes, July 14, 1953; "City Agency Approves Japanese Shop Center Plan," *Nichi Bei Times*, July 16, 1953.

68. Charles Kikuchi, "Japanese American Youth in San Francisco: Their Background, Characteristics, and Problems," report prepared for the National Youth Agency, 1941, 101, frame 10, reel 370, Japanese American Evacuation and Resettlement Records, Bancroft Library, University of California, Berkeley, 12–13; Tamotsu Shibutani, "The Initial Impact of the War on the Japanese Communities in the San Francisco Bay Region: A Preliminary Report," unpublished paper, 1942, 41, reel 10, frame 0515, Japanese American Evacuation and Resettlement Records.

69. WRA, *People in Motion: The Postwar Adjustment of the Evacuated Japanese Americans* (Washington, DC: GPO, 1947), 97–98, 112–13.

70. Harry H. L. Kitano, "Housing of Japanese-Americans in the San Francisco Bay Area," in *Studies in Housing and Minority Groups*, ed. Nathan Glazer and Davis McEntire (Berkeley: University of California Press, 1960), 178–97; Record, "Minority Groups," 22–23; Charlotte Brooks, *Alien Neighbors, Foreign Friends: Asian Americans, Housing, and the Transformation of Urban California* (Chicago: University of Chicago Press, 2009).

71. Hirota interview.

72. Lon Kurashige, *Japanese American Celebration and Conflict: A History of Ethnic Identity and Festival, 1934-1990* (Berkeley: University of California Press, 2002), 43-49. For a similar dynamic among Chicago interwar ethnics, see Lizabeth Cohen, *Making a New Deal: Industrial Workers in Chicago, 1919-1939* (New York: Cambridge University Press, 1990), 109-20.

73. Sidney Herschel Small, "San Francisco's Chinatown," *Holiday*, August 1954, 98, 101.

74. Quote from "Out of the Shadows," *Newsweek*, August 15, 1955, 19-20. Christopher Lowen Agee, *The Streets of San Francisco: Policing and the Creation of a Cosmopolitan Liberal Politics, 1950-1972* (Chicago: University of Chicago Press, 2014), 34-35. For more on Chinese Americans' political participation, see Brooks, *Between Mao and McCarthy*. For contemporary views of Chinatown assimilation, see "Hard Work at Hip Wo," *Life*, April 25, 1955, 71; and Jade Snow Wong, *Fifth Chinese Daughter* (Seattle: University of Washington Press, 1989). For more on Chinatown residents' efforts to portray themselves as assimilable into mainstream US culture, see Chiou-Ling Yeh, *Making an American Festival: Chinese New Year in San Francisco's Chinatown* (Berkeley: University of California Press, 2008); and Wu, *Color of Success*.

75. Nayan Shah, *Contagious Divides: Epidemics and Race in San Francisco's Chinatown* (Berkeley: University of California Press, 2001), 231-45; Amy L. Howard, *More Than Shelter*, 102-23.

76. "Redevelopment Agency Reveals Plans for Loans, Moving Costs," *Nichi Bei Times*, December 7, 1956.

77. Housing Act of 1949, Public Law 171, 81st Cong., 1st sess., chapter 338, S. 1070, 2.

78. "Negroes Build for Tomorrow, *San Francisco Chronicle*, December 31, 1964.

79. "Church Housing Project to Start," *San Francisco Chronicle*, January 17, 1964.

80. The Housing Act of 1954 changed redevelopment to "renewal," allowing for rehabilitation and the financial instruments necessary for low- and moderate-priced housing. It was in place well before the first Western Addition project began, but the project was implemented under the Housing Act of 1949, which lacked such innovations.

81. SFRA, *Annual Report to Mayor George Christopher for the Year July 1, 1960, to June 30, 1961* (San Francisco: SFRA, 1961), 2; SFRA, *Annual Report to Mayor George Christopher for the Year July 1, 1959, to June 30, 1960* (San Francisco: SFRA, 1960), 7-8; SFRA Commission, minutes, December 15, 1959; SFRA, *Annual Report, 1965-1966* (San Francisco: SFRA, 1966), 7; "Churches to Join in Race Conference," *San Francisco Chronicle*, July 21, 1963; "S.F. Negroes Split over Methods," *San Francisco Chronicle*, August 14, 1963; Jones Memorial Methodist Church, "The Jones History," jonesumc.com; "Rev. Hamilton T. Boswell—Led S.F. Blacks," *SFGate.com*, May 11, 2007.

82. "Negroes Build for Tomorrow, *San Francisco*, December 31, 1964.

83. Planning Committee of the JCCNC, "Report," 2.

84. Brian Masaru Hayashi, *Democratizing the Enemy: The Japanese American Internment* (Princeton, NJ: Princeton University Press, 2004), 95.

85. For example, see "S.F. Redevelopment Project to Be Started in October," *Nichi Bei Times*, January 19, 1953; and "US Lends $21,710,000 for S.F. Redevelopment," *Nichi Bei Times*, May 16, 1953. For the classic discussion of the Cold War family, see Elaine Tyler May, *Homeward Bound: American Families in the Cold War Era* (New York: Basic Books, 1999).

86. For example, "Nisei to Aid New Uptown SF AVC Unit," *Nichi Bei Times*, September 5, 1946; and "Arrest of Suspect Brings Down Burglaries in SF Uptown Area," *Nichi Bei Times*, April 4, 1954.

87. Planning Committee of the JCCNC, "Report," 1.

88. Ibid.

89. SFRA, *Annual Report to Mayor George Christopher for the Year July 1, 1957, to June 30, 1958* (San Francisco: SFRA, 1958), 1.

90. SFRA, *Report to Mayor Robinson 1952-1953*, 5-6, 17; Levine, *Urban Renewal of San Francisco*, 1; Board of Supervisors, *Public Hearing*, 1; "Ways to Speed Redevelopment Projects Studied," *San Francisco Chronicle*, September 9, 1952; Black, "Redevelopment in California," 472, 475;

Gelfand, *Nation of Cities,* 202-5; George Lefcoe, "Finding the Blight That's Right for California Redevelopment Law," *Hastings Law Journal* 52 (2001): 993; Daniel S. Maroon, "Redevelopment in the Golden State: A Study in Plenary Power under the California Constitution," *Hastings Constitutional Law Quarterly* 40 (2013): 454-55; Mel Scott, *American City Planning since 1890: A History Commemorating the Fiftieth Anniversary of the American Institute of Planners* (Berkeley: University of California Press, 1969), 489-93; Teaford, *Rough Road to Renaissance,* 107-9; von Hoffman, "Study in Contradictions"; Catherine Bauer, "Redevelopment: A Misfit in the Fifties," in *The Future of Cities and Urban Redevelopment,* ed. Coleman Woodbury (Chicago: University of Chicago Press, 1953).

91. Levine, *Urban Renewal of San Francisco,* 12-16; Black, "Redevelopment in California," 475; Chester Hartman with Sarah Carnochan, *City for Sale: The Transformation of San Francisco* (Berkeley: University of California Press, 2002), 15-20; John H. Mollenkopf, *The Contested City* (Princeton, NJ: Princeton University Press, 1983), 151-53, 159-62, 167-70. Chester MacPhee was forced to resign as chief administrator in city hall after it was found that the Redevelopment Agency paid well over the mortgaged value of a property he held an interest in; he also held other redevelopment-area properties through another real-estate company. An appraiser with the assessor's office also faced official investigations for similar dealings. "The Mayor Wrestles a Case of Anguish," *San Francisco Chronicle,* January 17, 1959; "Story of MacPhee's Troubled Months," *San Francisco Chronicle,* January 23, 1959; Paul T. Miller, *The Postwar Struggle for Civil Rights: African Americans in San Francisco, 1945-1975* (New York: Routledge, 2010), 110. On Diamond Heights and its legal controversy, see "S.F. Project Plans Nears Property Purchase Stage," *Nichi Bei Times,* December 8, 1954; Richard Brandi, "San Francisco's Diamond Heights: Urban Renewal and the Modernist City," *Journal of Planning History* 12 (2012): 134-47; SFRA in cooperation with the Department of City Planning, *Diamond Heights: A Report on the Tentative Redevelopment Plan* (San Francisco: SFRA, 1951); SFRA, *Report to Mayor Robinson 1952-1953,* 5-6; and SFRA, *Annual Report to Mayor Elmer E. Robinson for the Fiscal Year Ended June 30, 1955* (San Francisco: SFRA, 1955), 1.

92. "Further Study Planned for Japanese Shopping Center," *Nichi Bei Times,* November 6, 1953.

93. SFRA Commission, minutes, November 23, 1953.

94. "SF Businessmen Tourists to Japan to Be Feted Here," *Nichi Bei Times,* May 2, 1958.

95. Yuji Ichioka, *The Issei: The World of the First Generation Japanese Immigrants, 1885-1924* (New York: Free Press, 1988), 160.

96. Eiichiro Azuma, *Between Two Empires: Race, History, and Transnationalism in Japanese America* (New York: Oxford University Press, 2005), 43-44; Ichioka, *Issei,* 156-64; Lon Kurashige, *Japanese American Celebration,* 29-31, 53-54, 79-84, 114-15; Jere Takahashi, *Nisei/Sansei: Shifting Japanese American Identities and Politics* (Philadelphia: Temple University Press, 1997), 54-64, 125-26; Wu, *Color of Success,* 75-78.

97. "Improvements for Uptown Business Area Discussed," *Nichi Bei Times,* May 20, 1955.

98. As one *Sunset* article suggests, non-Japanese audiences were much smaller than those that attended Chinatown's famous New Year's celebrations, but growing. "Browsing and Shopping in San Francisco's Japantown," *Sunset,* July 1962, 52-53.

99. Agee, *Streets of San Francisco,* 154.

100. "Merits Support," *Sun-Reporter,* April 11, 1959.

101. "SF Issei-Nisei Merchants Protest Towaway Plan," *Nichi Bei Times,* December 10, 1954; "Parking Meters, Stop Signals Asked for Post-Buchanan Area," *Nichi Bei Times,* October 12, 1956; "NC JCC Asks SF Police for Extra Attention," *Nich Bei Times,* April 27, 1957.

102. Wu, *Color of Success,* chapter 3. See also Scott Kurashige, *The Shifting Grounds of Race: Black and Japanese Americans in the Making of Multiethnic Los Angeles* (Princeton, NJ: Princeton University Press, 2008), 189-92; Greg Robinson *A Tragedy of Democracy: Japanese Confinement*

in North America (New York: Columbia University Press, 2009), 207-10, 244; T. Fujitani, *Race for Empire: Koreans as Japanese and Japanese as Americans during World War II* (Berkeley: University of California Press, 2011), chapter 3.

103. "Planning for $1,000,000 Shopping Center Studied," *Nichi Bei Times*, October 12, 1956; "Uptown Shopping Center Planning Discussion Slated," *Nichi Bei Times*, November 24, 1956; "'Japan Village' May Be Built in SF Project Area," *Nichi Bei Times*, November 29, 1956; T. C. Bell to Eugene J. Riordan, memorandum, October 17, 1958, folder Block 700 #1, box D/RE 99, Project WA-A1, SFRA; Riordan, memorandum, November 24, 1958, folder Block 700 #1, box D/RE 99, Project WA-A1, SFRA; Victor S. Abe to Joseph Alioto, December 2, 1958, folder Block 700 #1, box D/RE 99, Project WA-A1, SFRA.

104. Masao Ashizawa, interview with the author, Ken Yamada, and Clement Lai, July 15, 2004, transcript, "Back in the Day" Collection, National Japanese American Historical Society, San Francisco, CA.

105. "Western Addition Project to Start Buying in January," *Nichi Bei Times*, August 11, 1957; Ashizawa interview; "Japanese Garden Center Incorporated for Project," *Nichi Bei Times*, March 26, 1959.

106. "Higher Settlement Asked by Nisei from SF Redevelopment Agency," *Nichi Bei Times*, February 6, 1959; "Japanese Garden Center Incorporated," *Hoku Bei Times*, March 26, 1959; "NB Dept. Store to Move into New Location Next Week," *Nichi Bei Times*, October 22, 1959; "Dave Tatsuno, 92, Whose Home Movies Captured History, Dies," *New York Times*, February 13, 2006. The late Dave Tatsuno gained fame as a home-film maker. While in the Topaz internment camp, he smuggled in a movie camera and recorded rare footage of life in the camps.

107. M. Justin Herman to Victor S. Abe, May 31, 1960, folder Block 700 #1, box D/RE 99, Project WA-A1, SFRA. According to one lawyer, in the 1950s there were about half a dozen other Japanese American lawyers in the entire Bay Area. Mas Yonemura, interview by author, March 29, 2006, Berkeley, CA.

108. Brooks, *Alien Neighbors, Foreign Friends*, 221-22. For more on the use of property values and rights to forward racial segregation, see, for example, Connolly, *World More Concrete*; Kevin M. Kruse, *White Flight: Atlanta and the Making of Modern Conservatism* (Princeton, NJ: Princeton University Press, 2005); and Matthew D. Lassiter, *Silent Majority: Suburban Politics in the Sunbelt South* (Princeton, NJ: Princeton University Press, 2006).

CHAPTER FOUR

1. Quote from "Redevelopment Set for S.F. Area North of Post St.," *Nichi Bei Times*, February 28, 1958. SFRA Commission, minutes, January 28, 1954; "New Western Addition Project Office Opens at 'Y,'" *Nichi Bei Times*, January 28, 1958; "100 Parcels of Land Already Bought for S.F. Redevelopment," *Nichi Bei Times*, February 21, 1958; "Japanese Shopping Center Proposal for S.F. Revealed," *Nichi Bei Times*, November 20, 1958; Masao Ashizawa, interview with the author, Ken Yamada, and Clement Lai, July 15, 2004, transcript, "Back in the Day" Collection, National Japanese American Historical Society, San Francisco, CA (hereafter NJAHS); Michael Dobashi, interview with Henry S. Francisco and Jovilynn T. Olegario, April 29, 2004, San Francisco, CA, transcript, NJAHS; Tomoye Takahashi, interview with the author and Judy Hamaguchi, July 16, 2004, San Francisco, CA, transcript, NJAHS; Japantown Task Force, *San Francisco's Japantown* (Chicago: Arcadia, 2005), 50, 79.

2. Among others, see Akira Iriye, *Across the Pacific: An Inner History of American-East Asian Relations* (Chicago: Imprint Publications, 1992), 379, 383; and Theodore H. White, "The Danger from Japan," *New York Times*, July 28, 1985.

3. "S.F. Nisei Group Reveals Plans for Shopping Center," *Nichi Bei Times*, November 28, 1958.

4. There were fourteen officers, directors, and charter investors at this point. Of these, one was not listed in the phone book; two lived out of town; four in the Richmond, Sunset, and Merced Heights districts; and seven in Japanesetown. R. L. Polk & Co., *Polk's San Francisco City Directory 1958* (Los Angeles: R. L. Polk, 1958), 3, 734, 951, 969, 1015, 1026, 1029, 1250, 1294, 1392, 1405, 1569; SFRA Commission, minutes, December 16, 1958; Japantown Task Force, *San Francisco's Japantown*, 22; Takahashi interview.

5. "Preliminary Plans Prepared for Japanese Garden Center," *Nichi Bei Times*, November 26, 1959.

6. "S.F. Nisei Group Reveals Plans."

7. "Japanese Garden Center Incorporated for Project," *Nichi Bei Times*, March 26, 1959.

8. "S.F. Nisei Group Reveals Plans."

9. Victor S. Abe to Everett Griffin, November 4, 1959, folder Block 700 #1, box D/RE 99, Project WA-A1, San Francisco Redevelopment Agency Central Records, San Francisco, CA (hereafter SFRA).

10. Nayan Shah, *Contagious Divides: Epidemics and Race in San Francisco's Chinatown* (Berkeley: University of California Press, 2001), 152-53.

11. "Preliminary Plans Prepared."

12. Abe to Griffin, November 4, 1959.

13. "Japanese Garden Center Plan Will Be Submitted," *Nichi Bei Times*, June 18, 1959; "Preliminary Plans Prepared"; "'Japan Town' Planned for S.F. Block," *San Francisco Chronicle*, December 17, 1958.

14. For more on the contemporary "Japan boom," see Meghan Warner Mettler, *How to Reach Japan by Subway: America's Fascination with Japanese Culture, 1945-1965* (Lincoln: University of Nebraska Press, 2018). For its 1980s iteration, see Andrew C. McKevitt, *Consuming Japan: Popular Culture and the Globalizing of 1980s America* (Chapel Hill: University of North Carolina Press, 2017).

15. "Program Set for SF Park Japan Week," *Nichi Bei Times*, July 21, 1957; "Japan Day Fete Planned by NCJCC," *Nichi Bei Times*, May 25, 1957; "Examiner Editorial Praises Japan Day at Golden Gate Park," *Nichi Bei Times*, August 20, 1957.

16. "Japan Village for SF," *San Francisco Chronicle*, December 19, 1958.

17. "President's Message," *Bay Region Business*, January 13, 1961, 1, 4; "S.F. Examiner Hails Chamber Executive HQ Campaign," *Bay Region Business*, May 25, 1962, 1.

18. "30 San Francisco Corporations Total $32.6 Billion in Assets," *Bay Region Business*, December 20, 1957, 4; "S.F. Executive HQ Program Geared by Chamber," *Bay Region Business*, August 10, 1962, 1.

19. "Chamber's '59 Annual Report," *Bay Region Business*, January 29, 1960, 1.

20. Aaron Levine, *The Urban Renewal of San Francisco* (San Francisco: San Francisco Planning and Housing Association of the Blyth-Zellerbach Committee, 1959), 19, 21.

21. Christopher Lowen Agee, *The Streets of San Francisco: Policing and the Creation of a Cosmopolitan Liberal Politics, 1950-1972* (Chicago: University of Chicago Press, 2014), 7, 29-31; John H. Mollenkopf, *The Contested City* (Princeton, NJ: Princeton University Press, 1983), 151, 167.

22. "HHFA Area Office to Be Opened Here," *San Francisco Chronicle*, April 28, 1951.

23. Quote from Chester Hartman with Sarah Carnochan, *City for Sale: The Transformation of San Francisco* (Berkeley: University of California Press, 2002), 16.

24. "SF Starting Search Nation's Top Developers," *San Francisco Chronicle*, April 20, 1959; "Slum Chief: US Aide Gets Job," *San Francisco Chronicle*, May 22, 1959; Hartman, *City for Sale*, 17-18; Mollenkopf, *Contested City*, 151-54, 167-69; "Leaving a Heart in San Francisco," *Business Week*, May 1969, 111.

25. SFRA, *The Golden Gateway Redevelopment Project* (San Francisco: SFRA, 1959), 2.

26. Quote from "President's Message," *Bay Region Business*, December 15, 1961, 1; "Construc-

tion in San Francisco Reaches Record High of $625 Million for 1959," *Bay Region Business*, January 15, 1960, 1; Gray Brechin, *Imperial San Francisco: Urban Power, Earthly Ruin* (Berkeley: University of California Press, 1999); Hartman, *City for Sale*, 7-11; Stephen J. McGovern, *The Politics of Downtown Development: Dynamic Political Cultures in San Francisco and Washington, D.C.* (Lexington: University Press of Kentucky, 1998), 61-64. For more on the urban redevelopment of downtowns, see, for example, Alison Isenberg, *Downtown America: A History of the Place and the People Who Made It* (Chicago: University of Chicago Press, 2004), 168-73; and David Schuyler, *A City Transformed: Redevelopment, Race, and Suburbanization in Lancaster, Pennsylvania, 1940-1980* (University Park, PA: Penn State University Press, 2002).

27. "Queen of the Pacific—or Just an Aging Matron?," *San Francisco Chronicle*, September 12, 1958.

28. "Expert Picking Art for Brundage Show," *San Francisco Chronicle*, April 9, 1960.

29. "Mayor Maps Fight for Art Collection," *San Francisco Chronicle*, September 11, 1958.

30. "A Is for Brundage," *San Francisco Chronicle*, June 6, 1960.

31. "Board Backs Brundage Bonds," *San Francisco Chronicle*, April 16, 1960; "Cooks Union Votes to Back Proposition A," *San Francisco Chronicle*, May 4, 1960; "Two State Measures, Five City Propositions Get Chamber Backing," *Bay Region Business*, June 3, 1960, 1; "Vote 'Yes' on Proposition A Urged," *Bay Region Business*, May 20, 1960, 1; "Top Officials Here Support Prop. A," *San Francisco Chronicle*, June 3, 1960.

32. "Mission Swept by Oriental Artistry," *San Francisco Chronicle*, May 24, 1960; "Lectures on Brundage Collection," *San Francisco Chronicle*, May 12, 1960.

33. "Record Crowds See Asian Art," *San Francisco Chronicle*, May 25, 1960.

34. "A Is for Brundage"; "Brundage Collection—a Public Treasure," *San Francisco Chronicle*, June 10, 1966.

35. "The Fabled Brundage," *San Francisco Chronicle*, May 8, 1960; "Unmentioned Gift," *San Francisco Chronicle*, May 25, 1960; "Prop. A Wins New Support," *San Francisco Chronicle*, June 7, 1960.

36. "Not a Gift but a Trust," *San Francisco Chronicle*, June 14, 1966.

37. "Brundage Art at Crossroads," *San Francisco Chronicle*, August 28, 1966.

38. "Museum History," Asian Art Museum, http://www.asianart.org/about/history.

39. "New Hotel Built on Historic Site," *Sun-Reporter*, February 10, 1968.

40. SFRA Commission, minutes, September 22, 1959.

41. As quoted in "Mayor Christopher Backs Japan Center Project Plan," *Nichi Bei Times*, September 24, 1959.

42. Robert L. Rumsey to Clifford J. Geertz, December 8, 1959, folder Block 700 #1, box D/RE 99, Project WA-A1, SFRA.

43. "Western Slum Lands on Sale in February," *San Francisco Chronicle*, October 14, 1959.

44. Kobe Consulate to the Department of State, Washington, DC, April 23, 1959, dispatch 202, folder 210.2 1959-1961, Assistance to American Trade, box 13, Classified General Records, 1952-1963, Kobe Consulate, Japan, RG 84, NARA.

45. "Japanese Stores Plan Foreign Units," *New York Times*, February 6, 1959; "Daimaru Planning to Set Up Stores in Southeast Asia," April 23, 1959, enclosure 3, dispatch 202, folder 210.2, box 13, Classified General Records, 1952-1963, Kobe Consulate, Japan, RG 84, NARA; "Daimaru May Build Stores in Kuala Lumpur," n.d., enclosure 4, folder 210.2, box 13, Classified General Records, 1952-1963, Kobe Consulate, Japan, RG 84, NARA; George M. Emory to the Department of State, April 23, 1959, dispatch 202, enclosure, 2, folder 510.2, box 13, General Records, 1936-1961, US Consulate Kobe, Japan, RG 84, NARA; "General Economic Matters, 1956-1958," folder 500, box 11, General Records, 1936-1961, US Consulate Kobe, Japan, RG 84, NARA; William S. Borden, *The Pacific Alliance: United States Foreign Economic Policy and Japanese Trade Recovery, 1947-1955* (Madison: University of Wisconsin Press, 1984), 189-90.

46. Quote from "Trade Topics," *Japan Times*, March 3, 1959, folder 510.2, box 13, Records of the Consular Posts, Subgroup 84.3, RG 84, NARA.

47. "Japanese Specialty Store to Open on Fifth Avenue," *New York Times*, June 27, 1958; "Japanese Department Store to Open on Fifth Avenue Saturday," *New York Times*, October 15, 1958.

48. "Buying Rush at Japanese Store in LA," *San Francisco Chronicle*, March 15, 1962; Amcongen Kobe-Osaka to the Department of State, Foreign Service Dispatch, Washington, August 11, 1959, folder 510.2, box 13, subgroup 3, RG 84, NARA.

49. "Leaving a Heart in San Francisco," 111.

50. "Cultural Center Plan 'Vetoed,'" *San Francisco Chronicle*, November 25, 1959; "Leaving a Heart in San Francisco," 111.

51. "HHFA Area Office to Be Opened Here."

52. M. Justin Herman to Kakuhei Matsui, June 5, 1959, folder Block 700 #1, box D/RE 99, Project WA-A1, SFRA; Herman to Matsui, September 21, 1959, folder Block 700 #1, box D/RE 99, Project WA-A1, SFRA.

53. Herman to Akira Nishiyama, September 29, 1959, and Tadamasu Tani to Herman, October 14, 1959; both in folder Block 700 #1, box D/RE 99, Project WA-A1, SFRA.

54. Tani to Herman, October 14, 1959.

55. Herman to Matsui, June 5, 1959.

56. MJH to file, July 6, 1961, folder Block 700 #2, box D/RE 99, Project WA-A1, SFRA.

57. Walter LaFeber, *The Clash: U.S.-Japanese Relations throughout History* (New York: W. W. Norton, 1997), 303–4. For more on the Japanese elite view of migrants, see Eiichiro Azuma, *Between Two Empires: Race, History, and Transnationalism in Japanese America* (New York: Oxford University Press, 2005); Andrea Geiger, *Subverting Exclusion: Transpacific Encounters with Race, Caste, and Borders, 1885–1928* (New Haven, CT: Yale University Press, 2011).

58. "Western Addition Slum Lands on Sale in February"; "Japanese Shopping Center Proposal."

59. "S.F. Japanese Shopping Center Proposal under Study Tuesday," *Nichi Bei Times*, October 20, 1959; Eugene J. Riordan, memorandum, March 24, 1958, folder Block 700 #1, box D/RE 99, Project WA-A1, SFRA; "More Development Projects," *Nichi Bei Times*, December 24, 1958; SFRA Commission, minutes, November 12, 1958, January 13, 1959; SFRA, Japan Cultural Center meeting, minutes, November 24, 1959, folder Block 700 #1, box D/RE 99, Project WA-A1, SFRA; "Three Show Interest in Japan Center," *Nichi Bei Times*, November 26, 1959.

60. "Japanese Center Plan," *San Francisco Chronicle*, November 30, 1959.

61. "Hung Wo Ching Pays $1,150,000 for 1,534 Acres in California," *Honolulu Star-Bulletin*, November 8, 1956; "New Aloha Airline Prexy Hopes to Find Black Ink," *Honolulu Star-Bulletin*, February 28, 1958; "Entrepreneur Soared with Aloha," *Honolulu Star-Bulletin*, September 24, 1999; "Hung Wo Ching," *Honolulu Advertiser*, July 2, 2006; "Hawaiian Executives Guests of the Chamber," *Bay Region Business*, September 25, 1959; George Cooper and Gavan Daws, *Land and Power in Hawaii: The Democratic Years* (Honolulu: University of Hawai'i Press, 1990), 46, 53, 63; Lawrence H. Fuchs, *Hawaii Pono: A Social History* (New York: Harcourt, Brace & World, 1961), 399, 400.

62. "Cultural Center Plan 'Vetoed.'"

63. SFRA, Japan Cultural Center meeting, minutes, November 24, 1959, folder Block 700 #1, box D/RE 99, Project WA-A1, SFRA.

64. M. Justin Herman to Vining Fisher, December 28, 1959, folder Block 700 #1, box D/RE 99, Project WA-A1, SFRA; SFRA, Japan Cultural Center meeting, minutes, November 24, 1964.

65. Ching to Herman, November 30, 1959, folder Block 700 #1, box D/RE 99, Project WA-A1, SFRA.

66. E. J. Burns to Herman, December 9, 1959, folder Block 700 #1, box D/RE 99, Project WA-A1, SFRA.

67. Ibid.

68. Hung Wo Ching to Herman, November 30, 1959, folder Block 700 #1, box D/RE 99, Project WA-A1, SFRA.

69. Christina Klein, *Cold War Orientalism: Asia in the Middlebrow Imagination, 1945–1961* (Berkeley: University of California Press, 2003).

70. Herman to Ching, November 1959, folder Block 700 #1, box D/RE 99, Project WA-A1, SFRA.

71. SFRA, *The San Francisco Redevelopment Agency Now Offers . . . 43 Acres of Prime Land in the Western Addition* (San Francisco: SFRA, 1960), 12.

72. Ching to Herman, October 19, 1959, folder Block 700 #1, box D/RE 99, Project WA-A1, SFRA; SFRA, "Western Addition Land Sales," press release, July 8, 1960, folder "WA A-1 Desk File, General Information," box D/RE #33, Project WA-A1 General Files, SFRA; "Nippon: Charted Survey of Japan, 1959," no date, folder "WA A-1 Desk File, General Information," box D/RE #33, Project WA-A1 General Files, SFRA.

73. "San Francisco Plans for Japanese Center," *Nichi Bei Times*, December 24, 1959.

74. Ching to Herman, October 19, 1959.

75. Gavan Daws, *Shoal of Time: A History of the Hawaiian Islands* (Honolulu: University of Hawai'i Press, 1989), 176–77; William Issel and Robert W. Cherny, *San Francisco 1865–1932: Politics, Power, and Urban Development* (Berkeley: University of California Press, 1986), 48; Jacob Adler, *Claus Spreckels: The Sugar King in Hawaii* (Honolulu: University of Hawai'i Press, 1966), 9; Ronald Takaki, *Pau Hana: Plantation Life and Labor in Hawai'i* (Honolulu: University of Hawai'i Press, 1983), 19–20; Cecil G. Tilton, "The History of Banking in Hawaii," University of Hawai'i Research Publications no. 3 (Honolulu: University of Hawai'i, 1927), 43–44; John Liu, "Race, Ethnicity and the Sugar Plantation System: Asian Labor in Hawai'i, 1850–1900," in *Labor Migration under Capitalism: Asian Workers in the United States before World War II*, ed. Lucy Cheng and Edna Bonacich (Berkeley: University of California Press, 1984).

76. According to one 1959 report, sugar was valued at $131 million (an unusual low due to a recent four-month strike) and pineapple at $127 million, while defense accounted for $338 million and tourists over $100 million. Bank of Hawaii, *Hawaii: The First Year of Statehood* (Honolulu: Bank of Hawaii, 1960), 12–13, 14, 25–26.

77. Edward Beechert, *Honolulu: Crossroads of the Pacific* (Columbia: University of South Carolina Press, 1991), 117.

78. *Hawaii-Alaska Statehood: Hearings before the United States House Committee on Interior and Insular Affairs*, 84th Cong., 1st sess., 88 (1955) (statement of Washington representative Russell V. Mack).

79. US Congress, House, 27th Cong., 3rd sess., House Doc. No 35, "Sandwich Islands and China: Message from the President of the United States," 1; David Igler, *The Great Ocean: Pacific Worlds from Captain Cook to the Gold Rush* (New York: Oxford University Press, 2013), 27.

80. Alfred Thayer Mahan, "Hawaii and Our Future Sea Power," *Forum*, March 1893, 4, 7.

81. *Statehood for Hawaii: Hearings before the Subcommittee on Territorial and Insular Affairs of the Committee of Interior and Insular Affairs, House of Representatives*, 85th Cong., 1st sess., 46 (1957) (statement of Hon. Hatfield Chilson, Under Secretary of the Interior); *Statehood for Hawaii: Hearings before the Senate Committee on Interior and Insular Affairs on S. 50: A Bill to Provide for the Admission of the State of Hawaii into the Union and S 36: A Bill to Provide for the Election of the Governor and Secretary of the Territory of Hawaii by the People of the Territory of Hawaii*, 85th Cong., 1st sess., 10 (1957) (statement of Hon. John A. Burns, a delegate in Congress from the Territory of Hawaii); US Congress, House, *Hawai'i Statehood: Report of the Committee of Internal and Insular Affairs*, 86th Cong., 1st sess. (1959), 12; Stuart Banner, *Possessing the Pacific: Land, Settlers, and Indigenous People from Australia to Alaska* (Cambridge, MA: Harvard University Press, 2007), 159–60; Walter LaFeber, *The New Empire: An Interpretation of American Expansion, 1860–1898* (Ithaca, NY:

Cornell University Press, 1963), 362; Bruce Cumings, *Dominion from Sea to Sea: Pacific Ascendency and American Power* (New Haven, CT: Yale University Press, 2009), 186-88; Daws, *Shoal of Time*, 285-92; Gary Okihiro, *Pineapple Culture: A History of the Tropical and Temperate Zones* (Berkeley: University of California Press, 2009), 117-27; Noenoe K. Silva, *Aloha Betrayed: Native Hawaiian Resistance to American Colonialism* (Durham, NC: Duke University Press, 2004), especially chapters 4 and 5.

82. "Hawaii—Beauty, Wealth, Amiable People," *Life*, March 23, 1959, 58.

83. *Statehood for Hawaii: Hearing before the Subcommittee on Territories and Insular Affairs of the Committee on Interior and Insular Affairs on S. 50, a Bill to Provide for the Admission of the State of Hawaii into the Union*, 86th Cong., 1st sess., 6 (1959) (questioning by Senator Henry M. Jackson, committee member).

84. "Hawaii—Beauty, Wealth, Amiable People," 58; US Congress, House, *Official Trip to Conduct a Study and Investigation of the Various Questions and Problems Relating to the Territory of Hawaii*, 79th Cong., 2nd sess. (1946), 9.

85. Ralph S. Kuykendall and A. Grove Day, "'Racial Aloha' in Hawaii," *Nation*, August 14, 1948, 185. For examples from social science, see Romanzo Adams, *Interracial Marriage in Hawaii: A Study of the Mutually Conditioned Processes of Acculturation and Amalgamation* (New York: AMS Press, 1937); C. K. Chan and Douglas S. Yamamura, "Interracial Marriage and Divorce in Hawaii," *Social Forces* 36 (October 1957): 77-84; Andrew Lind, *Hawai'i's People* (Honolulu: University of Hawai'i Press, 1955).

86. Data from US Bureau of the Census, *Statistical Abstract of the United States 1960* (Washington, DC: GPO, 1962), section 33, 991. For more on Native Hawaiian colonization and resistance, see, for example, Adria L. Imada, *Aloha America: Hula Circuits through the U.S. Empire* (Durham, NC: Duke University Press, 2012); Jonathan K. K. Osorio, *Dismembering Lahui: A History of the Hawaiian Nation to 1887* (Honolulu: University of Hawai'i Press, 2002); Sally Engle Merry, *Colonizing Hawai'i: The Cultural Power of Law* (Princeton, NJ: Princeton University Press, 2000); and Hauani-Kay Trask, "Settlers of Color and 'Immigrant' Hegemony: 'Locals in Hawai'i," in *Asian Settler Colonialism: From Local Governance to the Habits of Everyday Life in Hawai'i*, ed. Candace Fujikane and Jonathan Y. Okamura (Honolulu: University of Hawai'i Press, 2008). Native Hawaiians consistently challenged colonialism, historically and in present-day movements. See, for example, Noelani Goodyear-Kaopua, Ikaika Hussey, and Erin Kahunawaika'ala, *A Nation Rising: Hawaiian Movements for Life, Land, and Sovereignty* (Durham, NC: Duke University Press, 2014); Silva, *Aloha Betrayed*; Hauani-Kay Trask, *From a Native Daughter: Colonialism and Sovereignty in Hawai'i* (Honolulu: University of Hawai'i Press, 1999).

87. "Hawaii: A Melting Pot," *Life*, November 26, 1945, 103.

88. "Hawaii—Beauty, Wealth, Amiable People," 58.

89. "Welcome to Hawaii!," *San Francisco Chronicle*, March 13, 1959.

90. Gretchen Heefner, "'A Symbol of the New Frontier': Hawaiian Statehood, Anticolonialism, and Winning the Cold War," *Pacific Historical Review* 74 (2005): 545-74. For more on the Cold War and civil rights, see Mary L. Dudziak, *Cold War Civil Rights: Race and the Image of American Democracy* (Princeton, NJ: Princeton University Press, 2000).

91. *Statehood for Hawaii: Hearings on S. 50 and S. 36*, 10 (statement of Hon. John A. Burns).

92. "Enchanting 'State,'" *Newsweek*, February 23, 1959.

93. Roger Bell, *Last among Equals: Hawaiian Statehood and American Politics* (Honolulu: University of Hawaii Press, 1984), 136-39, 253-56; Ann K. Ziker, "Segregationists Confront American Empire: The Conservative White South and the Question of Hawaiian Statehood, 1947-1959," *Pacific Historical Review* 76 (2007): 439-66.

94. "Enchanting 'State'"; Heefner, "Symbol of a New Frontier," 545-48; Ellen D. Wu, *The Color of Success: Asian Americans and the Origins of the Model Minority* (Princeton, NJ: Princeton University Press, 2014), chapter 7.

95. *Statehood for Hawaii: Hearings before the House Committee on Interior and Insular Affairs*, 86th Cong., 1st sess., 24 (1959) (statement of Massachusetts representative John W. McCormack).

96. *Statehood for Hawaii: Hearing on S. 50*, 119–20 (1959) (statement of Hon. Fred A. Seaton, Secretary of the Interior).

97. "Hawaiian Section of Chamber Plans Tour of 50th State," *Bay Region Business*, April 10, 1959, 1.

98. "Hawaiian Statehood Stand Reaffirmed," *Bay Region Business*, February 27, 1959, 4; "It's Crowded in Hawaii, but No One Cares," *San Francisco Chronicle*, July 13, 1959; "S.F. May Back More Hawaii Trips," *San Francisco Chronicle*, November 19, 1959; "Stanton Delaplane's Postcard from Hawaii," *San Francisco Chronicle*, November 26, 1959; "Continental in Bid for Hawaii Trips," *San Francisco Chronicle*, December 4, 1959; "Hawaii Week," *Bay Region Business*, February 15, 1957, 2; "Trade Development Tour of Hawaii Set," *Bay Region Business*, March 13, 1959, 1; "S.F. Businessmen on Hawaiian Tour," *Bay Region Business*, October 20, 1961, 2; "Mayor of Honolulu and Hawaii Businessmen to Hold Conference Here," *Bay Region Business*, October 9, 1959, 2.

99. In 1920, only 44 percent of all Japanese Americans aged sixteen were still in school, and fewer at age seventeen or eighteen. By the college age of twenty, only 7 percent were. US Bureau of the Census, *14th Census of the United States, 1920, Vol. III: Population Composition and Characteristics of the Population by States* (Washington, DC: GPO, 1922), 1185.

100. Eileen H. Tamura, *Americanization, Acculturation, and Ethnic Identity: The Nisei Generation in Hawaii* (Urbana: University of Illinois Press, 1994), 10; Yamato Ichihashi, *Japanese in the United States: A Critical Study of the Problems of the Japanese Immigrants and Their Children* (Stanford, CA: Stanford University Press, 1932), 68; Yuji Ichioka, *The Issei: The World of the First Generation Japanese Immigrants, 1885–1924* (New York: Free Press, 1988), 40; George Engebretson, *A Century of Trust: The Story of Masayuki Tokioka* (Honolulu: Island Insurance, 1993), 7–16.

101. Dorothy Swaine Thomas and Richard Nishimoto, *The Salvage* (Berkeley: University of California Press, 1952), 67.

102. Takie Okumura and Umetaro Okumura, "Expatriation—Back to the Soil," *Mid-Pacific Magazine*, no. 173 (January–March 1935): 82, folder 2, box 1, Pan Pacific Union Records, Manuscript M003, Archives and Manuscripts Department, University of Hawai'i at Manoa, Honolulu.

103. Gary Okihiro, *Cane Fires: The Anti-Japanese Movement in Hawai'i, 1865–1945* (Philadelphia, PA: Temple University Press, 1991), 132–34, 141–42; Shimada Noriko, "Social, Cultural, and Spiritual Struggles of the Japanese in Hawai'i: The Case of Okumura Takie and Imamura Yemyo and Americanization," in *Hawai'i at the Crossroads of the US and Japan before the Pacific War*, ed. Jon Thares Davidann (Honolulu: University of Hawai'i Press, 2008), 147; Tamura, *Americanization*, 15, 129–32; Roland Kotani, *The Japanese in Hawaii: A Century of Struggle* (Honolulu: Hawaii Hochi, 1985), 53–69; Tyler Tokioka, telephone interview with author, June 18, 2013.

104. "He Grew with the Community," *Sunday Star-Bulletin & Advertiser*, November 14, 1965; Engebretson, *Century of Trust*, 14–16, 22–24.

105. "He Grew with the Community"; "From Small Beginnings . . . : The Story of International Savings and Loan Association, Limited, National Mortgage and Finance Company, Limited, Island Insurance Company, Limited, National Securities and Investment, Incorporated," 3, pamphlet, circa 1961, folder Block 700 #2, box D/RE 99, Project WA-A1, SFRA; Daws, *Shoal of Time*, 108, 313; "Thayer Resigns as Consulate Attorney," *Honolulu Star-Bulletin*, December 31, 1941; "Wade Thayer, Former Secretary of Hawaii, Dies," *Honolulu Advertiser*, June 5, 1959; Fuchs, *Hawaii Pono*, 198, 250; International Building and Loan Association, Limited, Affidavit, October 15, 1927, International Savings & Loan, file #50529 D1, Department of Commerce and Consumer Affairs, Honolulu, HI; Engebretson, *Century of Trust*, 14–16, 22–24, 29–32; Tilton, "Banking in Hawaii," 139.

106. Commissioner of Labor Statistics, *Labor Conditions in Hawaii* (Washington, DC: GPO, 1916), table 3; Ichioka, *Issei*, 40–51; Adam McKeown, *Chinese Migrant Networks and Cultural Change: Peru, Chicago, Hawaii, 1900–1936* (Chicago: University of Chicago Press, 2001), 33–35,

101-2, 229-30; Okihiro, *Cane Fires*, 23, 27; Silva, *Aloha Betrayed*, 48-54; Takaki, *Pau Hana*, 23-28.

107. Okihiro, *Cane Fires*, 41-43.

108. Masayo Umezawa Duus, *The Japanese Conspiracy: The Oahu Sugar Strike of 1920* (Berkeley: University of California Press, 1999), 55-58, 63-65, 67; Rick Baldoz, *The Third Asiatic Invasion: Migration and Empire in Filipino America, 1898-1946* (New York: New York University Press, 2011), 54-55; Melinda Tria Kerkvliet, *Unbending Cane: Pablo Manlapit, a Filipino Labor Leader in Hawaii* (Honolulu: University of Hawai'i Press, 2002), 26-28.

109. Duus, *Japanese Conspiracy*, 16, 82.

110. Moon-Kie Jung, *Reworking Race: The Making of Hawaii's Interracial Labor Movement* (New York: Columbia University Press, 2010), 88. See also Takaki, *Pau Hana*, 164-73; Tamura, *Americanization*, 214-15; John J. Stephan, *Hawai'i under the Rising Sun: Japan's Plans for Conquest after Pearl Harbor* (Honolulu: University of Hawai'i Press, 1984); Duus, *Japanese Conspiracy*.

111. *Japanese in Hawai'i: Hearings on S. 3206 before a Subcommittee of the Committee on Immigration*, 66th Cong. 19 (1920) (statement of Charles J. McCarthy, Governor of Hawai'i).

112. Mariko Takagi-Kitayama, "In the Strong Wind of the Americanization Movement: The Japanese-Language School Litigation Controversy and Okumura's Educational Campaign," in Davidann, *Hawai'i at the Crossroads*, 221, 226-28.

113. Romanzo Adams, *The Japanese in Hawaii: A Statistical Study Bearing on the Future Number and Voting Strength on the Economic and Social Character of the Hawaiian Japanese* (New York: National Committee on American Japanese Relations, 1924), 3.

114. See, for example, Tamura, *Americanization*; and Okihiro, *Cane Fires*.

115. Adams, *Japanese in Hawaii*, table A; Shiho Imai, *Creating the Nisei Market: Race and Citizenship in Hawaii's Japanese American Consumer Culture* (Honolulu: University of Hawai'i Press, 2010), 165; Fuchs, *Hawaii Pono*, 120, 123; Arnold T. Hiura with Glen L. Grant, *A History of Service: The Central Pacific Bank Story* (Honolulu: Central Pacific Bank, 1994), 10; "Rakasui" Tsuneichi Yamamoto, "About the Olden Times—until the Birth of the Japanese Chamber of Commerce" in *The Rainbow: A History of the Honolulu Japanese Chamber of Commerce*, ed. Tsuneichi Yamamoto (Honolulu: Honolulu Japanese Chamber of Commerce, 1970), 42; Tilton, "Banking in Hawaii," 130-33.

116. "He Grew with the Community." This trend grew with increased postwar wages and GI benefits. Matsuo Takabuki assisted by Dennis M. Ogawa with Glen Grant and Wilma Sur, *An Unlikely Revolutionary: Matsuo Takabuki and the Making of Modern Hawai'i* (Honolulu: University of Hawai'i Press, 1998), 81.

117. Tilton, "Banking in Hawaii," 83, 130-33; Takabuki, *Unlikely Revolutionary*, 81; Tamura, *Americanization*, 35-37.

118. International Savings and Loan, Articles of Association, January 16, 1925, International Savings and Loan, International Building & Loan Association, Ltd., Installment Capital Stock, September 30, 1927, and International Building & Loan Association, Ltd., Installment Capital Stock, September 30, 1927, all in file #50529 D1, Department of Commerce and Consumer Affairs, Honolulu, HI; Lionel Tokioka, telephone interview with author, June 28, 2013; "From Small Beginnings . . . ," 4, folder Block 700 #2, box D/RE 99, Project WA-A1, SFRA; Sydney G. Walton to Herman, September 9, 1960, folder Block 700 #1, box D/RE 99, Project WA-A1, SFRA; Robert Sasaki to Herman, July 18, 1960, folder Block 700 #1, box D/RE 99, Project WA-A1, SFRA; Hiura, *History of Service*, 4-6, 10.

119. "'Aloha' for the Fiftieth State," *New York Times Magazine*, April 19, 1959.

120. Takaki, *Pau Hana*, 66-78; Daws, *Shoal of Time*, 312-14; Fuchs, *Hawaii Pono*, 378-79.

121. Quote from Commission on Wartime Relocation and Internment of Civilians, *Personal Justice Denied: Report of the Commission on Wartime Relocation and Internment of Civilians* (Seattle: University of Washington Press, 1997), 118-33; Beth Bailey and David Farber, *The First Strange Place: The Alchemy of Race and Sex in World War II Hawaii* (New York: Free Press, 1992), 2-10;

Lon Kurashige, *Two Faces of Exclusion: The Untold History of Anti-Asian Racism in the United States* (Chapel Hill: University of North Carolina Press, 2016), 175–76; Franklin Odo, *No Sword to Bury: Japanese Americans in Hawai'i during World War II* (Philadelphia, PA: Temple University Press 2004), 112–13; Greg Robinson, *A Tragedy of Democracy: Japanese Confinement in North America* (New York: Columbia University Press, 2009), 113–21. Although there was no mass internment, there were two Department of Justice camps in Hawai'i. For one account, see Gail Honda, ed., *Family Torn Apart: The Internment Story of the Otokichi Muin Ozaki Family* (Honolulu: Japanese Cultural Center of Hawai'i, 2012).

122. Cooper and Daws, *Land and Power in Hawaii*, 5, 42–43; Tom Coffman, *The Island Edge of America: A Political History of Hawai'i* (Honolulu: University of Hawai'i Press, 2003), 100–159; Dennis M. Ogawa assisted by Claire Marumoto, *First among Nisei: The Life and Writings of Masaji Marumoto* (Honolulu: Japanese Cultural Center of Hawai'i, 2007), 158.

123. Fuchs, *Hawaii Pono*, 426.

124. Cooper and Daws, *Land and Power*, 4–5, 43–46; Fuchs, *Hawaii Pono*, 258–59, 397–400; Cumings, *Dominion from Sea to Sea*, 379–80.

125. Tokioka's son and the vice president of National-Braemar, Franklin Tokioka, was uncertain but suggested this as a possibility in an interview.

126. "Pagoda of the Dream Come True," *Honolulu Star-Bulletin*, March 2, 1968. I was not able to identify the men Tokioka described.

127. "Hope Fading for Building SF Uptown Japan Center," *Nichi Bei Times*, May 7, 1960; SFRA, *San Francisco Redevelopment Agency*, 12.

128. "Japanese Center Hopes Fade," *San Francisco Chronicle*, May 6, 1960.

129. "Japan Center Site Sold for $1 Million," *San Francisco Chronicle*, September 7, 1960.

130. Engebretson, *Century of Trust*, 65.

131. Rakusui Yamamoto, "Chronological Record of Events," *Rainbow*, 74–82; "Tokioka Takes Over Japanese Chamber Head," *Hawaii Hochi*, February 4, 1953; "Tokioka Named President of Economic Club," *Honolulu Advertiser*, January 24, 1952, folder C-57, box 3, subseries 4, Romanzo Adams Social Research Laboratory Clippings Files A1979:042b, Archives and Manuscripts Department, University of Hawai'i at Manoa, Honolulu; "Economic Study Club Marks 10 Years of Civic Work Here," *Honolulu Star-Bulletin*, January 31, 1958; Clarence L. Hodge, ed., *Hawaii Facts and Figures, 1948* (Honolulu: Chamber of Commerce of Honolulu, 1949), 52–53; Samuel Wilder King, *Annual Report of the Governor of Hawaii to the Secretary of the Interior* (Washington, DC: GPO, 1954), 31–32; Samuel Wilder King, *Annual Report of the Governor of Hawaii to the Secretary of the Interior* (Washington, DC: GPO, 1956), 71–72; "Growth of Pan American in the Pacific," folder "Transportation: Air: 1909–1964," box 8a, Robert C. Schmitt Papers, Hawai'i State Archives, Honolulu; "Japan Lines' First Plane Stops Here," *Honolulu Star-Bulletin*, November 24, 1953; table of overseas passengers, 1949–1969, folder "Transportation: Air: 1909–1964," box 8a, Schmitt Papers; Samuel Wilder King, *Annual Report of the Governor of Hawaii to the Secretary of the Interior* (Washington, DC: GPO, 1953), 16.

132. "Japanese Business, Civic Leaders Arrive for Visit," *Honolulu Star-Bulletin*, September 23, 1953.

133. "Japan Envoy Stops in Isles on Way to Defense Parley," *Honolulu Star-Bulletin*, September 30, 1953; Franklin Tokioka interview; Yamamoto, *Rainbow*, 146; Lionel Tokioka, telephone interview by author, June 18, 2013; "He Grew with the Community."

134. Minutes of the San Francisco Redevelopment Agency Commission meeting, August 30, 1960, 9.

135. M. Justin Herman memorandum, August 29, 1960, folder Block 700 #1, box D/RE 99, Project WA-A1, SFRA.

136. LaFeber, *Clash*, 321.

137. Japanese American Citizens League (JACL) 1960 Convention, minutes, 37–38, folder 32, box 35, series 8, JACL History Collection, JANL.

138. Tadashi Aruga, "Security Treaty Revision of 1960," in *The United States and Japan in the Postwar World*, ed. Akira Iriye and Warren I. Cohen (Lexington: University Press of Kentucky, 1989), 61-79; Iriye, *Across the Pacific*, 311-13; John Swenson-Wright, *Unequal Allies? United States Security and Alliance Policy toward Japan, 1945-1960* (Stanford, CA: Stanford University Press, 2005), 232-33.

139. "Causes of Japan Crisis," *Foreign Policy Bulletin* 39 (1960): 161-62; "Challenge to US by Riotous Japanese," *Life*, June 20, 1960, 26-35; "Ike's Defeat in Asia," *US News and World Report*, June 27, 1960, 37-41.

140. "Syndicate Has Plan for Japanese Center," *San Francisco Chronicle*, June 14, 1960.

141. SFRA, *The Peace Pagoda* (San Francisco: SFRA, 1965); "Crisis of the Pacific: Setback in Japan Forebodes Challenge to U.S. Strategic Foothold in Asia," *New York Times*, June 18, 1960.

142. "Concern on Japanese Center Delay," *San Francisco Chronicle*, May 24, 1962.

CHAPTER FIVE

1. Robert Ezra Park, "Human Migration and the Marginal Man," *American Journal of Sociology* 33 (1928): 888, 893.

2. Henry Yu, *Thinking Orientals: Migration, Contact, and Exoticism in Modern America* (New York: Oxford University Press, 2001), 109-12.

3. Eiichiro Azuma, *Between Two Empires: Race, History, and Transnationalism in Japanese America* (New York: Oxford University Press, 2005), 138-39, 145-47; Yuji Ichioka, *The Issei: The World of the First Generation Japanese Immigrants, 1885-1924* (New York: Free Press, 1988), 252-53.

4. US Department of War, *Final Report: Japanese Evacuation from the West Coast, 1942* (Washington, DC: GPO, 1943), 34.

5. Lon Kurashige, *Japanese American Celebration and Conflict: A History of Ethnic Identity and Festival, 1934-1990* (Berkeley: University of California Press, 2002), 48-51; Peggy Pascoe, "Miscegenation Law, Court Cases, and Ideologies of 'Race' in Twentieth-Century America," *Journal of American History* 83 (1996): 47-48; Lisa Lowe, *Immigrant Acts: On Asian American Cultural Politics* (Durham, NC: Duke University Press, 1996), 5; Mae M. Ngai, *Impossible Subjects: Illegal Aliens and the Making of Modern America* (Princeton, NJ: Princeton University Press, 2004), 2.

6. Alan Jacobson and Lee Rainwater, "A Study of Management Representative Evaluations of Nisei Workers," *Social Forces* 32 (1953): 35.

7. William Caudill, "Japanese-American Acculturation and Personality" (PhD diss., University of Chicago, 1952), 11.

8. Setsuko Matsunaga Nishi, "Japanese American Achievement in Chicago: A Cultural Response to Degradation" (PhD diss., University of Chicago, 1963), 26.

9. Caudill, "Japanese-American Acculturation," 14.

10. William Petersen, "Success Story, Japanese American Style," *New York Times Magazine*, January 9, 1966. See also "Disguised Blessing," *Newsweek*, December 29, 1958, 23; and Demaree Bess, "California's Amazing Japanese," *Saturday Evening Post*, April 30, 1955, 38-39.

11. Scott Kurashige, *The Shifting Grounds of Race: Black and Japanese Americans in the Making of Multiethnic Los Angeles* (Princeton, NJ: Princeton University Press, 2010), 192.

12. Frank J. Taylor, "Home Again," *Collier's*, February 15, 1946, 36.

13. Victor Boesen, "The Nisei Come Home," *New Republic*, April 26, 1948, 16.

14. "Disguised Blessing."

15. Ellen D. Wu, *The Color of Success: Asian Americans and the Origins of the Model Minority* (Princeton, NJ: Princeton University Press, 2014); Madeline Y. Hsu, *The Good Immigrants: How the Yellow Peril Became the Model Minority* (Princeton, NJ: Princeton University Press, 2015); "Success Story of One Minority Group in US," *US News and World Report*, December 26, 1966.

16. See, for example, Charlotte Brooks, *Alien Neighbors, Foreign Friends: Asian Americans, Housing, and the Transformation of Urban California* (Chicago: University of Chicago Press, 2009); Robert G. Lee, *Orientals: Asian Americans in Popular Culture* (Philadelphia, PA: Temple University Press, 1999), chapter 5; Wu, *Color of Success.*

17. Lisa Yoneyama, "Habits of Knowing Cultural Differences: *Chrysanthemum and the Sword* in the U.S. Liberal Multiculturalism," *Topoi* 18 (1999): 71-80.

18. Ruth Benedict, *The Chrysanthemum and the Sword: Patterns of Japanese Culture* (Boston: Houghton Mifflin, 1946), 1.

19. Clifford Geertz, *Works and Lives: The Anthropologist as Author* (Stanford, CA: Stanford University Press, 1988), 116-17.

20. Quote from John Morris, "*The Chrysanthemum and the Sword* by Ruth Benedict," *Pacific Affairs* 20 (1947): 208, 210. See also Gordon Bowles, "*The Chrysanthemum and the Sword* by Ruth Benedict," *Harvard Journal of Asiatic Studies* 10 (1947): 237-41; John A. Rademaker, "*The Chrysanthemum and the Sword: Patterns of Japanese Culture* by Ruth Benedict," *American Journal of Sociology* 53 (1947): 156-58; and Jesse F. Steiner, "*The Chrysanthemum and the Sword* by Ruth Benedict," *Far Eastern Quarterly* 6 (1947): 433.

21. Caudill, "Japanese-American Acculturation," 118.

22. Paul Broman to M. Justin Herman, October 7, 1960, folder Block 700 #1, box D/RE 99, Project WA-A1, San Francisco Redevelopment Agency Central Records, San Francisco, CA (hereafter SFRA).

23. Broman to Herman, February 6, 1961, 1, folder Block 700 #2, box D/RE 99, Project WA-A1, SFRA.

24. Herman to file, November 21, 1960, folder Block 700 #1, box D/RE 99, Project WA-A1, SFRA.

25. Herman to file, May 15, 1961, May 31, 1961, folder Block 700 #2, box D/RE 99, Project WA-A1, SFRA; National-Braemar, Inc., press release, March 17, 1965, 7-8, folder "Updates—WA Reports/Studies #85," box CR-156, file code 0170, Project Western Addition, SFRA.

26. SFRA, press release, August 10, 1961; Herman to Mark L. Gerstle, August 9, 1961; both in folder Block 700 #2, box D/RE 99, Project WA-A1, SFRA.

27. Douglas C. Myers to Robert L. Rumsey, memorandum, January 17, 1962, folder Block 700 #3, box D/RE 99, Project WA-A1, SFRA.

28. Tokioka to Herman, September 18, 1961, folder Block 700 #2, box D/RE 99, Project WA-A1, SFRA; Walter LaFeber, *The Clash: U.S.-Japanese Relations throughout History* (New York: W. W. Norton, 1997), 303-4.

29. National-Braemar, Inc., and Dream Entertainments, Inc., agreement, November 25, 1963, 10, folder Block 700 #6, box D/RE 99, Project WA-A1, SFRA.

30. Broman to Herman, March 19, 1962, folder Block 700 #3, box D/RE 99, Project WA-A1, SFRA.

31. "Meitetsu Cancels Contract with Japan Trade Center," *Nichi Bei Times*, February 24, 1963; Hideo Satoh to Herman, February 27, 1963, folder Block 700 #5, box D/RE 99, Project WA-A1, SFRA; SFRA, minutes of meeting on Japanese Cultural and Trade Center, April 5, 1963, folder Block 700 #5, box D/RE 99, Project WA-A1, SFRA; Broman to Herman, October 2, 1963, 2, folder Block 700 #6, box D/RE 99, Project WA-A1, SFRA; National-Braemar, Inc., and Kintetsu Enterprises Company of America, agreement, December 12, 1963, 11-12, folder Block 700 #6, box D/RE 99, Project WA-A1, SFRA; Herman to Roger Boas, January 6, 1964, folder Block 700 #7, box D/RE 99, Project WA-A1, SFRA.

32. Broman to Herman, October 2, 1963, 2.

33. Masayuki Tokioka to Herman, April 2, 1963, folder Block 700 #5, box D/RE 99, Project WA-A1, SFRA.

34. SFRA, *Japanese Cultural and Trade Center* (San Francisco: SFRA, February 1968), 11; Bro-

man to Herman, October 2, 1963; "List of Tenants," January 29, 1968, folder "Newsclip-WA, JCTC," box CR-77, file code 0165, SFRA.

35. SFRA Commission, minutes, August 30, 1960, 8–9.

36. SFRA, press release, April 10, 1961, folder Block 700 #2, box D/RE 99, Project WA-A1, SFRA.

37. William A. Kellar to Herman, March 15, 1965, folder Block 700 #8, box RE-977, Project WA-A1, SFRA.

38. SFRA, "Japanese Cultural and Trade Center Groundbreaking Ceremony—Participants' Instructions and Sequence of Ceremony," March 18, 1965, folder Block 700 #8, box RE-977, Project WA-A1, SFRA.

39. Everett Griffin, "Statement at Groundbreaking Ceremony, Japanese Cultural and Trade Center," March 18, 1965, folder Block 700 #8, box RE-977, Project WA-A1, SFRA.

40. San Francisco Board of Supervisors, Resolution No. 2096, adopted September 6, 1960, folder Block 700 #1, box D/RE 99, Project WA-A1, SFRA.

41. SFRA, *The San Francisco Redevelopment Agency Now Offers . . . 43 acres of prime land in the Western Addition* (San Francisco: SFRA, 1960).

42. Board of Supervisors, Resolution No. 2096.

43. Minoru Yamasaki to Frederick L. Holborn, July 19, 1962, folder #4, box D/RE-99, Project WA-A1, SFRA.

44. Two of Yamasaki's buildings had already won the American Institute of Architects' First Honor Award, and he was elected a fellow of the institute in 1960. "Minoru Yamasaki," in *Fellows of the American Institute of Architects* (Memphis, TN: Books LLC, 2010), 262; "Minoru Yamasaki," *American Institute of Architects Directory* (New Providence, NJ: R. R. Bowker, 1970).

45. Norman Murdoch to Herman, August 22, 1960, folder Block 700 #1, box D/RE 99, Project WA-A1, SFRA.

46. Quote from Herman to Broman, August 31, 1960, Murdoch to Herman, August 29, 1960, both in folder Block 700 #1, box D/RE 99, Project WA-A1, SFRA.

47. Murdoch to Herman, August 22, 1960.

48. The Pruitt-Igoe housing project was ambitious, high-rise public housing with ten thousand residents at its height. It quickly became, in the words of contemporary sociologist Lee Rainwater, "a community scandal" that "condensed into one 57-acre tract all of the problems and difficulties that arise from race and poverty and all of the impotence, indifference and hostility with which our society has so far dealt with these problems." It was completely demolished by 1976. Lee Rainwater, *Behind Ghetto Walls: Black Families in a Federal Slum* (Chicago: Aldine, 1970), 8, 9.

49. Minoru Yamasaki, *A Life in Architecture* (New York: Weatherhill, 1979).

50. Architectural historian Myungkee Min notes that there were as many Western publications about Japanese architecture from the mid-1950s through the 1960s as there had been in total before that period. Myungkee Min, "Japanese/American Architecture: A Century of Cultural Exchange" (PhD diss., University of Washington, 1999), 212.

51. "Museum of Modern Art Speeds Building of House from Japan," *New York Times*, June 12, 1954; Arthur Drexler, *The Architecture of Japan* (New York: Museum of Modern Art, 1955), 6.

52. Minoru Yamasaki, "History and Emotional Expression," *Journal of Architectural Education* 12 (Summer 1957): 8.

53. "The Road to Xanadu," *Time*, January 18, 1963, 54; see also Minoru Yamasaki, "Toward an Architecture of Enjoyment," *Architectural Record*, August 1955, 142–49.

54. "Yamasaki's Serene Campus Center," *Architectural Forum*, August 1958, 79.

55. "A Conversation with Yamasaki," *Architectural Forum*, July 1959, 111.

56. "Road to Xanadu," 61.

57. "A Handsome Outpost in Japan," *Architectural Forum*, February 1958, 71.

58. Sanae Nakatani, "Blueprints for New Designs: Japanese American Cultural Ambassadorship during the Cold War," *Journal of Asian American Studies* 19 (2016): 358.

59. "Handsome Outpost," 70-71, 78-79; "U.S.A. Abroad," *Architectural Forum*, December 1957, 114-23.

60. Jane Loeffler, *Architecture of Diplomacy: Building America's Embassies* (New York: Princeton Architectural Press, 1998), 78, 164.

61. "A Compliment to Traditional Japanese Architecture," *Architectural Record*, February 1958, 157, quoted in Bert Winther-Tamaki, "Minoru Yamasaki: Contradictions of Scale in the Career of the Nisei Architect of the World's Largest Building," *Amerasia* 26 (Winter 2000-2001): 170.

62. Penny von Eschen, *Satchmo Blows Up the World: Jazz Ambassadors Play the Cold War* (Cambridge, MA: Harvard University Press, 2004); Wu, *Color of Success*, chapter 4.

63. Loeffler, *Architecture of Diplomacy*, 8, 148.

64. Winther-Tamaki, "Minoru Yamasaki," 170-71.

65. "American Architect Yamasaki," *Architectural Forum*, August 1958, 85, 166.

66. "A Humanist Architecture for America and Its Relation to the Traditional Architecture of Japan," *Zodiac* 8 (1961): 141, 143.

67. Isamu Noguchi, whose fluid midcentury wooden, paper, ceramic, glass, and stone pieces remain iconic in interior design, went through a similar midcareer transformation that included a new design vocabulary as well as marriage to a Japanese woman and Japanese residence. He and Yamasaki might illustrate a shared interest or a similar arc of individual life history, but, as Matthew Frye Jacobson shows, these seemingly personal affiliations (often explained by generational position) are better understood in larger social and structural contexts. Masayo Duus, *The Life of Isamu Noguchi: Journey without Borders*, trans. Peter Duus (Princeton, NJ: Princeton University Press, 2004), 288; Matthew Frye Jacobson, *Roots Too: White Ethnic Revival in Post-Civil Rights America* (Cambridge, MA: Harvard University Press, 2006).

68. SFRA, *Japanese Cultural and Trade Center* (San Francisco: SFRA, n.d.), 4.

69. National-Braemar, *Japanese Cultural and Trade Center* (San Francisco: SFRA, n.d.), 5; "Japanese Temple-Style Structures Not Planned for New SF Japanese Center," *Nichi Bei Times*, January 27, 1961.

70. Yamasaki to Herman, February 7, 1964, folder Block 700 #7, box RE-977, Project WA-A1, SFRA.

71. Nakatani, "Blueprints for New Designs," 363.

72. Noboru Nakamura, interview with author, April 11, 2006, Oakland, CA, tape recording in author's possession; Matsuo Kunizo to Paul Broman, March 21, 1967, folder Block 700 #11, box RE-977, Project WA-A1, SFRA.

73. Nakamura interview.

74. Francis Sill Wickware, "The Japanese Language," *Life*, September 7, 1942, 58.

75. Some Nisei with especial fluency took accelerated three-month courses, and those truly fluent were recruited as instructors. James C. McNaughton, *Nisei Linguists: Japanese Americans in the Military Intelligence Service during World War II* (Washington, DC: Department of the Army, 2006), 33-34, 96.

76. Sociologist Dorothy Thomas estimated that "in some localities all, in most others the overwhelming majority" of Japanese American children attended their local Japanese school, if there was one; there were 261 in California in 1940. Dorothy Swaine Thomas and Richard Nishimoto, *The Salvage* (Berkeley: University of California Press, 1952), 63; Masaharu Ano, "Loyal Linguists: Nisei of World War II Learned Japanese in Minnesota," *Minnesota History* 45 (Fall 1977): 276, 277.

77. Kazue Masuyama, "Foreign Language Education," in *Transnational Competence: Rethinking the U.S.-Japan Educational Relationship*, ed. John N. Hawkins and William K. Cummings (Albany: State University of New York Press, 2000), 49; McNaughton, *Nisei Linguists*, 103, 108; Eiichiro Azuma, "Brokering Race, Culture, and Citizenship: Japanese Americans in Occupied Japan and Postwar National Inclusion," *Journal of American-East Asian Relations* 16 (Fall 2009): 183-211; Commission on Wartime Relocation and Internment of Civilians, *Personal Justice Denied: Report of*

the Commission on Wartime Relocation and Internment of Civilians (Seattle: University of Washington Press, 1997), 254-55; McNaughton, *Nisei Linguists*.

78. Miyako Hotel, press release, circa 1968, folder 5, box 14, series 2, SFH 5, San Francisco History Center, San Francisco Public Library (hereafter SFPL).

79. Japanese Cultural and Trade Center, advertisement, March 26, 1968.

80. Quote from Minoru Yamasaki to Edward E. Carlson, Executive VP to Western Hotels, Inc., November 10, 1960, 2, folder Block 700 #1, box RE-977, Project WA-A1, SFRA; Murdoch to Herman, February 9, 1965, folder Block 700 #8, box RE-977, Project WA-A1, SFRA.

81. "Welcome to San Francisco's Miyako," *Holiday*, March 1970, 86.

82. Murdoch to Herman, memorandum, February 28, 1964, folder Block 700 #7, box RE-977, Project WA-A1, SFRA.

83. Murdoch to Herman, July 24, 1964, folder Block 700 #7, box RE-977, Project WA-A1, SFRA.

84. National-Braemar, Inc., "Groundbreaking Ceremony, Japanese Cultural and Trade Center," March 17, 1965, 6, folder Block 700 #7, box RE-977, Project WA-A1, SFRA.

85. SFRA, "Japanese Cultural and Trade Center Status Report," February 15, 1963, folder Block 700 #5, box D/RE 99, Project WA-A1, SFRA; "Japanese Center 'No Chinatown,'" *San Francisco Chronicle*, May 27, 1962.

86. Yoshiro Taniguchi to Herman, July 27, 1965, folder Block 700 #9, box D/RE 99, Project WA-A1, SFRA; "Nisei, Japanese Architects Will Design Local Center," *Nichi Bei Times*, September 14, 1960.

87. Herman to Tange Kenzō, June 16, 1960, folder Block 700 #1, box D/RE 99, Project WA-A1, SFRA.

88. "Japanese Center 'No Chinatown.'"

89. Yamasaki to Herman, February 7, 1964, folder Block 700 #7, box D/RE 99, Project WA-A1, SFRA; "Yamato Was Imported in 7000 Tiny Pieces," *San Francisco Chronicle*, December 6, 1959.

90. Herman to Broman, May 7, 1965, folder Block 700 #9, box RE-977, Project WA-A1, SFRA.

91. Herman to Broman, February 26, 1962, folder Block 700 #3, box D/RE 99, Project WA-A1, SFRA.

92. "Japan Plans Gift to Trade Center," *San Francisco Chronicle*, December 7, 1960; SFRA, press release, 27 April 1962, folder Block 700 #3, box D/RE 99, Project WA-A1, SFRA.

93. Amcongen Kobe-Osaka to the Department of State, Foreign Service Dispatch, Washington, DC, April 23, 1959, folder 510.2, box 13, Records of the Consular Posts, Subgroup 84.3, Records of the Foreign Service Posts of the Department of State, Record Group 84, NARA; Tomoko Tamari, "Rise of the Department Store and the Aestheticization of Everyday Life in Early 20th Century Japan," *International Journal of Japanese Sociology* 15 (2006): 99-118; Penelope Francks, *The Japanese Consumer: An Alternative Economic History of Modern Japan* (New York: Cambridge University Press, 2009).

94. For example, Drexler's *Architecture of Japan* focused on a particular shoin style of home that he argued was characteristic of Japanese traditional architecture, to the exclusion of other quite different forms. Min, "Japanese/American Architecture," 218-19.

95. Joseph M. Henning, *Outposts of Civilization: Race, Religion, and the Formative Years of American-Japanese Relations* (New York: New York University Press, 2000), 101; Kristin L. Hoganson, *Consumer's Imperium: The Global Production of American Domesticity, 1865-1920* (Chapel Hill: University of North Carolina Press), 36-37.

96. Quotes from Broman to Herman, February 6, 1961, 1, folder Block 700 #3, box D/RE 99, Project WA-A1, SFRA; "$2 Million Japanese Theater Here," *San Francisco Chronicle*, December 3, 1961; "A Japanese One-Up on Disneyland," *San Francisco Chronicle*, December 4, 1961; Barbara E. Thornbury, "America's Kabuki-Japan, 1952-1960: Image Building, Myth Making, and Cultural

Exchange," *Asian Theatre Journal* 25 (2008): 193–230; and Kevin J. Wetmore Jr., "1954: Selling Kabuki to the West," *Asian Theatre Journal* 26 (2009): 78–93.

97. "Agreement on Japanese Center Deal," *San Francisco Chronicle*, August 11, 1961.

98. *Japanese Cultural and Trade Center* (San Francisco: National-Braemar, circa 1962); Herman to Jane P. Hale, November 8, 1965, folder Block 700 #9, box RE-977, Project WA-A1, SFRA; SFRA, *1967–1968 Annual Report* (San Francisco: SFRA, 1968), 16; "YW-Wives Council Plans Holiday Bazaar in Ignacio," *San Rafael Daily Independent Journal*, November 29, 1963; "'World's Fair' in a Camera Store," *San Francisco Examiner*, February 6, 1966; "Previews of YW Classes in Art, Music for Public," *San Francisco Examiner*, January 8, 1967; "Shinto Dedication for S.F. Japanese Center," *San Francisco Chronicle*, March 19, 1965; Civic Committee for the Formal Opening of the Japanese Cultural and Trade Center, memorandum, March 18, 1968, folder 4, box 14, Joseph L. Alioto Papers (SFH 5), SFPL.

99. "Japanese Center 'No Chinatown'"; "Japanese Temple-Style Structures."

100. Minoru Yamasaki to Frederick L. Holborn, 19 July 1962, folder Block 700 #4, box D/RE-99, Project WA-A1, SFRA.

101. For example, "After Night Falls," *San Francisco Chronicle*, August 25, 1956; "After Night Falls," *San Francisco Chronicle*, August 29, 1953; Herb Caen, "Friday Fish Fry," *San Francisco Chronicle*, February 13, 1959; and David Hulburd, "Talk around Town," *San Francisco Chronicle*, August 23, 1956.

102. "New Tokyo Sukiyaki Is the Perfect Rendezvous," *San Francisco Chronicle*, October 20, 1954.

103. Sidney Herschel Small, "Sukiyaki in San Francisco," *Holiday*, July 1955, 65.

104. Bob de Roos, "Now Hear This," *San Francisco Chronicle*, April 25, 1950.

105. "After Night Falls," *San Francisco Chronicle*, August 20, 1955; Yamato Sukiyaki advertisement, *San Francisco Chronicle*, August 28, 1955.

106. Tokyo Sukiyaki advertisement, *San Francisco Chronicle*, July 3, 1951.

107. "After Night Falls," *San Francisco Chronicle*, February 13, 1954; Marjorie Trumbull, "Exclusively Yours," *San Francisco Chronicle*, October 20, 1954; Hal Schaefer, "After Night Falls," *San Francisco Chronicle*, April 9, 1955.

108. Trumbull, "Exclusively Yours."

109. "Sukiyaki Restaurant," *San Francisco Chronicle*, November 8, 1958; "New Sukiyaki House to Open in S.F. in August," *Nichi Bei Times*, June 4, 1958; "'Birthday' for Nikko Sukiyaki," *San Francisco Chronicle*, November 19, 1960.

110. "New Nikko Restaurant Is Open," *San Francisco Chronicle*, November 15, 1958.

111. Bob Keely, "The Owl Steps Out," *San Francisco Chronicle*, February 11, 1961.

112. Macy's advertisement, *San Francisco Chronicle*, September 24, 1959. See also "Meet the Chef," *San Francisco Chronicle*, November 3, 1958.

113. "Japanese Foods to Be Demonstrated at Trade Center," *Nichi Bei Times*, July 26, 1956; Japan Trade Center advertisement, *San Francisco Chronicle*, February 3, 1960.

114. See, for example, "Gourmet Guide: Yamato Sukiyaki," *San Francisco Chronicle*, January 6, 1957.

115. Lucy Seligman, "The History of Japanese Cuisine," *Japan Quarterly* 41 (April 1994): 172; Katarzyna J. Cwiertka, *Modern Japanese Cuisine: Food, Power and National Identity* (London: Reaktion Books, 2006), 24–34.

116. Mas Yonemura, interview by author, March 29, 2006, Berkeley, CA, tape recording held by author; Mas Yonemura obituary, *San Francisco Chronicle*, August 10, 2008; "Legal Directory," *Crisis*, June–July 1952, 404.

117. Yonemura interview; "Japan Seaman Held for Fatal Stabbing of Lodi Man," *Nichi Bei Times*, September 28, 1954; "Seaman to Be Arraigned for Murder of Lodi Man," *Nichi Bei Times*, October 7, 1954; "Bail Refused Seaman Slaying," *Nichi Bei Times*, October 10, 1954.

118. Yonemura interview; "Oakland Link to Japan City Studied," *Oakland Tribune*, October 9, 1959; "Oakland Seeking to Establish Sister-City Ties with Fukuoka," *Nichi Bei Times*, May 11, 1962; "Mayor of Oakland to Attend Sister-City Tieup Fete in Fukuoka," *Nichi Bei Times*, September 19, 1962; "Oakland May Adopt Sister-City in Japan," *Oakland Tribune*, May 8, 1962; "Many Nisei Attend White House Dinner for Premier Eisaku Sato," *Nichi Bei Times*, January 17, 1965; "Political Notes, Comment of Bay Area Interest," *Oakland Tribune*, April 1, 1955; "Political Notes," *Oakland Tribune*, October 4, 1962.

119. Quote from "Peter Ohtaki Made Commercial Sales Manager for JAL," *Nichi Bei Times*, August 24, 1954. "Local Man Named to Japanese Department of P&O Orient Lines," *Nichi Bei Times*, March 15, 1960; "Japanese Department," *Nichi Bei Times*, February 14, 1954.

120. "Company Profile: Oakland's American President Lines," *California Business*, September 1985, 54; "New Air-Sea Combination Reduced Rate for Japan Trips Announced by APL, JAL," *Nichi Bei Times*, August 27, 1954; "APL Nisei Purser Now on Tour of US, Canada," *Nichi Bei Times*, January 22, 1960.

121. "Japan Language Proved Handy, Declare Pan Am Stewardesses," *Nichi Bei Times*, October 21, 1956; Christine Yano, *Airborne Dreams: "Nisei" Stewardesses and Pan American World Airways* (Durham, NC: Duke University Press, 2011), 5–7, 13.

122. "Northwest Airlines Seeks More Nisei Stewardesses for US Lines," *Nichi Bei Times*, January 11, 1959; "Pan American Seeks to Double Number of Nisei Stewardesses," *Nichi Bei Times*, February 12, 1963; Yano, *Airborne Dreams*, 3, 58–59.

123. Yano, *Airborne Dreams*, 13, 75.

124. Phil Tiemeyer, *Plane Queer: Labor, Sexuality, and AIDS in the History of Male Flight Attendants* (Berkeley: University of California Press, 2013), chapters 2 and 3; Kathleen Barry, *Femininity in Flight: A History of Flight Attendants* (Durham, NC: Duke University Press, 2007), chapters 1 and 2.

125. Mike Masaoka to Roy Nishikawa, memorandum, September 21, 1956, folder "Committee on Japanese-American Affairs, 1954, 1956–1958, 1960 1 of 2," box 6, series 1, Japanese American Citizens League (JACL) History Collection, JANL.

126. "New Nisei Role Urged to Aid U.S.-Japan Relations," *Nichi Bei Times*, September 5, 1956.

127. Masao Satow to Northern California–Western Nevada District Council Chapters, memorandum, April 14, 1958, folder "Committee on Japanese-American Affairs, 1954, 1956–1958, 1960 2 of 2," box 6, series 1, JACL History Collection, JANL.

128. JACL 1954 Convention, minutes, 69, folder 29, box 35, series 8, JACL History Collection, JANL.

129. Haruo Ishimaru to George Inagaki, memorandum, folder "Committee on Japanese-American Affairs, 1954, 1956–1958, 1960 1 of 2," box 6, series 1, JACL History Collection, JANL.

130. JACL 1958 Convention, minutes, 93, folder "15th Biennial Convention," box 8, accession 0217-006, Japanese American Citizen's League, Seattle Branch records, Special Collections, University of Washington Libraries, Seattle.

131. Quote from Haruo Ishimaru to George Inagaki, memorandum, March 10, 1954, folder "Committee on Japanese-American Affairs, 1954, 1956–1958, 1960 1 of 2," box 6, series 1, JACL History Collection, JANL. Masaoka to Ishimaru, memorandum, March 22, 1954, folder "Committee on Japanese-American Affairs, 1954, 1956–1958, 1960 1 of 2," box 6, series 1, JACL History Collection, JANL; JACL 1954 Convention, minutes, 69; Wu, *Color of Success*, 100–108.

132. Masaoka and Hosokawa, *They Call Me Moses*, 314, 254.

133. Quotes from Masaoka and Hosokawa, *They Call Me Moses*, 239; "United States Dancing to the Tune of Foreign Money," 134th Cong. Rec H 4703, House, June 23, 1988, Cong. 100, sess. 2, speaker Helen Delich Bentley; "Washington's Most Successful Lobbyist," *Reader's Digest*, May 1949, 125–29. For a laudatory account of Masaoka and the JACL, see Mike Masaoka and Bill Hosokawa, *They Call Me Moses Masaoka: An American Saga* (New York: William Morrow, 1987); and Bill Hosokawa, *JACL: In Quest of Justice* (New York: William Morrow, 1982).

134. Fred Hirasuna, circa 1956, folder "Committee on Japanese-American Affairs, 1954, 1956–1958, 1960 1 of 2," box 6, series 1, JACL History Collection, JANL.

135. Harold Gordon to Masaoka, memorandum, April 16, 1958; Gordon to Abe Hagiwara, August 13, 1958; both in folder "Committee on Japanese-American Affairs, 1954, 1956–1958, 1960 2 of 2," box 6, series 1, JACL History Collection, JANL.

136. Gordon to Masaoka, May 13, 1958, ibid.

137. "N.C. Chapters to Discuss CL's Japan Relations Role," *Nichi Bei Times*, January 7, 1958. See also Roy Nishikawa to Satow and Masaoka, memorandum, September 27, 1957, folder "Committee on Japanese-American Affairs, 1954, 1956–1958, 1960 1 of 2," box 6, series 1, JACL History Collection, JANL; and Saburo Kido, "Why I Am against JACL Involvement in U.S. Japan Relations," *Pacific Citizen*, July 11, 1958.

138. Nishikawa to Masaoka, memorandum, October 3, 1956, folder "Committee on Japanese-American Affairs, 1954, 1956–1958, 1960 1 of 2," box 6, series 1, JACL History Collection, JANL.

139. JACL 1958 Convention, minutes, 93, 90.

140. "Liberalize JACL Attitude on U.S.-Japan: 'Tokuzo' Gordon," *Pacific Citizen*, July 18, 1958. Wu discusses the formation of a similar but separate committee in 1956. The controversy continued, however, leading to another decision in 1958. Wu, *Color of Success*, 107.

141. JACL 1958 Convention, minutes, 92.

142. JACL 1966 Convention, minutes, 45, folder 34, box 35, series 8, JACL History Collection, JANL.

143. Ibid., 51.

144. JACL 1968 Convention, minutes, 39, 92, folder 33, box 35, series 8, JACL History Collection, JANL.

145. Jacobson, *Roots Too*.

146. JACL 1960 Official Convention Minutes, 56, folder 29, box 34, series 8, JACL History Collection, JANL.

147. California Department of Industrial Relations, Division of Fair Employment Practices, *Californians of Japanese, Chinese, and Filipino Ancestry: Population, Education, Employment, Income* (Sacramento, CA: California Department of Industrial Relations 1965), 33; Charles Kikuchi, "Japanese American Youth in San Francisco: Their Background, Characteristics, and Problems," report prepared for the National Youth Agency, 1941, 101, frame 10, reel 370, Japanese American Evacuation and Resettlement Records, Bancroft Library, University of California, Berkeley.

148. T. Fujitani, *Race for Empire: Koreans as Japanese and Japanese as Americans during World War II* (Berkeley: University of California Press, 2011), 211–36; Wu, *Color of Success*; Richard D. Alba, *Blurring the Color Line: The New Chance for a More Integrated America* (Cambridge, MA: Harvard University Press, 2009), 21–27; Richard D. Alba and Victor Nee, *Remaking the American Mainstream: Assimilation and Contemporary Immigration* (Cambridge, MA: Harvard University Press, 2005); California Department of Industrial Relations, *Californians of Japanese, Chinese, Filipino Ancestry*; US Bureau of the Census, *16th Census of the United States, 1940: Characteristics of the Nonwhite Population by Race* (Washington, DC: GPO, 1943), 109.

149. Albert Broussard, *Black San Francisco: The Struggle for Racial Equality in the West, 1900–1954* (Lawrence: University Press of Kansas, 1993), 212, 216–17.

150. Alexander Yoshikazu Yamato, "Socioeconomic Change among Japanese Americans in the San Francisco Bay Area" (PhD diss., University of California, Berkeley, 1986), 196; Jere Takahashi, *Nisei/Sansei: Shifting Japanese American Identities and Politics* (Philadelphia. PA: Temple University Press, 1997), 114.

151. Business figures from a Japanese Chamber of Commerce survey. "Many Japanese Firms Now Maintain Offices in SF," *Nichi Bei Times*, August 4, 1960.

CHAPTER SIX

1. All descriptions of the ceremony and courtyard are from "The Civic Committee for the Formal Opening of the Japanese Cultural and Trade Center," memorandum, circa 1968, folder 4, box 14, SFH 5, San Francisco History Center, San Francisco Public Library (hereafter SFPL); "Ceremonies Dedicating the Japanese Cultural and Trade Center," circa March 1968, folder 4, box 14, SFH 5, SFPL; "Japan Center's Dedication Ceremony," archival news film, March 28, 1968, CBS5 KPIX-TV, Japanese American Collection, San Francisco Bay Area Television Archive, J. Paul Leonard Library, San Francisco State University, San Francisco, CA; "SF Japanese Center Dedicated," *San Francisco Chronicle*, March 29, 1968.

2. "Alioto Gives His Blessing to Herman," *San Francisco Examiner*, March 20, 1968.

3. "Ceremonies Dedicating the Japanese Center."

4. Quote from caption, San Francisco Convention & Visitors Bureau photograph, image AAB-9223, San Francisco Historical Photograph Collection, SFPL.

5. Project Services Company for the SFRA, *The Population of Western Addition Area 2: A Sample Survey of the Residents and Their Relocation Needs* (San Francisco: SFRA, 1962), 2, 11, box CR-3, file 0170, Western Addition Reports and Studies, SFRA.

6. Allen Okamoto, interview with Princess Bustos and Jerwyn Sendaydiego, April 26, 2004, 4, San Francisco, CA, transcript, NJAHS.

7. Tracts J-2, J-6, and J-8 in 1950; and 152, 155, and 159 in 1970. US Bureau of the Census, *17th Census of the United States, 1950: Census of the Population, Vol. III, Part IV: Census Tracts Statistics* (Washington, DC: GPO, 1952); US Bureau of the Census, *19th Census of the United States, 1970: Census of the Population and Housing, Series PHC (1): Census Tracts, Part 19*.

8. US Bureau of the Census, *1950 Census of the Population, Census Tracts Statistics, San Francisco–Oakland California and Adjacent Area*, 29; Real Estate Research Corporation for the SFRA, "Marketability Analysis: Nihonmachi Area Western Addition Redevelopment Area A-2," unpublished report, December 1971, file 0170, box CR-1, Western Addition Reports and Studies, SFRA; US Bureau of the Census, *1970 Census Tracts*, 44–45; San Francisco Department of City Planning, *San Francisco 1970: Population by Ethnic Groups* (San Francisco: Department of City Planning, 1975), 4. Comparisons are not exact: the 1970 city report uses census data accounting for ethnicity. The 1950 tract-level census only refers to "other," although anecdotal evidence suggests this was primarily Japanese Americans.

9. Harry H. L. Kitano, "Japanese Americans in the Bay Area," in *Studies in Housing and Minority Groups*, ed. Nathan Glazer and Davis McEntire (Berkeley: University of California Press, 1960), 186.

10. Tract J-8. Rents rose by 56 percent in J-2 and 47 percent in J-6. These increases were actually less than for the city as a whole, but could force out low-income families. US Bureau of the Census, *1950 Census Tracts*, 55–56; US Bureau of the Census, *18th Census of the United States, 1960: Census of the Population and Housing, Vol. X, series PHC (1): Census Tracts* (Washington, DC: GPO, 1962), 367.

11. Department of City Planning, *San Francisco 1970*, 6–9; Paul T. Miller, *The Postwar Struggle for Civil Rights: African Americans in San Francisco, 1945-1975* (New York: Routledge, 2010); John H. Mollenkopf, *The Contested City* (Princeton, NJ: Princeton University Press, 1983), 202–3; Christina Renee Jackson, "Black San Francisco: The Politics of Race and Space in the City" (PhD diss., University of California, Santa Barbara, 2014), 47–57; Chester Hartman with Sarah Carnochan, *City for Sale: The Transformation of San Francisco* (Berkeley: University of California Press, 2002), 374.

12. Herb Caen, *The San Francisco Book* (Boston: Houghton Mifflin, 1948), 112.

13. The 1950 tracts were J-2, J-3, J-4, J-6, and J-8. US Bureau of the Census, *1950 Census Tracts*, 11–13. The 1960 tracts were J-3, J-6, J-7, J-8, J-10, and J-12. US Bureau of the Census, *1960 Census Tracts*, 38–39.

14. Project Services Company, *Western Addition Area 2*, 29; California Department of Industrial Relations, Department of Fair Employment Practices, *Negro Californians: Population, Employment, Income, Education* (San Francisco: Department of Industrial Relations, 1963); Thomas J. Sugrue, *The Origins of the Urban Crisis: Race and Inequality in Postwar Detroit* (Princeton, NJ: Princeton University Press, 1996).

15. "A War on Home Sale Bias Here," *San Francisco Examiner*, August 1, 1961.

16. Wilson Record, "Minority Groups and Intergroup Relations in the San Francisco Bay Area," in *The San Francisco Bay Area: Its Problems and Future*, ed. Stanley Scott (Berkeley: University of California, Berkeley Institute for Governmental Studies, 1966), 14.

17. Quote from Kitano, "Japanese Americans in the Bay Area," 191, see also 188. "Renewal Experts Start S.F. Survey," *San Francisco Chronicle*, July 16, 1959; Project Services Company, *Western Addition Area 2*, 29, 30.

18. "New Plans for SF Area Approved," *Nichi Bei Times*, December 24, 1961.

19. Masao Ashizawa, interview by author, Ken Yamada, and Clement Lai, San Francisco, CA, July 15, 2004, transcript, 7, National Japanese American Historical Society, San Francisco, CA (hereafter NJAHS).

20. Nobuo Mihara, interview by Andrew Kleinhenz and Brian Fok, San Francisco, CA, May 7, 2004, transcript, 11, NJAHS.

21. Sam Seiki, interview with author, San Francisco, CA, July 13, 2004, transcript, NJAHS; Ashizawa interview with author; Sumi Honnami, interview with Ken Yamada, San Francisco, CA, July 21, 2000, transcript, NJAHS; Japantown Task Force, *San Francisco's Japantown* (San Francisco: Arcadia, 2005), 22, 54, 74, 83.

22. Masao Ashizawa, interview with author, San Francisco, CA, March 28, 2006; "Togasaki to Head Local Area Group," *Nichi Bei Times*, March 24, 1962.

23. SFRA, *Nihonmachi* (San Francisco: SFRA, 1968), 2.

24. Emphasis mine. "Redevelopment Director Backs Nihonmachi for SF," *Nichi Bei Times*, January 31, 1962.

25. "Renewing the Heart of San Francisco," *San Francisco Chronicle*, October 15, 1959.

26. US Commission on Civil Rights, *Hearings before the United States Commission on Civil Rights*, held in San Francisco, CA, January 27, 28, 1960 (Washington, DC: GPO, 1960), 534 (hereafter *1960 Hearings*).

27. Comptroller General of the United States, *Report to the Congress of the United States: Review of Slum Clearance and Urban Renewal Activities of the San Francisco Regional Office, Housing and Home Finance Agency* (Washington, DC: General Accounting Office, 1959), 40.

28. Quote from Commission on Civil Rights, *1960 Hearings*, 701. Beth Sattler, *Family Properties: How the Struggle over Race and Real Estate Transformed Chicago and Urban America* (New York: Picador, 2009), 135–37; SFRA, "Report on the Redevelopment Plan for the Western Addition," 1964, folder "Redevelopment A-2, 1 of 2," Western Addition, San Francisco Districts, San Francisco Ephemera Collection (hereafter SUB COLL), SFPL; Wilton S. Sogg and Warren Wertheimer, "Legal and Governmental Issues in Urban Renewal," in *Urban Renewal: The Record and the Controversy*, ed. James Q. Wilson (Cambridge, MA: MIT Press, 1966), 157; SFRA, *Summary of Project Data and Key Elements* (San Francisco: SFRA, 1971), 4; "Western Addition #2 Housing Problems Discussed," *Sun-Reporter*, October 12, 1963; International Longshore and Warehouse Union et al., "Relocation in Redevelopment Area A-2," circa 1964, 3, folder 12, box 4, subseries C, series 1, SFH 10, SFPL.

29. Commission on Civil Rights, *1960 Hearings*, 705; SF NAACP, "Western Addition Redevelopment Survey," 1957, folder 61, carton 103, series 6.2, National Association for the Advancement of Colored People, Region I, Records, BANC MSS 78/180c (hereafter NAACP), Bancroft Library, University of California, Berkeley (hereafter BANC).

30. "Redevelopment A-2 Plan Hit," *Sun-Reporter*, April 18, 1964.

31. Commission on Civil Rights, *1960 Hearings*, 552.

32. Comptroller General, *Review of Slum Clearance*, 37-42; "Uprooting of Negros Hit," *San Francisco Chronicle*, November 22, 1961.

33. "7 in Path of Slum Job Won't Move," *San Francisco Chronicle*, March 18, 1959.

34. "Western Addition #2 Housing Problems Discussed."

35. James Howard Pye Jr., interview with Clement Lai and Ken Yamada, August 20, 2004, video recording, NJAHS; "Wide Looting in Slum Project," *San Francisco Chronicle*, August 5, 1959; "Relocation Major Problem in Western Addition Project," *San Francisco Chronicle*, February 1, 1959; "Goodbye Slums, Hello Corruption," *Sun-Reporter*, July 16, 1960; Mollenkopf, *Contested City*, 182-83; SFRA, *Annual Report to Mayor George Christopher for the Year July 1, 1958, to June 30, 1959* (San Francisco: SFRA, 1959), 10-11.

36. Quote from Herbert J. Gans, "The Failure of Urban Renewal," in James Q. Wilson, *Urban Renewal*, 537-57. Alison Isenberg, *Downtown America: A History of the Place and the People Who Made It* (Chicago: University of Chicago Press, 2004), 173.

37. Quote from Robert P. Groberg, "Urban Renewal Realistically Reappraised," in James Q. Wilson, *Urban Renewal*, 518. Commission on Civil Rights, *Hearings before the United States Commission on Civil Rights*, held in San Francisco, CA, May 1-3, 1967, and Oakland, CA, May 4-6, 1967 (Washington, DC: GPO, 1967), 932 (hereafter *1967 Hearings*).

38. Commission on Civil Rights, *1967 Hearings*, 932. For redevelopment as "Negro removal," see, for instance, Robert A. Caro, *The Power Broker: Robert Moses and the Fall of New York* (New York: Alfred A. Knopf, 1974), 968-69, 972; Mark I. Gelfand, *A Nation of Cities: The Federal Government and Urban America, 1933-1965* (New York: Oxford University Press, 1975), 359; Arnold R. Hirsch, *Making the Second Ghetto: Race and Housing in Chicago, 1940-1960* (New York: Cambridge University Press, 1983), 250; and June Manning Thomas, *Redevelopment and Race: Planning a Finer City in Postwar Detroit* (Baltimore: Johns Hopkins University Press, 1997), 107.

39. See, for example, Martin Anderson, *The Federal Bulldozer: A Critical Analysis of Urban Renewal, 1949-1962* (Cambridge, MA: MIT Press, 1964); Herbert J. Gans, *Urban Villagers: Group and Class in the Life of Italian-Americans* (New York: Free Press, 1962); Jane Jacobs, *The Death and Life of Great American Cities* (New York: Random House, 1961); Christopher Klemek, *The Transatlantic Collapse of Urban Renewal: Postwar Urbanism from New York to Berlin* (Chicago: University of Chicago, 2011), 79-127; and Jon C. Teaford, *The Rough Road to Renaissance: Urban Revitalization in America, 1940-1985* (Baltimore: Johns Hopkins University Press, 1990), 153-60.

40. SFRA, *Report on the Redevelopment Plan for the Western Addition Approved Redevelopment Project Area A-2* (San Francisco: San Francisco Redevelopment Agency, 1964), 215.

41. SFRA, *Annual Report, 1966-1967* (San Francisco: SFRA, 1967).

42. "Renewal Plan for Japan Town," *San Francisco Chronicle*, March 20, 1968.

43. Twentieth Century Fund Task Force on Community Development Corporations with Geoffrey Faux, *CDCs: New Hope for the Inner City* (New York: Twentieth Century Fund, 1971), 3. For more on Community Development Corporations, see Brian Goldstein, *The Roots of Urban Renaissance: Gentrification and the Struggle over Harlem* (Cambridge, MA: Harvard University Press, 2017); and Michael Woodsworth, *Battle for Bed-Stuy: The Long War on Poverty in New York City* (Cambridge, MA: Harvard University Press, 2016).

44. Quote from SFRA, "Nihonmachi Community," pamphlet, no date, folder "History of Japantown," Japantown History Collection, Japantown Task Force, San Francisco, CA. "Background and Purpose of the Nihonmachi Community Development Corporation," mimeograph, no date, folder "History of Japantown," Japantown History Collection, Japantown Task Force, San Francisco, CA.

45. Project Services Company, *Population of Western Addition Area 2*, 3.

46. SFRA, *Japanese Cultural and Trade Center* (San Francisco: SFRA, 1968).

47. Quotes from SFRA, "More Collaborative Planning—Nihonmachi," press release, March 19, 1968; "A Gentle Groundbreaking," *San Francisco Chronicle*, July 31, 1970; and Real Estate Research Corporation, "Marketability Analysis," 4.

48. SFRA, "Redevelopment Plan"; Allen Okamoto, interview with Princess Bustos and Jerwyn Sendaydiego, April 26, 2004, San Francisco, CA, transcript, NJAHS.

49. The first reference I found to Nihonmachi in the *Nichi Bei Times* was in 1956, while the last for Uptown was in 1960.

50. Masao Ashizawa to Shareholders, March 2, 1970, Nihonmachi Community Development Corporation file, CANE box, JANL.

51. "Browsing and Shopping in San Francisco's Japantown," *Sunset*, July 1962, 52–53.

52. "For Christmas Shopping . . . or Just a Visit," *Sunset*, December 1969, 44–46.

53. "Alioto Gives His Blessing."

54. Emphasis in original. "But Is It Legal?," *Sun-Reporter*, June 25, 1960.

55. Thomas J. Sugrue, *Sweet Land of Liberty: The Forgotten Struggle for Civil Rights in the North* (New York: Random House, 2008), 301–2. For a conceptualization of civil rights struggle as one against displacement, see Greta de Jong, "Staying in Place: Black Migration, the Civil Rights Movement, and the War on Poverty in the Rural South," *Journal of African American History* 90 (2005): 387–409.

56. Quote from "San Francisco Answers Birmingham," *Dispatcher*, May 31, 1963; "S.F. Rights March," *San Francisco Chronicle*, May 27, 1961; "30,000 Hear SCLC Leader," *Sun-Reporter*, June 1, 1963.

57. "Response to Plan for S.F. Racial Study," *San Francisco Chronicle*, June 18, 1963; "San Francisco Answers Birmingham."

58. "Mayor Harbors Gradualistic Approach to End Racial Strife," *Sun-Reporter*, July 27, 1963; "Bi-racial Group 'Helpless'; CORE Wants Power Meeting," *Sun-Reporter* July 27, 1963.

59. "Mass Meeting for Negro Rights Here," *San Francisco Chronicle*, July 30, 1963.

60. David Thomas Wellman, "Negro Leadership in San Francisco" (MA thesis, University of California, Berkeley, 1966), 49.

61. "Mass Meeting for Negro Rights Here."

62. Wellman, "Negro Leadership," 24.

63. "SF Civil Rights Groups Unite," *Sun-Reporter*, July 27, 1963; Paul T. Miller, *Postwar Struggle*, 65; Wellman, "Negro Leadership," 23–25; Daniel Crowe, *Prophets of Rage: The Black Freedom Struggle in San Francisco, 1945–1969* (New York: Garland, 2000), 119, 123–25, 127–28.

64. "USFFM Asks for Power Structure Meet," *Sun-Reporter*, August 10, 1963; "Come, Let Us Reason Together," *Sun-Reporter*, August 17, 1963; "S.F. Rights Pledge: 'We Will Try,'" *San Francisco Chronicle*, September 12, 1963; "Human Relations Job in Housing," *San Francisco Chronicle*, March 6, 1964; Christopher Lowen Agee, *The Streets of San Francisco: Policing and the Creation of a Cosmopolitan Liberal Politics, 1950–1972* (Chicago: University of Chicago Press, 2014), 185.

65. "1500 Pickets Turn Out—500 Sit in the Lobby," *San Francisco Chronicle*, March 7, 1964.

66. "Negro Civil Rights Groups Unite for Mass Picketing at Hotel," *San Francisco News-Call Bulletin*, March 6, 1964; "Hundreds Arrested at Sheraton-Palace Sit-In," *Sun-Reporter*, March 7, 1964; "New Mass Arrests—Shelley Arranges Pact," *San Francisco Chronicle*, March 8, 1964; "Shelley Gets No-Bias Pact on Hotel Jobs," *San Francisco Examiner*, March 8, 1964; Paul T. Miller, *Postwar Struggle*, 77–83; James Richardson, *Willie Brown: A Biography* (Berkeley: University of California Press, 1996), 86–87.

67. NAACP San Francisco Branch, "Negro Revolt Finally Comes to San Francisco," newsletter, April 1964, 1, folder 1, box 94, series 4.1, NAACP, BANC.

68. Ibid., 2.

69. "Pickets, 'Shop-ins' Mark Renewed Drive for Jobs for Negroes in SF Bay Area," *People's World*, February 29, 1964; "CORE Will Picket Bank Tomorrow," *San Francisco Chronicle*, May 21, 1964; Jo Freeman, *At Berkeley in the Sixties: The Education of an Activist, 1961–1965* (Bloomington: Indiana University Press, 2004), 94–106; Crowe, *Prophets of Rage*, 125–27; Paul T. Miller, *Postwar Struggle*, 77–87; Sugrue, *Sweet Land of Liberty*, 145–47; Frederick Wirt, *Power in the City: Decision Making in San Francisco* (Berkeley: University of California Press, 1974), 256–57.

70. "Certainly Not a Japanese Realtor?," *Sun-Reporter*, July 14, 1962.

71. "Nakamura Refutes CORE Charges of Discrimination by Realty Firm," *Nichi Bei Times*, July 11, 1962.

72. "CORE Calls Off Lucky Shop-ins," *Sun-Reporter*, February 29, 1964; "Does the End Justify the Means?," *Sun-Reporter*, February 29, 1964.

73. Quote from "Feud within the NAACP—2 Slates," *San Francisco Chronicle*, November 17, 1964. Rev. George L. Bedford et al. to NAACP members, December 15, 1964, folder 8, box 93, series 4.1, NAACP, BANC; "Vote NAACP Election," circa December 1964, folder 8, box 93, series 4.1, NAACP, BANC; Albert Broussard, *Black San Francisco: The Struggle for Racial Equality in the West, 1900-1954* (Lawrence: University Press of Kansas, 1993), 222-25.

74. "San Francisco's Negro Leaders," *San Francisco Examiner*, December 3, 1961; Richard Young, "The Impact of Protest Leadership on Negro Politicians in San Francisco," *Western Political Quarterly* 22, no. 1 (1969): 99, 96, 111; "Redevelop Aide Quits NAACP Post," *San Francisco Chronicle*, June 15, 1966; "End of Hearings on Western Addition," *San Francisco Chronicle*, April 29, 1964; "New Supervisor Rights Fighter," *San Francisco Chronicle*, August 25, 1964; "The Shelley Legacy," *Sun-Reporter*, January 6, 1968; Mrs. Richard Barnes to Roy Wilkins, September 10, 1965, folder 8, box 93, series 4.1, NAACP, BANC; SF NAACP, press release, April 10, 1966, folder 9, box 93, series 3.1, NAACP, BANC; Charles Wollenberg, *All Deliberate Speed: Segregation and Exclusion in California Schools, 1855-1975* (Berkeley: University of California Press, 1976), 141-42.

75. Quote from Mrs. Marie Denevi to Jack Shelley, March 5, 1964, folder 7, box 1, series 1, John F. "Jack" Shelley Papers (SFH 10), SFPL. Shelley's papers include three bulging folders of citizens' responses. See, for example, Phillip B. Stapp to Shelley, telegram, March 9, 1964; and Mrs. William Paul Butler to Shelley, March 6, 1964; both in folder 7, box 1, series 1, SFH 10, SFPL.

76. C. Petersen to Mayor Shelley, March 7, 1964, folder 7, box 1, subseries A, SFH 10, SFPL.

77. "The Negro in San Francisco, No. 1—Population," *San Francisco News-Call Bulletin*, December 4, 1961. Historian Christopher Agee notes a similar emphasis on the southern origins of many black San Franciscans by observers, which in his story shaped perceptions of disproportionate black crime rates. Agee, *Streets of San Francisco*, 155-57.

78. Quote from "A New Look for San Francisco's Bit of Japan," *San Francisco Chronicle*, October 23, 1963. Agee, *Streets of San Francisco*, 155-57.

79. Quote from M. Justin Herman to Percy Moore, January 21, 1964, folder 12, box 4, subseries C, series 1, SFH 10, SFPL. SFRA, "Summary of Community Relations Program," August 2, 1963, folder 10, box 4, subseries C, series 1, SFH 10, SFPL; SFRA, *Relocation Program*; SFRA, *The Change in Rehousing and Relocation Aids* (San Francisco: SFRA, 1965), 1-2; Gelfand, *Nation of Cities*, 260, 371, 375.

80. "Western Addition Plan Means Exodus of 13,500," *San Francisco News-Call Bulletin*, March 31, 1964, folder 10, box 4, subseries C, series 1, SFH 10, SFPL.

81. "City Reveals Its Renewal Plan—and Stirs Furor," *San Francisco Chronicle*, April 15, 1964.

82. "Redevelopment A-2 Plan Hit," *Sun-Reporter*, April 18, 1964.

83. "Housing Program Proposed by Freedom Movement," *Sun-Reporter*, January 23, 1964.

84. Freedom House, "San Francisco Area 2 Summer Project," pamphlet, circa 1964, folder 11, box 4, subseries C, series 1, SFH 10, SFPL.

85. Robert O. Self, *American Babylon: Race and the Struggle for Postwar Oakland* (Princeton, NJ: Princeton University Press, 2003), 179, 191.

86. Ibid., 178-214.

87. "New Civil Rights Move," *Sun-Reporter*, June 6, 1964; Freedom House, "Area 2 Summer Project"; Mollenkopf, *Contested City*, 186.

88. John H. Mollenkopf, "Community Organization and City Politics" (PhD diss., Harvard University, 1974), 276; "Rights Group Outlines Needs," *San Francisco Chronicle*, October 7, 1963; Charles M. Payne, *I've Got the Light of Freedom: The Organizing Tradition and the Mississippi Freedom Struggle* (Berkeley: University of California Press, 1995), 293.

89. Young, "Impact of Protest Leadership," 105.

90. "Doctor of Civil Rights," *Sun-Reporter*, June 27, 1964; "Joye Goodwin—Early Proponent of Civil Rights," *San Francisco Chronicle*, April 8, 2009; California State Employment Service, *The Economic Status of Negroes in the San Francisco–Oakland Bay Area* (San Francisco: California State Employment Service, 1963), 4, table 8; R. L. Polk & Co., *San Francisco City Directory 1963* (Monterey Park, CA: R. L. Polk, 1963), 1577; R. L. Polk & Co., *Polk's San Francisco City Directory 1966* (Monterey Park, CA: R. L. Polk, 1966), 198, 586, 1487; R. L. Polk & Co., *San Francisco City Directory 1964–1965* (Monterey Park, CA: R. L. Polk, 1965), 1561.

91. Freedom House, "Area 2 Summer Project."

92. Young, "Impact of Protest Leadership," 104–5; Natalie Becker and Marjorie Myhill, *Power and Participation in the San Francisco Community Action Program, 1964–1967* (Berkeley: Institute of Urban and Regional Development, University of California, Berkeley, 1967), 10–12. For movement distinctions, see Sattler, *Family Properties*, 213; and Payne, *I've Got the Light*, 3–4.

93. "S.F. Slums and Their People," *San Francisco Chronicle*, April 14, 1964.

94. "100 Area-2 Residents Gagged," *A-2 Stand*, circa July 1964, 1, folder 11, box 4, subseries C, series 1, SFH 10, SFPL.

95. M. Sharon, "Some Errors Cleared Up," *A-2 Stand*, circa July 1964, 2, folder 11, box 4, subseries C, series 1, SFH 10, SFPL.

96. Quote from International Longshore and Warehouse Union et al., "Relocation in Redevelopment Area A-2," circa 1964, folder 12, box 4, subseries c, series 1, SFH 10, SFPL. "Big Debate on Slum Clearance," *San Francisco Chronicle*, September 22, 1964; Council for Civic Unity, *Among These Rights . . . ,* newsletter, March 1964, 3, folder 10, box 4, subseries C, series 1, SFH 10, SFPL.

97. "New Probe on Western Addition," *San Francisco Chronicle*, July 24, 1964.

98. "New Attack on Redevelopment," *San Francisco Chronicle*, August 26, 1964.

99. "Redevelopment Hearing," *Sun-Reporter*, April 11, 1964.

100. "Renewal Plans Rejected," *San Francisco Chronicle*, July 31, 1964; "Board OKs 2nd Western Addition Plan," *San Francisco Chronicle*, October 6, 1964; "A Socially Oriented Redevelopment Plan," *San Francisco Chronicle*, October 7, 1964; "Controls on Western Addition," *San Francisco Chronicle*, October 10, 1964.

101. "New Civil Rights Move."

102. "WA Starts Clean Up Plan," *Sun-Reporter*, May 9, 1964; "Freedom House: Area-2 Residents Working to Beat Redevelopment," *Sun-Reporter*, June 13, 1964.

103. "Fund Drive for Rights Defendants," *Sun-Reporter*, August 21, 1964; "Dramatic Bid for a Slum Cleanup," *San Francisco Chronicle*, October 2, 1964; "Freedom House Begins Tutorial Project," *Sun-Reporter*, October 10, 1964; "Freedom House Training Program," *Sun-Reporter*, November 21, 1964; Young, "Impact of Protest Leadership," 105. This local emphasis was characteristic of contemporary activism. Sugrue, *Sweet Land of Liberty*, 314–15.

104. Fred Hayden, "If You Need Help," *A-2 Stand*, September 26, 1964, 5; and "Having Trouble with Rats or Roaches?," *A-2 Stand*, September 26, 1964, 5; both in folder 11, box 4, subseries C, series 1, SFH 10, SFPL.

105. "Western Addition Renewal Row," *San Francisco Chronicle*, May 4, 1967.

106. Quote from "WACO Speaks Out," *Sun-Reporter*, February 10, 1968. "WACO Speaks," *Sun-Reporter*, November 4, 1967; "Western Addition Community Organization," circa 1967, folder "Redevelopment—WACO," Western Addition, San Francisco Districts Collection, SFPL; Clement Lai, "Between 'Blight' and a New World: Urban Renewal, Political Mobilization, and the Production of Spatial Scale in San Francisco, 1940–1980" (PhD diss., University of California, Berkeley, 2006), 220–25; Mollenkopf, *Contested City*, 186–90; "Prop. 14 Vote," *San Francisco Chronicle*, November 4, 1964; SFRA, *Annual Report, 1965–1966* (San Francisco: SFRA, 1966), 3; Daniel Martinez HoSang, *Racial Propositions: Ballot Initiatives and the Making of Postwar California* (Berkeley: University of California Press, 2010), chapter 3.

107. Quotes from "Western Addition Renewal Row," *San Francisco Chronicle*, May 4, 1967; and

Mary Rogers, interview with Clement Lai and Ken Yamada, August 4, 2004, San Francisco, CA, transcript, 3, NJAHS.

108. Mollenkopf, "Community Organization," 279, 281; "Mayor Misses a Redevelopment Protest," *San Francisco Chronicle*, May 10, 1967; "Walkout on Shelley over Renewal Housing," *San Francisco Chronicle*, May 18, 1967; "Redevelopment Foes Plan Fight," *San Francisco Chronicle*, July 22, 1967; Reverend Arnold Townsend, interview by Alex Momirow and Molly Miranker, May 12, 2007, San Francisco, CA, "Fillmore Redevelopment/Dislocation" Oral History Archives Project, Urban School of San Francisco, http://www.tellingstories.org/fillmore/arnold_townsend/index .html; Clement Lai, "Between 'Blight' and a New World," 222–23.

109. "Mayor Misses a Redevelopment Protest."

110. "Renewal Official Responds to Citizen Participation Statements," *Journal of Housing* 26 (1969): 602.

111. Quote from "Walkout on Shelley over Renewal Housing." "Weaver's Relocation Pledge," *San Francisco Chronicle*, June 23, 1967; "Redevelopment Project Blocked," *San Francisco Chronicle*, April 26, 1968.

112. "Redevelop Chief's Feud with WACO," *San Francisco Chronicle*, July 21, 1967.

113. Dody to John Anderson, memorandum, July 18, 1967, folder 18, box 5, subseries D, series 1, SFH 10, SFPL.

114. John Anderson to T. J. Kent Jr., memorandum, August 3, 1967, folder 18, box 5, subseries D, series 1, SFH 10, SFPL.

115. "First Choice for Renewal Residents," *San Francisco Chronicle*, July 26, 1967; "Unique Offer in Area 2 Renewal," *San Francisco Chronicle*, July 28, 1967.

116. "Relocation Plan Delayed by Board," *San Francisco Chronicle*, June 27, 1967.

117. City and County of San Francisco, *Journal of Proceedings, Board of Supervisors* (San Francisco: City and County of San Francisco, 1967), October 23, 1967, 849.

118. "Flurry over New Vote on Renewal," *San Francisco Chronicle*, October 27, 1967; "Morrison's Stand on Renewal," *San Francisco Chronicle*, October 10, 1967; City and County of San Francisco, *Journal of Proceedings, Board of Supervisors*, October 9, 1967, 817.

119. City and County of San Francisco, *Journal of Proceedings, Board of Supervisors*, November 6, 1967, 870.

120. "Herman Back, Won't Budge on A-2 Plan," *San Francisco Chronicle*, October 28, 1967.

121. E. R. Hambrick, letter to the editor, *Sun-Reporter*, December 6, 1958.

122. "Urban Affairs Beat," *San Francisco Chronicle*, May 13, 1973; Fillmore Community Development Corporation, "Application for an EDA Technical Assistance Grant," April 15, 1968, 1, folder 28, box 9, series 2B, SFH 5, SFPL.

123. Fillmore Community Development Corporation, "Application for an EDA Technical Assistance Grant."

124. Quote from "Fleishhacker Housing Plan Attacked," *San Francisco Chronicle*, August 25, 1966. "Rent Subsidy Co-op Plan Urged Here," *San Francisco Chronicle*, July 16, 1966; "Fillmore Plan for Low Rents," *San Francisco Chronicle*, August 2, 1967; Goldstein, *Roots of Urban Renaissance*, 53–54.

125. John E. Dearman to Joseph Alioto, May 6, 1968, folder 30, box 9, series 2B, SFH 5, SFPL.

126. Quote from "Housing Squabble Divides the Western Addition," *San Francisco Progress*, May 8, 1968. "WACO Hecklers Disturb Renewal Groundbreaking," *San Francisco Chronicle*, April 22, 1968.

127. M. Justin Herman to Joseph Alioto, May 24, 1968, 1, folder 30, box 9, series 2B, SFH 5, SFPL.

128. Family Service Agency of San Francisco, "Statement on Martin Luther King Square," press release, May 3, 1968, folder 30, box 9, series 2B, SFH 5, SFPL.

129. Western Addition Project Area Committee (WAPAC), "Resolution Endorsing the Fillmore

Community Development Association's Application for an E.D.A. Technical Assistance Grant," March 12, 1969, folder 28, box 9, series 2B, SFH 5, SFPL; "Why WACO Now Likes King Square," *San Francisco Chronicle*, July 19, 1968.

130. "King Square Doors Open," *San Francisco Chronicle*, July 29, 1969.

131. "FHA Helps City's King Project," *San Francisco Examiner*, June 28, 1968; "Urban Affairs Beat," *San Francisco Chronicle*, May 13, 1973. Martin Luther King Square itself continued under federal receivership.

132. "New Charge by Redevelop Foes," *San Francisco Chronicle*, August 10, 1967.

133. "Relocation Issue Goes to Court," *San Francisco Chronicle*, December 16, 1967; "A New Crisis for Western Addition," *San Francisco Chronicle*, December 27, 1968; "WACO's Tough Stand on Relocation Plan," *San Francisco Chronicle*, December 18, 1968.

134. San Francisco Neighborhood Legal Assistance Foundation, "Federal Judge Orders Relocation Stopped in Western Addition," press release, December 26, 1968, WACO folder, Western Addition, San Francisco Districts, SUB COLL, SFPL.

135. "WAPAC Funded," *Sun-Reporter*, September 27, 1969.

136. "WACO Suit Stops Western Addition Displacement," *Sun-Reporter*, January 4, 1969.

137. Eva Brown, "WACO Speaks Out," *Sun-Reporter*, February 10, 1968; "Western Addition Residents Confer with Mayor Alioto," *Sun-Reporter*, March 9, 1968; "Western Addition Power and Unity," *Sun-Reporter*, January 11, 1969.

138. "WACO Attacks Redevelopment," *Sun-Reporter*, November 4, 1967.

139. Quote from "A Congress Organizes in Fillmore," *San Francisco Chronicle*, July 30, 1967.

140. John H. Anderson to Peter G. Trimble, memorandum, July 28, 1967, folder 18, box 5, subseries D, series 1, SFH 10, SFPL; Mollenkopf, "Community Organization," 284; Self, *American Babylon*, 206-7.

141. SFRA, "Proposed Agreement between Western Addition Project Area Committee and the San Francisco Redevelopment Agency," February 3, 1969, 1-2, folder 28, box 9, series 2B, SFH 5, SFPL.

142. "Mayor Misses a Redevelopment Protest."

143. "Redevelopment Foes Plan Fight"; "The Sansei," *San Francisco Examiner*, August 14, 1966.

144. "Congress Organizes in Fillmore"; "Mayor Misses a Redevelopment Protest."

145. Committee against Nihonmachi Evictions (CANE), "What Is CANE?," no date, report, folder 23, box 7, series 2.4, Social Protest Collection, BANC MSS 86/157c, BANC.

146. Benh Nakajo, interview with Ruth Meas and Sarah Lew, May 11, 2004, San Francisco, CA, transcript, NJAHS, 49, 54.

147. Karen Umemoto, "'On Strike!' San Francisco State College Strike, 1968-1969: The Role of Asian American Students," *Amerasia* 15 (1989): 21-22, 34; Diane C. Fujino, "Who Studies the Asian American Movement? A Historiographical Analysis," *Journal of Asian American Studies* 11 (2008): 128; Daryl J. Maeda, *Chains of Babylon: The Rise of Asian America* (Minneapolis: University of Minnesota Press, 2009); Laura Pulido, *Black, Brown, Yellow, and Left: Radical Activism in Los Angeles* (Berkeley: University of California Press, 2006), 77-85, 105-15, 134-42; William Wei, *The Asian American Movement* (Philadelphia, PA: Temple University Press, 1993).

148. Quotes from Nakajo interview, 54, 20; and Umemoto, "On Strike!," 21. Clement Lai, "Saving Japantown," 472, 187-204.

149. "J-Town Collective," *Rodan*, September 1971, 8, CANE box, JANL.

150. "Victory!," *Rodan*, November 1971, 3, ibid.

151. CANE, "Relocation, Redevelopment, C.A.N.E.," flier, circa 1973, CANE box, JANL; J-Town Collective, "Historic Struggle Emerging in J-Town," *New Dawn*, March 1973, 1.

152. Glenn K. Omatsu, "Listening to the Small Voice Speaking the Truth: Grassroots Organizing and the Legacy of Our Movement," in *Asian Americans: The Movement and the Moment*, ed.

Steve Louie and Glenn K. Omatsu (Los Angeles: UCLA Asian American Studies Center Press, 2001), 313; CANE, "Low Cost Housing Is the Right of Nihonmachi," circa 1975, 1, CANE box, JANL.

153. CANE, "Low Cost Housing," 1.

154. Nakajo interview, 42; "Japantown Redevelop Sit-in Protest," *San Francisco Chronicle*, February 14, 1975.

155. CANE, "One Year of Struggle: United, Together, We'll Never Be Defeated!" (San Francisco: CANE, 1974), 5, Asian American Studies Collection, Ethnic Studies Library, University of California, Berkeley; "Old Victorians Win a Skirmish," *San Francisco Chronicle*, August 13, 1975; "Group in Japantown Fights Big New Hotel," *San Francisco Chronicle*, December 13, 1973; "CANE Holds Rally," *CANE Newsletter*, March–April 1974, 2, folder 23, box 7, series 2.4, Social Protest Collection, BANC MSS 86/157c, BANC; "A Hot Meeting over Japantown Housing," *San Francisco Chronicle*, June 22, 1977; Clement Lai, "Saving Japantown, Serving the People: The Scalar Politics of the Asian American Movement," *Environment and Planning D: Society and Space* 31 (2013): 471–72.

156. CANE, "CANE Examines the Motives and Interests of NCDC," circa 1974, 2, folder 23, box 7, series 2.4, BANC MSS 86/157c, BANC.

157. CANE, *CANE Meets with Little Tokyo Anti-eviction Task Force* (San Francisco: CANE, 1974), 2.

158. CANE, "CANE Examines the Motives."

159. "Community Confronts RDA: Redevelopment Refuses CANE Demands," *New Dawn*, May 1973, 3; CANE, "What Is Kintetsu?," circa 1974, 1, CANE box, JANL; "Hotel Wins Fight in Japantown," *San Francisco Chronicle*, December 18, 1973.

160. Michael Dobashi, interview with Henry S. Francisco and Jovilynn T. Olegario, April 29, 2004, San Francisco, CA, transcript, 11, NJAHS.

161. Peggy Kanzawa, "Why Is There a Committee against Nihonmachi Eviction?," circa 1973, 1, folder 23, box 7, series 2.4, Social Protest Collection, BANC MSS 86/157c, BANC; CANE, press release, September 4, 1974, 6, CANE box, JANL.

162. Richard Hashimoto, interview with the author and Ken Yamada, July 2, 2004, San Francisco, CA, transcript, 19, NJAHS.

163. "Destruction of a Community?," *Rodan*, November 1971, 7.

164. Quotes from "Historic Struggle Emerging in J-Town," 7; and "Committee against Nihonmachi Eviction," *New Dawn*, April 1973, 3. Helen Jones, "Redevelopment for Whom?," *New Dawn*, August 1974, 3; "City Wide Support at CANE Demonstration," *New Dawn*, November 1974, 3; Clement Lai, "Saving Japantown," 472–74; "Gen. Mtg. Summary," *CANE Newsletter*, May–June 1974, 1.

165. "CANE Opposes Tourism in Nihonmachi," *New Dawn*, February 1974, 3; Clement Lai, "Saving Japantown," 479–80.

166. Quote from "One Year of Struggle," 32. Hartman, *City for Sale*, 70; Ocean Howell, *Making the Mission: Planning and Ethnicity in San Francisco* (Chicago: University of Chicago Press, 2015), 271; Eduardo A. Contreras, "The Politics of Community Development: Latinos, Their Neighbors, and the State in San Francisco, 1960s and 1970s" (PhD diss., University of Chicago, 2008); Tomás F. Summers Sandoval Jr., *Latinos at the Golden Gate: Creating Community and Identity in San Francisco* (Chapel Hill: University of North Carolina Press, 2013), 125–33.

167. Dai-Ming Lee, "Chinatown Needs Fresh Thoughts and Ideas," *Chinese World*, March 8, 1960.

168. Charlotte Brooks, *Between Mao and McCarthy: Chinese American Politics in the Cold War Years* (Chicago: University of Chicago Press, 2015), 77, 191, 298; Him Mark Lai, "China and the Chinese American Community: The Political Dimension," *Chinese America: History and Perspectives* (1999): 11; Him Mark Lai, *Becoming Chinese American: A History of Communities and Institu-*

tions (Walnut Creek, CA: AltaMira Press, 2004), 64-65; Chiou-Ling Yeh, *Making an American Festival: Chinese New Year in San Francisco's Chinatown* (Berkeley: University of California Press, 2008), 138, 249n67.

169. City and County of San Francisco, *Journal of Proceedings, Board of Supervisors*, April 13, 1964, 248.

170. Quotes from "Chinese Center a Boon to City," *San Francisco Chronicle*, April 6, 1964; and SFRA, "Famous Chinese Artist to Design Pedestrian Bridge," press release, December 2, 1968, 2, folder 14, box 8, series 2B, SFH 5, SFPL.

171. J. K. Choy, "Proposed Financing of the Chinese Cultural & Trade Center," January 13, 1964, 1, folder 34, box 3, series 1C, SFH 10, SFPL.

172. J. K. Choy to Jack Shelley, January 20, 1965, 1, 3, folder 34, box 3, series 1C, SFH 10, SFPL; Chinese Culture Foundation of San Francisco, "Chinese Cultural Center, San Francisco," pamphlet, circa 1965, folder 14, box 8, series 2B, SFH 5, SFPL.

173. "The Chinatown Center," *San Francisco Chronicle*, May 19, 1964; "Chinese Center a Boon."

174. Quote from "Chinese Cultural Center—East-West Bridge of Hope," *San Francisco Chronicle*, August 21, 1965. "Taiwan Artist to Work on Chinese Cultural Bridge Design," *East/West*, December 11, 1968; "Chinese Art Gift to the City," *San Francisco Chronicle*, January 29, 1970; City and County of San Francisco, File No. 7-69-3, Resolution No. 862-69, *Journal of Proceedings, Board of Supervisors* (San Francisco: City and County of San Francisco, 1969), 1002, December 15, 1969.

175. "Choy Blasts Six Co. for Cop Out," *East/West*, October 7, 1970; "Chinese Culture Finds a Home—Finally," *San Francisco Chronicle*, January 28, 1973; Yeh, *Making an American Festival*, 141.

176. "'New' Chinatown Urged to Boost Asian Trade," *San Francisco Chronicle*, June 26, 1965.

177. Even domestic funding was hindered by local and transnational politics. The pro-Nationalist Six Companies refused to lend their formidable fund-raising prowess to the center. See "Cultural Meeting a Puzzle," *East/West*, August 26, 1970; "Choy Blasts Six Co. for Cop Out"; Yeh, *Making an American Festival*, 138-41.

178. Allan Temko, "Dr. Fu Manchu's Plastic Pagoda," *S.F. Magazine*, May 1971, 22, folder "S.F. Bldgs: Chinese Cultural Center," Buildings, SUB COLL, SFPL.

179. "Chinatown Center"; "Gold Mixed into Chinatown's New Bridge," *San Francisco Chronicle*, February 4, 1970.

180. "Capturing a Culture on Kearny Street," *San Francisco Chronicle*, January 25, 1970.

181. Alvin H. Baum Jr., "What Should Be Done about Chinatown," *San Francisco Chronicle*, June 2, 1965.

182. "'New' Chinatown Urged."

183. "San Francisco's Chinatown," *Atlantic*, March 1970, 32; Charlotte Brooks, *Alien Neighbors, Foreign Friends: Asian Americans, Housing, and the Transformation of Urban California* (Chicago: University of Chicago Press, 2009), 227-29; Iris Chang, *The Chinese in America: A Narrative History* (New York: Viking, 2003), 262; Peter Kwong and Dušanka Mišĉevič, *Chinese America: The Untold Story of America's Oldest New Community* (New York: New Press, 2005), 317-18.

184. San Francisco Chinese Community Citizens' Survey and Fact Finding Committee, *Report* (San Francisco: San Francisco Chinese Community Citizens' Survey and Fact Finding, 1969) 40, 46; Iris Chang, *Chinese in America*, 261, 265-68; Madeline Y. Hsu, *The Good Immigrants: How the Yellow Peril Became the Model Minority* (Princeton, NJ: Princeton University Press, 2015), chapters 5 and 6; Madeline Y. Hsu and Ellen D. Wu, "'Smoke and Mirrors': Conditional Inclusion, Model Minorities, and the Pre-1965 Dismantling of Asian Exclusion," *Journal of American Ethnic History* 34 (2015): 43-65; Kwong and Mišĉevič, *Chinese America*, 227-36, 336-40; Buck Wong, "Need for Awareness: An Essay on Chinatown, San Francisco," in *Roots: An Asian American Reader*, ed. Amy Tachiki, Eddie Wong, and Franklin Odo with Buck Wong (Los Angeles: UCLA Asian American Studies Center, 1971), 265, 267.

185. "San Francisco's Chinatown," 32; Kenneth Rexroth, "Chinatown's Deterioration," *San*

Francisco Examiner, February 17, 1965, folder 34, box 3, series 1C, SPH 10, SFPL; Department of City Planning, "An Action Program for Chinatown," report, 1969, 3, folder 3, box 15, subseries A, series 3, SFH 5, SFPL.

186. Department of City Planning, "Action Program," 1–2; C. W. Fullilove to Supervisor Jack Ertola, February 27, 1969, in Chinese Community Committee, *Report*, 51.

187. Quote from "The Threat to Chinatown," *San Francisco Chronicle*, November 4, 1968. Department of City Planning, "Action Program," 3.

188. Office of the Mayor, press release, March 13, 1969, folder 17, box 8, series 2B, SFH 5, SFPL.

189. "San Francisco's Chinatown," 32.

190. The suggested project would have celebrated "vibrant emigrants from the Philippines" and commerce "from this courageous and democratic republic which amounts to some $130 million annually, second only to Japan among all countries." Filipino community leaders, however, could not secure the promised transpacific investment, nor were local Filipino incomes adequate for financing. Ruperto M. Baliao to Joseph Alioto, January 8, 1969, folder "S.F. Associations—Filipino—Cultural and Trade Center, 1968–1969," Associations, SUB COLL, SFPL; SFRA Commission, minutes, December 10, 1974, 2; Board of Supervisors, "Draft—Endorsing the Concept of a Cultural and Trade Center Representing the Republic of the Philippines," circa 1969, folder "S.F. Associations—Filipino—Cultural and Trade Center, 1968–1969," Associations, SUB COLL, SFPL; SFRA Commission, minutes, October 15, 1974, 4; Bill Ong Hing, *Making and Remaking Asian America through Immigration Policy, 1850–1990* (Stanford, CA: Stanford University Press, 1993), 106–9. In 1970, there were almost 60,000 Chinese Americans, about 25,000 Filipinos, and not quite 12,000 Japanese Americans. Department of City Planning, *San Francisco 1970*, 3.

CONCLUSION

1. Shenglin Chang, *The Global Silicon Valley Home: Lives and Landscapes within Taiwanese American Trans-Pacific Culture* (Stanford, CA: Stanford University Press, 2006).

2. Hilary Jenks, "'Home Is Little Tokyo': Race, Community, and Memory in Twentieth-Century Los Angeles" (PhD diss., University of Southern California, 2008), 226–32; CANE, *CANE Meets with Little Tokyo Anti-eviction Task Force* (San Francisco: CANE, 1974).

Index

Note: Page numbers followed by *f* refer to figures.

A-2 Stand, 195, 196

Abbey barbershop, 208

Abe, Victor S., 88, 94-97, 98-99, 101, 111, 117, 119, 132, 136, 147-48, 181-82; law practice, 103; as representative for the merchant-planners, 74-75, 86-87, 105, 109

Abels, Ulma A., 84

Ad Hoc Committee to End Discrimination, 190

African Americans: businesses, 27, 31, 39, 195, 200-203; civil rights activism, 84, 94-95, 176, 188-92; community organizing, 193-96, 197; compared to Asian Americans, 27, 32-33, 94-95, 100-101, 103, 105, 143, 171, 176-81, 184, 192, 200, 202, 204, 205, 208, 212, 215; discrimination against, 27, 32-33, 37, 184; employment of, 32, 180, 191; housing and residential segregation, 31-33, 37, 38, 84, 179-80, 193; housing quality of, 32, 84; in Japanesetown, 26-27, 31-33, 176; migration to San Francisco, 31-32, 192; population in San Francisco, 31, 89, 177-78; rejection as "outsiders," 178, 192; relations with Japanese Americans, 34, 36-39, 188-89, 191, 200, 202, 204, 208; in the Western Addition, 84-85, 176-80, 189, 194

Ajinomoto Company, 135

"alien citizenship," 10

Alien Land Law (CA), 16-17, 33

Alioto, Joseph, 174, 188, 211

American Municipal Association, 49

American President Lines, 58-59, 165

anti-Japanese movement, 3, 9, 19, 22-23, 24-25, 28

Anti-Jap Laundry League, 22

Ashizawa, Masao, 102-3, 109, 148, 182

Asia, US foreign relations with, 7, 10, 16, 20, 40, 45-47, 49, 104, 108, 120, 123-24, 133, 138, 165, 169

Asia Foundation, 57*f*

Asian American Movement, 205, 208, 216

Asian Art Museum, 114

Asian exclusion, 9, 10, 17, 18, 22, 29, 80, 88, 139, 140, 192

Asiatic Exclusion League, 22

Bank of Hawaii, 130

Bank of Tokyo, 134, 144, 147

Baptist Ministers Union, 203

Bayview-Hunters Point. *See* Hunters Point

Benedict, Ruth, 144

Benkyodo, 182

Berkeley, CA, 1, 191, 216

Berman v. Parker, 98

Birmingham campaign, 189

Black Panthers, 205

Black Power, 185, 200, 205

blight, 12, 85, 98, 100, 118, 183; definition of, 80; as empirical and objective, 8, 75, 77–79; as guide for redevelopment and renewal, 8, 75–80, 83, 104; as stigma, 74, 85, 87, 97, 180; as subjective concept, 81

Blyth-Zellerbach committee, 89

Board of State Harbor Commissioners, 39–40

Bolles, John, 50–51

Booker T. Washington Community Center, 37, 38, 204

Boston, MA, 52, 64

Boswell, Hamilton T., 94–95

Bowron, Fletcher, 36

Boxer Rebellion (China), 18

Broman, Paul, 133, 145–46

Brown, Eva, 197, 203

Brown, Ken, 197, 203

Brundage, Avery, 113–14

Brundage art collection, 114

Buchanan Street, 30, 91

Buchanan YMCA, 204

Buckley, John T., 58, 59

Burbridge, Thomas, 188–92

Caen, Herb, 31, 180

California, 18, 26, 32, 33, 88, 96, 119, 120, 122, 149, 170–71, 191, 216; urban redevelopment in, 76, 97–98, 133, 197

California Midwinter Fair of 1894, 20

California Supreme Court, 98, 197

Camp Funston, 34

CANE (Committee against Nihonmachi Evictions), 204–8, 206f; interracial alliances, 207–8

Capitol Laundry, 106

Carson, Pleasant, 188–92

Castle & Cooke, 130

central business district, 81, 111–13

Chambers, E. Snowden, 50, 51

Cherry Blossom Festival, 186, 208

Chiang Kai-Shek, 209

Chicago, IL, 35, 37, 113, 140–41

China, 7, 16; Japanese militarism toward, 22, 26, 128; trade with San Francisco, 20–21f, 165; US foreign relations with, 18, 46–47, 80, 209–11

Chinatown, 16–19, 26, 91, 92–93, 209–12; architecture, 19, 20, 109; compared to

Japanesetown, 2, 17, 20, 23, 24, 40, 93, 109, 187–88, 205, 209–10, 211, 215; Japanese businesses in, 26; migration and, 210–11, 215; new immigrants, 210–11; redevelopment in, 209–11; role in San Francisco life, 16, 17, 19–20, 40, 93, 211; and San Francisco cosmopolitanism, 16–17, 19–20; segregation of, 19, 25; as slum, 80–81, 210–11; tourism, 2, 1–17, 19, 26, 81, 90–92, 109, 188; transpacific connections, 211; Yamato Sukiyaki House, 162

Chinese Americans, 2, 15–17; anti-Chinese animus and movement, 19, 22–23; housing and segregation, 19, 25, 27, 178, 210; as model minorities, 141–43, 211, 215; popular views about, 80–81, 93, 124, 214–15; population, 23, 211; relations with Japanese Americans, 26, 91, 208; during World War II, 80–81

Chinese Culture Center, 209–10

Chinese migrants, 16–17, 19, 210–11, 215; end of exclusion, 80

Chinese Six Companies, 209

Chinese World, 219

Ching, Hung Wo, 132

Choy, J. K., 209–10

Christopher, George, 3, 55–56, 66f, 70–71, 107, 112, 115–17, 189

Chrysanthemum and the Sword, The (Benedict), 144

Chrysanthemum Festival. *See* Kiku-Matsuri

Cold War, 7, 213, 217; in Asia, 42, 54, 124, 210; and Asian Americans, 10, 104, 170; cultural programs, 59–60; and Japanese American integration, 13, 104, 139, 170

Commerce High School, 73

Committee against Nihonmachi Evictions. *See* CANE

community development corporations: African American involvement, 185, 200; Fillmore Community Development Corporation, 200–202; Nihonmachi Community Development Corporation, 200, 204, 206–7

community organizing: CANE (*see* CANE); Freedom House (*see* Freedom House); WACO (*see* WACO)

Community Redevelopment Act (1945), 97

Community Redevelopment Law (1951), 97

Congress of Racial Equality. *See* CORE

Cooke, Clarence H., 127
CORE (Congress of Racial Equality), 189–91
Council for Civic Unity, 72–73, 195
Crisis, 37
Crocker-Citizens bank, 112
Crown-Zellerbach Corporation, 112
Custus, Lawrence, 196

Daiwa Securities Company, 133
Davies, Phillips S., 52
defense industries, 31–32, 45, 141
Department of Justice camps, 30
DeWitt, John, 29, 140
Diamond Heights, 98
dispersal and assimilation, 34–36
Dobashi, Frank, 163
Dodge, Joseph, 47
Dole, Inc., 130
Dream Entertainments, 154–55, 159, 175

Eichler, Edward P., 195
Eisenhower, Dwight D., 49, 64, 135
Emanuel Church of God in Christ, 39
"enemy aliens," 29
escheat cases, 33
ethnic studies, 205
Europe, postwar economic collapse, 46–47
Evacuation Claims Act (1948), 88
Executive Order 9066, Franklin Roosevelt, 28

442nd Regimental Combat Team, 30, 101, 170
fair employment policies, 171, 182
fair housing ordinance, 171
Fairmont Hotel, 52, 112
Family Service Agency, 202
Filipinos: enclave, 26, 176; in Hawai'i, 128; in Japanesetown, 25–26, 28, 91, 225n59, 230n; in San Francisco, 18, 74, 176, 208, 211, 272n190; relations with Japanese Americans, 26, 28, 128, 208
Fillmore Community Development Corporation, 200–202
Fillmore neighborhood, 31, 177, 179f, 195, 200; community organizing, 195, 200–204; during World War II, 31; pre-World War II, 27
Fillmore Street, 24, 27, 195
Fisherman's Wharf, 162
Fleishhacker, Mortimer, Jr., 200–201
Fong, Hiram, 131

Freedom House, 194–98
Fuji Hotel, 96–97, 106

Gateway to the Pacific: and Chinatown, 209; "Gateway to the Orient," 17; identity and ambitions, 3, 11–12, 42, 75, 105, 113, 123, 137, 143, 213; and Japanese Americans, 143, 161, 163, 172, 188; and the Japanese Cultural and Trade Center, 107–8, 111, 113, 148, 175
Geary Street, Boulevard, and Expressway, 24, 106, 174, 177, 178f
Getz Bros. & Co., 46, 52
Glazer, Nathan, 195
Go for Broke!, 170
Golden Gate Park, 20, 110, 148
Golden Gateway, 113
Gold Rush, 18–19
Goodlett, Carlton B., 84
Goodman, Lester L., 52, 58, 68, 71
Goodwin, Joye, 194
Grant Avenue, 90, 92
Greater East Asia Co-prosperity Sphere, 47
Griffin, Everett, 112, 196

Hagiwara family, 20, 39
Hawai'i, 7, 41, 107–8; anti-Japanese hostility in, 127–29, 130–31; Asian American economic and political rise in, 130–32; as "Crossroads of the Pacific," 132–34; economics, 130–31; Economic Study Club, 133; Filipino population, 128; Japanese banks, 129; Japanese in, 126–31; mainland investments, 132; marginalization of Native Hawaiians in, 123–24; mass incarceration of Japanese Americans, absence of, 130–31; militarism in, 123, 130; overthrow of monarchy, 123; plantation strikes, 127–28; "racial aloha," 123–24; relationship with San Francisco, 120, 122–23, 125; statehood, 108, 122–25, 133; sugar plantations, 126, 127–28; tourism, 122, 125, 130, 134; World War II in, 130–31
Hawaiian banks, 129–30
Hawbecker, David, 204
Hayashino, Carole, 205
Hedani, Tokuji, 109
Herman, M. Justin, 112–13, 115, 117–22, 133, 146, 148–50, 153, 158–59, 182–83, 196, 198–203; background, 112, 117
Hirano, James, 88–89

Hirota, Carl, 88–89, 91
Honnami, Kikoroku, 88–89, 98, 101, 182
Honnami, Sumi, 34
Honnami shop, 88, 182
Honolulu Japanese Chamber of Commerce, 134
Hoshiyama, Fred Y., 204
Housing Act of 1949 (federal), 76, 97–98
Housing and Home Finance Agency, 112
Howden, Edward, 83
Hunters Point, 31, 34, 179–80, 203

Ickes, Harold, 34–35
Ikeda Hayato, 135
Immigration Act (1924), 22
Immigration Act (1965), 211
incarceration during World War II, 4, 11, 12,
    13, 22, 89, 99, 104–5; and assimilation,
    34–36, 95–96, 142; conditions of, 29–30;
    as a "disguised blessing," 142; economic
    impact of, 131; immediate lead-up to
    in San Francisco, 28–39; influence on
    Japanesetown urban planning, 85, 95–96;
    Japanese American opposition to, 23;
    resettlement in San Francisco, 33–39, 85
individual property rights and ownership,
    103–4
Ingleside, 179
Inouye, Daniel, 131
Inouye, Mike, 109
Interior Department, 34–35
International Hospitality Center, 57f
International Hotel, 208
International Longshore and Warehouse
    Union, 183
International Savings and Loan, 126–27, 129
interracial organizations and activity, 37, 73,
    130–31, 183, 190, 204, 208
Ishizaki, Edward S., 162
Ishizaki Taizo, 135
Issei, 28, 109, 131, 144, 146. See also Japanese
    migrants
Italians in San Francisco, 19, 28, 29
Ito, Willie K., 38

JACL (Japanese American Citizens League),
    33, 37, 99, 101, 110, 170; and Japanese-
    town urban planning, 85–86, 99, 105,
    147, 183; San Francisco chapter, 86; and
    US-Japanese relations, 166–70
James, Joseph, 31, 37
Japan: colonialism, 21, 41, 51, 124, 128; depic-
    tions in US media and law, 43, 53–55,

61–62; economics, 46–47, 70; gendered
    representations, 43, 53–54, 65; General
    Agreement on Tariffs and Trade, 53; inter-
    preted for US audiences, 156–58, 160–61,
    162–63, 165; perceptions of Japanese
    Americans, 118; San Francisco, relations
    with, 20–23, 39–40, 54, 58–59, 60–61,
    64–65, 135; in San Francisco imagination,
    4, 17–18, 23, 39–40; Self-Defense Force,
    61, 62–64; United Nations membership,
    53; US academic understandings of, 144;
    US popular fascination with, 53, 54, 110,
    150–51, 159, 162; US relations, 7, 47,
    51, 135 (see also sister-city program); US
    stereotypes of, 43, 53–56, 57f, 62, 65, 68–
    69, 143–44, 166
Japan Air Lines, 52, 60, 70, 133, 134, 164, 165,
    169, 175
Japan-American Conference of Mayors and
    Chamber of Commerce Presidents, 71, 215
Japan Day Fête (1957), 110
Japan Expo in Osaka (1970), 51
Japan Export Trade Promotion Agency, 135
Japan External Trade Organization, 57f, 58
Japan International Trade Fair, 48, 51
Japan Society, 48, 165
Japan Times, 48
Japan Trade Center, 57f, 163, 167
Japanese American Christ United Presbyter-
    ian Church, 39, 204
Japanese American Citizens League. See JACL
Japanese American Merchants and Property
    Owners Association, 100
Japanese American Religious Federation,
    204–6
Japanese Americans: activism, 23, 204–8,
    206f; ascendance in California agricul-
    ture, 17; in Chicago, 35, 37, 141; conflated
    with Japan, 11, 13, 208; contestation of
    municipal redevelopment, 85, 102–3;
    cultural fluency as an asset, 138–39; as
    cultural interpreters, 11, 138–43, 144–46,
    152–54, 155–56, 160–61, 162–67; dis-
    crimination against, 16–17, 22–23, 24–25,
    27, 28–30, 33, 65, 88, 89, 91–92, 180–81;
    educational attainment, 126, 141, 170–71;
    homeownership, 34, 38, 89–90, 180,
    186; housing and residential segregation
    of, 24–25, 33–35, 40, 65, 85, 89, 91, 96,
    104, 179–81; as a "marginal man," 140; as
    "model minorities," 10, 141–43, 188, 211,
    214, 207–8; population in San Francisco,

10, 23, 25, 33–34, 89, 177–78; racism toward African Americans, 37, 100; residential dispersal and integration, 36, 91–92, 176, 182; as transpacific bridge, 139–40, 143; work, 25–26, 28, 30, 34, 36, 54, 89, 109, 126–27, 150–53, 155, 160, 162–66, 167, 170–72, 182; World War II military service, 30, 101, 155–56, 164, 170

Japanese American women: as cultural transmitters, 54, 160–61, 162–63; sexualized and racialized notions of, 53–54, 68, 111, 160, 165–66, 170, 174–75; work of, 54, 140, 160, 165–66, 170–72

Japanese Association. *See* Nichi Bei Kai

Japanese business: investing in US, 121, 139; Japanese capital regulation, 145; Japanese Chamber of Commerce of Honolulu, 133–34

Japanese Chamber of Commerce of Northern California, 86, 88, 99, 100, 110, 147, 164, 165

Japanese consumer strength, 116–17

Japanese Cultural and Trade Center, 5, 115–16, 132–34, 136, 138–39, 188, 209; architecture and design, 148–49, 153–54, 154*f*, 156–60, 157*f*; delays, 145–46, 158, 174; eminent domain, challenges to, 189; expression of transpacific networks, 107–8, 111, 113, 148, 175; hotel, 150, 153, 156–57, 164*f*; Japanese American tenants, 146–48; Japanese tenants and investors, 144–46, 158–60, 175; opening celebration, 174–75; Peace Pagoda, 135–36, 174, 175*f*, 178*f*; photos, 154*f*, 157*f*, 175*f*, 178*f*; restaurants, 145–46; role of Japanese Americans, 133–36, 138, 143–50, 153–56, 160–61, 174–76, 187–88

Japanese Exclusion League, 22

Japanese Federation of Economic Organizations, 135, 175

Japanese Garden Center (Japanese Village), 90–111, 119, 136; design ideas, 90, 109, 110*f*; and incarceration, 95–96

Japanese general trading companies, 51

Japanese migrants, 22–23; exclusion of, 23, 29, 80, 88, 139, 140; ineligibility for citizenship, 17, 98, 111, 140; post–World War II migration, 54, 211. *See also* Issei

Japanese restaurants and cuisine, 54, 92, 107, 110, 162–63

Japanese Tea Garden, 20, 21*f*, 23, 39, 40, 114, 148

Japanesetown, 2, 4, 5, 10, 12, 13, 166; businesses, 25–26, 87, 88–93; ethnic and racial diversity, 26–27, 34, 36–39, 91, 176; after incarceration, 33–34; postincarceration resettlement, 33–39, 96; prewar, 23–27; redevelopment, 2, 4–5, 10, 12–13 (*see also* urban redevelopment); substandard housing, 32; varied names for, 23–24, 31, 186–87; during World War II, 27–33

Jimbo's Bop City (Waffle House), 39

Jimmy's Barber Shop, 39

Johnson, Lyndon B., 7, 169, 216

Jones, Helen, 208

Jones, Percy, 193

Jones Memorial Homes, 94–95

Jones Memorial Methodist Church, 94

Kaiser, Henry, 132

Kakuhei Matsui, 118

Kanrin Maru centennial celebration, 64–66, 135

Kansai Steamship Company, 52

Kazue, Togasaki, 30, 85

Kido, Saburo, 168

Kik's Smoke Shop, 106

Kikuchi, Charles, 28, 31

Kiku-Matsuri, 99–100

Killion, George, 165

Kinmon Gakuen, 37

Kintetsu, 145, 146, 164, 175, 207–8

Kishi Nobusuke, 53

Kitano, Harry H. L., 181

Korean War, impact on Japanese economy, 47

Kosaka Zentaro, 135

Kurosawa, Akira, 65

Kusano, Hatsuko, 159

Lash, James, 87

Latinos, 9, 91, 212

Lee, Dai-Ming, 209

Lili'uokalani, Queen, 123

Little, Helen O., 197

Little Tokyo, 36, 38, 92, 216

Los Angeles, CA, 133, 215; Bronzeville, 36; as San Francisco rival, 3–4, 44–45, 48, 111, 117, 216

Luzon Coffee Shop, 39

Makiki Christian Church, 126

Maloney, E. D., 57*f*

Manilatown, 26, 176
Manzanar, 30
Mao Zedong, 47
Martin Luther King Square, 200–202, 269n131
Masaoka, Mike, 166–69
Matsuo Kunizo, 144, 146, 154–55, 159–60
Matsuo Sentochi Theatrical Company, 144, 154. *See also* Dream Entertainments
McCarran-Walter Immigration Act (1952), 88
McGinty, Ellen, 69
Meitetsu Department Store, 144–46, 158
merchant-planners group, 86–91, 93–99, 101–11, 115, 117–19, 132, 136, 147–48, 181–82, 184–89; background of members, 88–90, 99, 102–3, 109, 182; demonstrating participatory citizenship, 103–5; inspirations of, 106–7, 110–11; urban planning as community self-fashioning, 74–75, 87, 97, 99, 101, 105
Mexican Americans. *See* Latinos
Michener, James, 120, 130
Mike's Richfield Service Station, 109
Military Intelligence Service (MIS), 155–56, 164, 165
Miller, Mike, 194
Mission district, 9, 176, 180, 208, 212
"model minorities," Asian Americans as, 10, 141–43, 188, 207–8, 211, 214, 215; vis-à-vis African Americans, 143, 188, 214
Montgomery, Ulysses J., 200–202
Moriwaki, Yoshiaki, 88–89
Morning Call, 22
Morrison, Jack, 199
Murdoch, Norman, 149–51, 157

NAACP (National Association for the Advancement of Colored People), 37, 183
NAACP, San Francisco chapter, 31–32, 37; cooperation with other civil rights organizations, 37, 190–92, 194, 195; critique of San Francisco redevelopment and renewal, 84, 105, 183, 194
Nakagawa Apartments, 106
Nakai Mitsuji, 51, 55–56
Nakajo, Benh, 39
Nakamura, Noboru, 149–50, 154–56, 182, 187
Nakayama Goro, 127
National Association for the Advancement of Colored People. *See* NAACP

National-Braemar, 133–35, 138, 145, 146, 148, 149
National Construction Industry Association of Japan, 117
National Securities and Investment, 130
Native Hawaiians, 123–24, 126, 127
NB Department Store, 30, 102–3
New Deal, 19, 75–76, 142
New Orleans, LA, 52
New York City, NY, 20, 117, 149, 150, 151, 217
Nichi Bei Kai, 93, 99, 110
Nichi Bei Times, 85, 96, 106, 119, 181
Nihonmachi Community Development Corporation, 183, 184–88, 193, 200, 202, 204; as demonstration of nonwhite and resident redevelopment cooperation, 185–86; members, 181–82; and model minority image, 188; and neighborhood control, 185–86; opposed by CANE, 206–8
Nikko Sukiyaki, 162–63
Nippon Pool Hall, 39
Nisei Fishing Club, 101
Nisei Lions, 101
Nixon, Richard, 211
Northwest Orient Airlines, 165

Oakland, CA, 89, 149, 163–64, 194; as San Francisco competitor, 45; sister-city program, 164, 215
Obata, Chiura, 162
Obon festival, 100, 187
Ogawa, Frank, 164
Okubo, Miné, 29
Okumura, Takie, 126, 128
Omura, James, 23
Osaka, 5, 42–43, 46, 49, 50–53, 55–56, 58–62, 59f, 62–72, 159, 116, 159, 165, 175, 215 (*see also* sister-city programs); economic conditions of, 51; economic relationship with San Francisco, 70–71
Osaka Chamber of Commerce and Industry, 56
Osaka Shinbun, 64

Pacific Citizen, 35, 37, 85
Pacific Festival, 41–42, 44, 62–63, 67, 113, 115–16
Pacific Heights, 81, 177, 179f; the Pacific world, 6–7, 40, 111, 121

Pan Am, 165–66
Panama Canal, 15
Panama-Pacific International Exposition of 1915, 15–17, 41
peace treaty conference (1951), 54
Pearl Harbor, attack on, 23, 28, 101, 130, 155
People's World, 80
People-to-People Program, 49, 52, 60, 67, 215
Phelan, James Duval, 22
Philippine-American war, 18
Philippine Cultural and Trade Center, 211, 272n190
Ping Yuen public housing project, 93
Pittman, Tarea Hall, 183
Pope, Gilbert, 196
Post Pool Hall, 106
Post Street, 39, 85, 86, 91, 92, 94, 100, 102, 106
Powell, Freddie, 208
Proposition 14 (CA), 197
Proposition 15 (CA), 88

Rabb, Maxwell M., 166–67
racial housing covenants, 38, 171
racial liberalism, 152, 172, 191, 211
racism and discrimination. *See* African Americans, discrimination against; Chinese Americans, anti-Chinese animus and movement; Japanese Americans, discrimination against
Richmond district, 2, 27, 89, 91–92, 178, 180
Rogers, Mary, 198, 199, 203
Roosevelt, Franklin, 28, 39
Roosevelt, Theodore, 7, 22
Rossi, Angelo, 28, 33
Rowe, Thomas, 61

Sakai, Tamotsu, 109
Sakamoto Grocery, 106
Sakura Matsuri, 208
San Diego, CA, 45, 50, 52, 215
San Francisco: 1906 earthquake and fire, 15, 19, 24, 77, 103; cosmopolitanism, 3, 16, 17, 18–20, 40, 71, 81, 105, 111, 170, 209; cultural exchanges with Osaka, 42–43, 50–51, 58, 60–62, 64–65, 67–72; de Young Memorial Museum, 110, 113–15; economic and political role in the Pacific region, 15, 17–18, 63, 41–42, 49, 113–15; fair employment, 171, 182; immigration

from China, Cold War era, 210–11; Japanese cultural activities, popularity, 107, 110; *Kanrin Maru* centennial celebration, 64–65, 135; "Metropolis of the Pacific," 18, 45; as a "new city," 74–75; population, 31; racial liberalism in, 172, 191, 211; strategic importance, 17–19; transpacific transportation hub, 3, 46–48, 133, 165; during World War II, 27–28, 63; and the World War II eviction of Japanese Americans, 22–23, 29–30
San Francisco, commerce and economy, 112, 113, 115; commercial importance, 17, 111; economic challenges, postwar, 44–45; finance, 45; importance of ports, 18; international trade promotion, 48; Japanese banks, 22; loss of primacy in Japanese commerce, 44–45; trade and commerce with Japan, 20–22, 39–40, 46–48, 58–59; trade with Japan, 20–21, 39–40, 45–47, 51, 60–61, 70
San Francisco Board of Supervisors, 22, 25, 56, 69, 70–71, 74, 83–84, 100, 146, 183, 191, 195, 199, 203, 207, 209
San Francisco Building Trades Council, 83
San Francisco Chamber of Commerce, 70, 111; business tour of Asia, 48–49, 72; Chamber of Commerce, 57f; Hawaiian Affairs Section, 125; Industrial Development Committee, 50; Japanese Affairs Subcommittee, 125; World Trade Department, 49, 70
San Francisco Chronicle, 18, 22, 24, 25, 44, 47, 111, 113, 119, 163
San Francisco Citizens' Advisory Committee, 112
San Francisco City Planning Commission, 196
San Francisco City Planning Department, 76, 98, 211
San Francisco Examiner, 28
San Francisco Housing Authority, 76, 93, 94, 183
San Francisco Housing Authority Commission, 94
San Francisco Human Rights Commission, 190
San Francisco Juvenile Justice Commission, 94
San Francisco Neighborhood Legal Assistance Foundation, 202
San Francisco News-Call Bulletin, 192

San Francisco–Osaka sister-city affiliation, 4, 11, 48–52, 61; celebrations, 42–43, 62–66, 72; challenges, 55–56; challenge to popular perceptions of Japan, 53, 55–58, 57f, 60–66, 69, 70, 72; civic benefits, 50–51, 56, 61; demonstrating San Franciscans' support for Japanese relations, 58, 63, 65–66, 70–71; economic and commercial benefits, 49–52, 56–60, 59f, 70; implications for San Francisco civic identity, 42–44, 58; Japanese American participation, 63–64, 65, 72; "Miss Osaka" exchange, 68–69; Miss Sister City, 69; youth involvement, 58, 62, 67–70

San Francisco Planning and Urban Renewal Association, 112

San Francisco Police Community Relations Bureau, 196

San Francisco Port Authority, 52

San Francisco Redevelopment Agency, 2, 73, 85–86, 88, 94, 95–96, 101, 103, 105, 106–8, 112, 115–16, 121, 133–34, 136, 138, 145, 147–48, 158, 180, 181, 202 (see also Herman, M. Justin); criticism of, 84, 97, 182–84, 188–89, 193, 194–98, 204–7; early years, 84, 97–98, 112–13

San Francisco Youth Association, 61–62, 67–68

Sato, Robert Y., 130–31

Sayonara, 53

Seattle, WA, 44–45, 48, 50, 52, 134, 150, 215

Seibu, 117, 119

Seiichi, Harada, 127

Seiki, Richard M., 88–89

Seiki, Sam, 182

Sekino Hirai, Alice Setsuko, 37

Serlis, H. G., 59f

Sheets, Millard, 149–50, 153

Shelley, John F., 190, 195–96, 199

Sheraton Palace Hotel, protest of, 190–91, 192

Shichisaburo, Hideshima, 93–94, 99

Shimizu Construction Company, 158

sister-city programs, 43, 49–50, 63, 67–80, 72, 169, 215 (see also San Francisco–Osaka sister-city affiliation); Oakland-Fukuoka, 164, 215; risks, 52; San Diego–Yokohama, 50; San Francisco–Taipei, 210; San Jose–Okayama, 56; Seattle-Kobe, 50

Smith, Serena, 193

SNCC (Student Nonviolent Coordinating Committee), 190, 194, 195

Soko Hardware, 39, 109, 182

South of Market district, 24, 113, 208

Soviet Union, 7, 10, 46

Spanish-American War, 18

St. Francis Square cooperative, 183

State Board of Harbor Commissioners, 40, 52

Stratten, James, 37

Student Nonviolent Coordinating Committee (SNCC). See SNCC

Sugi Michisuke, 56–58, 57f

sukiyaki, 31, 54, 92, 110, 115, 158, 162–63, 166

Sullivan, Charles, 39

Sullivan, James J., 57f

Sumitomo Bank, 127, 129, 167

Sun-Reporter, 100, 189, 191

Sunset district, 180, 194

Sunset Magazine, Japantown feature, 187–88

Taiwan, 210–11, 215

Takahashi, Tomoye and Henri, 1–2, 3, 4, 5, 13

Takahashi Trading Company, 1, 106

Takenaka and Associates, 158

Takeshimaya, 117, 119

Tanforan racetrack "assembly center" (CA), 29–30

Tange Kenzō, 158

Tatsuno, Dave, 30, 102–3

Tatsuno, Masateru, 102–3

Tenants and Owners in Opposition to Redevelopment in the South of Market, 208

Tenants Union, 197–98

Teruya, Albert, 130

Thayer, Wade Warren, 127

Third World strikes at San Francisco State College and University of California, Berkeley, 197, 205

Todaiji Temple, 175

Togasaki, Kazue, 30, 85

Tokioka, Franklin, 132

Tokioka, Masayuki, 5, 12, 143, 148, 150, 191 (see also Japanese Cultural and Trade Center; National-Braemar); in context of Hawai'i, 119, 121–27, 132–37, 214; as intermediary, 107–8, 124–27, 129–30, 133–37, 138, 143–46, 158, 163, 214; Japanese Cultural and Trade Center, 108, 132–33, 134–36, 144–47, 156

Tokyo, Japan, 5, 17, 48, 51, 133, 135, 145–46, 158, 160, 166

Tokyo Chamber of Commerce, 135

Tokyo Kaikan, 145
Tokyo Sukiyaki, 162
Topaz, UT, 1, 3, 30, 89, 162
tourism, 19, 56, 90, 122, 125, 130, 134, 165, 207
Toyooka, Charles, 208
transpacific exchanges, 5-6, 12, 20, 43-44, 54-55, 56, 58-60, 68-69, 72, 117
Tsuji Keiko, 68-69
Tule Lake, CA, 30

Underwood, S. David, 149
Union Square, 13, 41, 62, 81
United San Francisco Freedom Movement, 190-95, 197
United States-Japan Centennial Year, 64-66, 66f
University of California, Berkeley, 1, 28, 141, 162, 163, 190, 205
Uoki K. Sakai Grocery, 109, 182
Uratsu, Marvin, 165
urbanism, definition, 3
urbanity, definition, 3
Urban League, 203
urban redevelopment, 8-9, 185f; in California, 97-98; criticism of, 148, 176-78, 184, 193-94, 195-96, 199, 208, 212; Golden Gateway, 113; laws, 76, 75-76; metaphors, 76-77, 82, 104; nationally, 5, 9, 75-76, 93-94, 97-98, 146, 148; South of Market, 113
urban redevelopment in the Western Addition, 2, 4-5, 8, 10, 13, 73, 102, 104, 107, 176, 186; A-2, 176-77, 180, 184, 185f, 186, 193, 195-96, 201; African American participation, 94-96, 200-202; Chinese American participation, 209-10; community organizing against San Francisco's program, 176, 183-84, 193-200, 204-7, 208, 212 (see CANE; Freedom House; WACO); criticism of, 83-85, 183-84, 188-89, 193-94, 195-96, 197-99, 201-2, 205-8; displacement and rehousing, 182-83, 193-94, 202-3, 205-6; ideas behind general program, 8, 74-75, 76-84, 149; Japanese American participation, 12, 74-75, 85-93, 95-97, 104, 108-11, 119, 181-88; process of, 83-84, 85-87, 98, 106, 108, 112-13, 115, 119-20, 136, 174, 179f; support for, 73, 83-84, 111-13, 115-17
urban rivalry, 42, 44-45, 47-48, 111, 216

US consulate in Kobe, 151-53
US Customs, 52
US Information Agency, 49-50
US Pacific in the Cold War, making of, 7
US-Japanese relations, 10, 22, 135, 167-68; cultural, 50, 60, 117
Ussery, Wilfred T., 189-90, 194
Uyeda, Masaru, 130

Van Bourg/Nakamura, 149-50, 153, 156
Veterans of Foreign Wars, 83
von Loewenfeldt, Charles, 60

WACO (Western Addition Community Organization), 197-200, 201-3; and direct-action protest, 197-98, 200-201, 201f; diverse political strategies, 198-99; Homeowners' Association, 197; Japanese American inclusion, 204; members' incorporation into the Redevelopment Agency, 203-4; "predominantly Negro" frame, 204; successful lawsuit halting renewal, 202-3
Wada, Yori, 38, 204
Wagner-Steagall Housing Act of 1937, 75-76
Wani (deputy mayor, Osaka), 59f
WAPAC (Western Addition Project Area Committee), 203-4
War Brides Act, 43
War Relocation Authority. See WRA
Waterhouse, John, 127
Watts uprising, 203
Weaver, Robert C., 198
Western Addition, 3, 4-5, 13, 92, 94, 118, 177-81, 189 (see also Fillmore neighborhood; Japanesetown; urban redevelopment in the Western Addition); prewar, 23-27, 31, 33, 39; redevelopment of, 102, 106, 115, 176-77, 183-84, 193-208; redevelopment planning, 77-87, 98, 104, 107-8, 111, 116, 136
Western Addition Community Organization. See WACO
Western Addition Project Area Committee. See WAPAC
Western Defense Command, 22, 29, 33
Williams, Hannibal, 197-99, 201, 203-4
Wilson, James P., 49, 50, 70
Wilson, Woodrow, 17
Wong's Bait Shop, 39
World Trade Association. See WTA

WRA (War Relocation Authority), 34–36, 89, 95, 142
WTA (World Trade Association), 56, 58; business tour of Asia, 48–49; Japanese Affairs Subcommittee, 48

Yamamoto, Masae, 160, 169
Yamasaki, Minoru, 149–54, 155, 156, 158, 161, 163, 164, 174

Yamato Auto Repair, 106, 186
Yasuda, Shotaro, 162
YMCA, 41, 58, 204
Yonemura, Mas, 163–64, 169
Yoshida Hidemi, 62–63
Yoshida Shigeru, 54, 65

zaibatsu, 51